The Science and Engineering of Mechanical Shock

Carl Sisemore • Vít Babuška

The Science and Engineering of Mechanical Shock

Springer

Carl Sisemore
Albuquerque, NM, USA

Vít Babuška
Albuquerque, NM, USA

This book describes objective technical material. Any subjective views or opinions that might be expressed are those of the authors and do not necessarily represent the views of Sandia National Laboratories, the US Department of Energy, or the US Government.

ISBN 978-3-030-12102-0 ISBN 978-3-030-12103-7 (eBook)
https://doi.org/10.1007/978-3-030-12103-7

Library of Congress Control Number: 2019933292

This Springer imprint is published by the registered company Springer Nature Switzerland AG.
The registered company address is: Gewerbestrasse 11, 6330 Cham, Switzerland

To
Sara, Seth, Hosea, Josiah, and Hanna;
Pavel and Tomáš;
and the next generation of engineers.

Preface

Mechanical shock is often the misunderstood and neglected half of system environmental testing and analysis. Engineers design their systems to pass vibration exposure and hope that the systems pass shock testing. It has been said that the vibration laboratory is where components go for testing and the shock laboratory is where they go to die. The primary reason for this is that mechanical shock is generally not taught in either the undergraduate or graduate curriculum nor is it particularly straightforward. There are relatively few books dedicated to the study of mechanical shock, and most of those are engineering discipline-specific.

While some undergraduate texts on vibrations may include a chapter on mechanical shock, thorough treatments of the topic are lacking. Nevertheless, shock is a mechanical environment that affects almost every device—from ships to spacecraft and from cell phones to shipping packages. Civil engineers worry about blast-induced shocks and earthquakes. Mechanical engineers are concerned with impact-induced shocks, while naval engineers must consider the effects of underwater explosive shocks on ships. Satellites must be robust to pyroshocks.

The goal of this book is to bring together engineering information about mechanical shock from various disciplines into a cohesive text. The science of mechanical shock involves understanding of the basic physics of the event: how the dynamics of the system interplay with the high-speed transient nature of the excitation and the numerical methods for quantifying the shock event and its damage potential to various structures, systems, and components. The book includes information from naval and military applications, earthquake engineering, aerospace, and spacecraft in one reference.

While the book is intended to provide basic information about the field of mechanical shock, the book and the topic are not themselves simplistic. It is assumed that the reader has a firm understanding of mechanical vibrations and differential equations. The book presents a historical context to the field of mechanical shock including numerous practical examples and anecdotes. Mathematical derivations of the fundamental principles and data processing methods are included and elucidated. Development of the single degree-of-freedom oscillatory theory and the theory of the shock response spectra are covered in detail in Chaps. 3 and 4. There is a section on nonspectral techniques for analyzing shock test data. Various types of shock test machines and their capabilities are covered in Chap. 9. In addition, a lengthy discussion on the methods for deriving shock test specifications is included in Chap. 11. The analysis of multi-degree-of-freedom systems with shock response spectra is discussed in Chap. 8. The energy spectrum, which is another type of response spectrum that is used primarily in the earthquake engineering community, is discussed at the end of the book, in Chap. 12. Throughout the book, example problems illustrate the main concepts. Problems are included at the end of the chapters to further the reader's understanding.

Both authors have worked in the field of shock and vibration analysis and testing for many years. Our experience has been predominately with naval, aerospace, and spacecraft systems, although there has also been some brief work with earthquake survivability. Throughout our careers, we have found that a considerable amount of information about mechanical shock is available if one is willing to search. Unfortunately, the information is widely dispersed, and not every source is necessarily a good source. The finer points of the science are often buried in "black box" routines and analysis codes. A combined, handy reference, grounded in structural dynamic theory with wide applicability, has not been readily available. It is into this arena that the authors place this work. It is our hope that the reader will find the text useful and interesting. After all, mechanical shock is a fun business.

Albuquerque, NM, USA Carl Sisemore
Albuquerque, NM, USA Vít Babuška
May 2019

Contents

Acronyms

ABS	Absolute
ARMA	Autoregressive Moving Average
CL	Close
CQC	Combined Quadratic Combination
DFT	Discrete Fourier Transform
ERS	Energy response spectra
FAA	Federal Aviation Administration
FFT	Fast Fourier Transform
FOH	First-order hold
FSP	Floating Shock Platform
HBX	High Blast Explosive
IFSP	Intermediate Floating Shock Platform
IIR	Infinite Impulse Response
KSF	Keel shock factor
LFSP	Large Floating Shock Platform
LFT&E	Live Fire Test and Evaluation
LWSM	Light-weight shock machine
MCE	Maximum Considered Earthquake
MDOF	Multiple Degree of Freedom
MMAA	Maxi-max absolute acceleration
MPE	Maximum Predicted Environment
MWSM	Medium-weight shock machine
NASA	National Aeronautics and Space Administration
NATO	North Atlantic Treaty Organization
NRL	Naval Research Laboratory
PV	Pseudo-Velocity
PVSRS	Pseudo-Velocity Shock Response Spectrum
RMS	Root-Mean-Square
SDOF	Single degree of freedom
SI	Système International
SRS	Shock response spectra

SRSS	Square Root of the Sum of Squares
SSTV	Submarine Shock Test Vehicle
SVIC	Shock and Vibration Information Center
TNT	Trinitrotoluene
UERD	Underwater Explosion Research Division
UNDEX	Underwater Explosion
USCS	United States Customary System
USGS	United States Geological Survey
ZOH	Zero-order hold

Chapter 1
Introduction

The opening shots of the War of 1812 were fired by USS *President*, a heavy frigate under the command of Commodore John Rodgers on June 23, 1812. Commodore Rodgers himself fired the opening shot of the war from the starboard forecastle bow-gun striking the stern of HMS *Belvidera* under the command of Captain Richard Byron. This was followed by a hit from the corresponding main-deck gun and a second hit by the Commodore. When the main-deck gun was fired a second time it ruptured, blowing up the forecastle deck, killing or wounding 16 men including Commodore Rodgers, whose leg was broken. Gun failures in the early days are generally attributed to poor metallurgy but the loading is most assuredly a shock loading. The pressure pulse resulting from the burning propellant generates a shock to the gun barrel. This shock is repeated every time the gun is fired. Weapon failures have been a significant contributor to the study of mechanical shock and military applications have been one of the primary sources for research and development in the field. The need has arisen out of numerous examples similar to the encounter of the frigates USS *President* and HMS *Belvidera*. As a result of the main-deck gun explosion on the *President*, *Belvidera* managed to escape [1].

The study of mechanical shock from a military perspective has generally focused on the need to keep weapons and equipment functional and reliable under harsh operating conditions. This need follows closely with the need for operator safety since damaged military equipment can quickly lead to more serious consequences. Similarly, the other early contributor to the field of mechanical shock is from the civil engineering field of seismic resistant design. In the early days of earthquake engineering, seismic events were generally referred to as transient vibrations since the term shock was not commonly used until the middle of the twentieth century.

Earthquakes can be extremely damaging and cause significant loss of life due to the unpredictable nature of the event and the fact that many structures have historically not been designed to be shaken. Buildings are essentially hollow structures that are largely focused on resisting the ever present and very predictable downward force of gravity. In contrast, earthquakes can impose upward and lateral

C. Sisemore, V. Babuška, *The Science and Engineering of Mechanical Shock*,
https://doi.org/10.1007/978-3-030-12103-7_1

motion on a structure simultaneously. Ironically, minor earthquakes happen every day at locations throughout the world although most of these are so small that they are not even noticed by people. However, large earthquakes are common and deadly enough that significant resources are applied to designing earthquake resistance into new structures and retrofitting existing structures.

Earthquake deaths and property damage are largely confined to occupied buildings or other structures such as bridges and dams. This is because seismic motions themselves are not particularly harmful to people or the earth. Rather, the proximity of people to collapsing structures is the threat.

Shocks are common in transportation systems as well. Rail cars experience shocks during coupling. Everyone knows well the jolt of an automobile hitting a pot hole. Commercial aircraft are designed to withstand some crashes, which are in effect severe shocks. NASA and the FAA conduct research to understand the damage potential of aircraft crashes and develop crash-worthiness design standards.

With the dawn of the space age, the field of mechanical shock expanded to include missiles, rockets, and spacecraft. Space systems are complex systems with many shock-sensitive parts, primarily electronics. Stage separation systems in rockets and missiles use pyrotechnic devices that induce high-intensity, high-frequency shocks, known as pyroshocks, which can damage sensitive electronics. They can also cause functional failures such as the inadvertent disconnection of cables and connectors.

1.1 Introduction to Mechanical Shock

The earliest substantial work on the theory of shock and vibration was *The Theory of Sound* written by Lord Rayleigh and first published in 1877 [2]. At that time the subjects of sound, vibration, and transient vibrations were considered as a single field of study. The term "shock" was not used until much later. Lord Rayleigh's text is still in publication today and contains several current and very applicable design ideas despite or perhaps attributing to its longevity. It was about 50 years before acoustics, vibration, and eventually shock begin to diverge off into separate specialities. The field of shock was even later specializing due to its treatment as a transient vibration.

A shock event is generally described as a dynamic loading whose duration is short relative to the excited system's natural frequency. While not required per the definition, shock events have traditionally been described as having relatively high accelerations. High acceleration is of course a relative measure based on the system's capabilities. Table 1.1 provides a list of some common shock acceleration magnitudes and durations. The data in this table are approximate and are simply intended to provide an overview of the various severity ranges that could be readily expected from mechanical shock loads.

From the simple data presented in Table 1.1 it is clear that the peak acceleration from a shock can vary significantly as can the duration of the shock event. However,

Table 1.1 Approximate common acceleration loads

Operation	Acceleration (g)	Duration (s)
Elevator in fast service	0.1–0.2	1–5
Automobile comfortable stop	0.25	3–5
Automobile crash	20–100	0.1
Aircraft ordinary takeoff	0.5	10–40
Aircraft catapult takeoff	2.5–6	1.5
Aircraft seat ejection	10–15	0.25
Magnitude 6.5 earthquake	0.5	20–30
Underwater shock	40–2000	0.1
Pyrotechnic shock	1000–100,000	0.001

shock loadings usually follow a trend of higher accelerations corresponding to shorter durations, while longer durations have lower peak accelerations. Thus, pyrotechnic shocks have tremendous peak accelerations but for a very short time period. In contrast, earthquakes have relatively low peak accelerations but can last for a relatively long period of time. This is better expressed by the principle of work. Work is energy transferred to or from an object. All mechanical shock events of interest transfer energy to an object, otherwise the study would be purely academic. The definition of work is the scalar product of a force and the distance through which the force acts

$$W = \mathbf{F} \cdot \mathbf{d}. \tag{1.1}$$

Force is given by Newton's second law as the product of mass and acceleration

$$\mathbf{F} = m\mathbf{a}. \tag{1.2}$$

Thus work can be defined in terms of mass, acceleration, and displacement

$$W = m\mathbf{a} \cdot \mathbf{d}. \tag{1.3}$$

From Eq. 1.3 it is apparent that acceleration and displacement are inversely proportional for a given shock input. If the work done to the system is constant and the acceleration is high, then the resulting displacements will be proportionately low. Likewise, if the displacements are large, the accelerations must be significantly lower for work to remain constant. This is a fundamental principal of mechanical shock and is crucial to the design of shock attenuation mechanisms. Reducing peak acceleration into a system is bought at the price of increased deflection. Increased deflection naturally leads to increased durations since the longer the distance, the more time is naturally required to traverse that distance. The only other option available in Eq. 1.3 is to increase the system mass. While this is a very effective solution, it may not be appropriate in many situations.

1.2 History of Shock Engineering

Advances in mechanical shock engineering are derived primarily from military applications and seismic engineering. In military applications, shock failures were generally treated with the dual approaches of making the part stronger and separating the delicate parts from the damage source. It was not until the last 100 years or so that a thorough study of shock environments in military applications was undertaken.

Likewise, the field of civil engineering also suffered from an inability to thoroughly understand the forces working against structures. This was compounded by the desire for economical buildings and buildings made with local materials. This mentality yielded extremely diverse construction methodologies across the globe. However, even in antiquity engineers knew that few things lasted as long as well-fitted stone. It is no accident that most of the ancient structures still standing are made of large slabs of fitted stone stacked one on top of another—most notably the pyramids of Egypt and Central America. These are naturally resistant to seismic events since they have some minor flexibility, are extremely heavy, and are not hollow which prevents them from collapsing upon themselves.

As the understanding of shock phenomena progressed, so did the efforts to better designs. As equipment and systems became more complex, they also became more sensitive to shock loadings. Further, the complex equipment was now being called upon to operate in ever increasingly harsh environments. For example, a heavy frigate during the War of 1812 might carry 44 main cannon plus a complement of smaller weapons. If a cannon failed, another one was readily available. In contrast, USS *Connecticut,* BB-18, the flagship of Roosevelt's Great White Fleet carried a main armament of four 12-in (305 mm) guns, eight 8-in (203 mm) guns, and twelve 7-in (178 mm) guns along with some smaller weapons. While *Connecticut* class battleships carried a formidable array of weaponry, the primary armament had been reduced from 44 guns to 24. Thus, each of those guns was required to be more reliable. This trend is further seen with the USS *Iowa,* BB-61, carrying nine 16-in (406 mm) main guns and nothing else larger than a 127 mm gun. Again, more is being demanded of the equipment. This trend has played out completely with USS *Arleigh Burke,* DDG-51, class destroyers carrying a single fully automatic 127 mm (5-in) main gun complemented by a few smaller caliber weapons and a host of missiles. Admittedly, the naval threat has changed significantly in the last 200 years and battleships no longer fight it out in blue-water engagements; however, the application to mechanical shock is clear. As the number of systems decreases and the complexity increases, it is increasingly necessary to understand the environments in which the systems are required to operate.

Likewise, the civil engineering community faces a similar challenge. If a farmer builds a simple house in the middle of an open field and it collapses as a result of an earthquake, the ramifications are minor to everyone except the farmer and his family. In contrast, if a modern skyscraper were to collapse as a result of an earthquake, the loss of life and property can be disastrous for the whole community. Just like the concentrations in the field of naval gunnery, the concentration of people into large cities made up of high-rise buildings increases risk substantially and hence the need to better understand the environments.

In more recent times, the rise of the aircraft and spacecraft industries has significantly furthered the study of mechanical shock. The aircraft industry is naturally focused on the safety of passengers and the reliability of its equipment. After all, poor reliability in the aircraft industry is usually associated with a plane crash. Aircraft are exposed to numerous shock environments with every takeoff and landing, yet their parts must withstand the repeated shock loads.

The spacecraft industry has further pushed the envelope of the field due to the extreme difficulty of escaping the Earth's atmosphere. Multi-stage rocket motors, stage separation pyrotechnics, thermal shock, and Mach transitions are among the many sources of mechanical shock in spacecraft. Here again, the cost of space travel is very high; therefore, the system reliability needs to be proportionately high. Spacecraft systems often have the added complexity that once they leave the launch pad, there is frequently no person there to intervene if something goes wrong and even if there is, the reaction times are usually too short for meaningful interventions.

1.2.1 Naval Shock

Prior to World War II, the major causes of shock damage to military ships were direct hits from enemy guns, torpedoes, and the shock resulting from firing their own guns. The latter source of damage is quite unacceptable since this damage is self-inflicted and generally occurs when full functionality is critically needed. As a result, the early days of U.S. Navy shock testing were designed to harden equipment to withstand the effects of their own weapons. The earliest solutions were very practical and generally involved relocating critical equipment as far from the guns as practical. Figure 1.1 shows a photograph of USS *Missouri* firing main guns during the Korean War. Notice the size of the disturbance on the water surface despite the elevated firing angle. The blast over pressure from the main gun discharge is obviously quite severe.

After World War I, the U.S. Navy designed and built a simple shock test machine and began a program of shock hardening ship systems. In those days, the mechanics of shock were poorly understood as the field was very new. The methodology used at the time was to collect damaged equipment from the fleet that engineers thought should have remained functional and test new equipment on the shock machine until the damage was replicated. At this point, new designs were sought and tested using the same inputs until the upgraded system survived the shock test. This experimental approach to shock design can be costly, but it is also effective.

Around the time of World War II, two developments converged necessitating a greater study of the mechanical shock problem. The first was the development of the non-contact underwater mine. These mines often exploded some distance from the ship creating little to no structural damage but frequently incapacitating the ship.

Fig. 1.1 USS *Missouri,* BB-63, firing 16-in main guns during the Korean War (U.S. Navy photograph)

This was a direct result of the second major development—the increased reliance on electronic and other complex equipment for the ship's operation. Mine explosions apply considerable force to the ship due to the large hull surface area exposed to the pressure pulse. This in turn transmits tremendous energy to the ship and its various systems and components. Incapacitation of the ship is often a result of non-structural failures such as power failures or dislocation of critical equipment. These functional failures can be as damaging as structural failures since an incapacitated warship is helpless to defend itself against further assaults. However, the problem was not necessarily limited to shocks caused by enemy weapons. It was not uncommon for shock damage to result from firing of their own guns. This was especially true with the larger caliber weapons, specifically the 16-in guns on the battleships.

As a result of shock hardening efforts during World War II, the U.S. Navy sponsored the first organization dedicated to shock and vibration. This organization was known as the Shock and Vibration Information Center (SVIC). While the organization has changed somewhat over the years, including the name, the organization continues today working to fulfill its original mission.

The U.S. Navy still tests smaller shipboard equipment on what has become known as the light-weight shock machine. With the desire to test larger systems, the medium-weight shock machine was developed. Larger still are the standard and large floating shock platforms, which are simply barges to which equipment is mounted with depth charges used to apply shock loads to the equipment. The purpose being to test every critical piece of ship machinery to a defined shock input. One interesting result of the use of these specialized test machines is that the applied shock loading does not necessarily resemble the shock from any particular wartime scenario. In reality, it would be impossible to test every possible loading scenario for a given component in a particular installation location and orientation. Therefore, the primary goal of the navy shock test program is to demonstrate a minimum level of robustness more so than to qualify to a particular environmental scenario. This often leads to a confusion among system developers claiming that the shock loads are unreasonable. In some cases they may be, but in other cases they may not be. The primary goal of navy shock testing is to ensure that if the ship's hull is not substantially breached and a reasonable number of the ship's crew remains capable of fighting, then the systems that they depend on to carry-on the fight should still be operational after a shock event.

The U.S. Navy commonly refers to the non-contact underwater mine explosion as a near-miss shock event. The term is intended to distinguish events that penetrate the ship's hull from events that typically do not penetrate the hull but nevertheless impose considerable excitation on the ship. One of the primary reasons for the near-miss shock requirement is that the U.S. Navy has suffered significant casualties as a result of underwater mines. Since World War II, 15 of the 19 U.S. Navy ships sunk or seriously damaged were the victims of sea mines. From those 15 U.S. ships, there were 183 people killed and another 81 wounded. Most of these casualties occurred during the Korean War, but there were two incidents during the Vietnam War, one during the Iran–Iraq War, and two during the first Gulf War. Table 1.2 gives a list of all U.S. Navy ships sunk or seriously damaged by sea mines since World War II.

Table 1.2 U.S. Navy sea mine casualties, Korean War to present

Ship	Date	Casualties
USS Brush (DD-745)	26 Sept 1950	19
USS Magpie (AMS-25)	29 Sept 1950	21
USS Mansfield (DD-728)	30 Sept 1950	53
USS Pirate (AM-275)	12 Oct 1950	6
USS Pledge (AM-277)	12 Oct 1950	7
USS Partridge (AMS-31)	2 Feb 1951	14
USS Walke (DD-723)	12 June 1951	61
USS Ernest G. Small (DDR-838)	7 Oct 1951	27
USS Sarsi (ATF-111)	27 Aug 1952	2
USS Barton (DD-722)	16 Sept 1952	11
USS Westchester County (LST-1167)	1 Nov 1968	26
USS Warrington (DD-843)	16 July 1972	0
USS Samuel B. Roberts (FFG-58)	14 April 1988	10
USS Tripoli (LPH-10)	18 Feb 1991	4
USS Princeton (CG-59)	18 Feb 1991	3

The primary standard for U.S. Navy shock design is MIL-DTL-901E [3]. This military specification has gone through several revisions since its inception shortly after World War II. The most recent revision, Revision E, was released in 2017. Most U.S. Navy shipboard systems are required to satisfy the shipboard shock requirement defined by MIL-DTL-901E although in practicality, the systems are usually required to meet the specification in force at the time of the contract definition or deployment. Therefore some systems in the fleet may be found qualified to Revision E, although many older systems are still qualified to the MIL-S-901D or MIL-S-901C revisions. In general, the differences between the three revisions are relatively minor.

1.2.2 Civil Engineering Shock

While there is an ever growing interest in civil engineering for the development of blast resistant structures, the primary focus of shock in civil engineering has historically been on earthquake resistant design. In the early days of construction, building materials were locally sourced and high-strength steel girders were not available so many aspects of seismic engineering were simply not practical. Unfortunately, this has led to undesirable situations throughout history. For example, a major earthquake occurred in 1976 near T'ang-shan, China. The coal-mining and industrial city was almost completely destroyed by the early morning earthquake with an estimated death toll exceeding 240,000 people. Most of the casualties were due to the collapse of unreinforced masonry homes where the people were asleep [4]. Earthquake damage can be significant whenever or wherever it occurs. Figure 1.2 shows a photograph of collapsed buildings in Imperial, California, as a result of the May 18, 1940 El Centro earthquake. Four people died in these buildings.

Fig. 1.2 Collapsed buildings in Imperial, California, after the May 18, 1940 El Centro earthquake. (U.S. Coast and Geodetic Survey photograph)

The focus of earthquake resistant design is always on the next big earthquake and how to ensure that structures survive adequately to prevent loss of life and property. One of the most recent well-known earthquakes is the March 11, 2011 Tōhoko earthquake which occurred off the coast of Japan. This magnitude 9.0 earthquake resulted in significant structural damage, unleashed a tsunami, resulted in the Fukushima nuclear power plant meltdown, and caused considerable loss of life and property destruction. Certainly the damage from this earthquake was not just associated with the earthquake since the resulting tsunami also did considerable damage. While earthquake secondary effects can result in significant damage through landslides, tsunamis, fires, and fault ruptures, the greatest losses have always been through the failure of man-made structures during ground shaking.

The list of catastrophic earthquakes is long and well distributed around the globe. Unfortunately, earthquake resistant design is predominately focused in the wealthiest countries. This is a direct result of risk calculations. If people do not have money for basic necessities, then they will not have money for what is often perceived to be an unlikely or unfortunate accident. In contrast, wealthier nations have more time and money to devote to managing risks and trying to guard against unlikely or unfortunate events.

The greatest advances in seismic resistant civil engineering have come with the understanding that earthquakes impart both upward and lateral loads on a structure. While it is impossible to prevent the most severe forms of earthquake damage, adding braces to thwart building wall collapse goes a long way to increasing survivability. Cross-bracing, diagonal bracing, shear walls and such are all important aspects of earthquake engineering. Anything that can be incorporated to help keep structural walls upright during an earthquake greatly increases the chances that the building and occupants will survive.

One of the greatest advances in the field of earthquake engineering is the use of structural steel members as the primary load-bearing elements. Structural steel is an ideal building material since it typically exhibits significant strain capacity beyond its yield point prior to rupture. Thus, even if the building is a total loss, the large deformations of the structural steel members will often absorb considerable energy during their deformation and also prevent total collapse of the structure, allowing occupants more opportunity to escape.

Failure of the primary structure is not the only risk that civil structures face in an earthquake. The structural integrity of modern structures has been improved to the point that buildings can withstand fairly significant shaking. However, just like in the naval examples, a building may be compromised if mechanical systems fail. The 1994 Northridge earthquake caused failures in sprinkler systems. Non-structural water damage in three hospitals was severe enough to close them for a week after the earthquake [5].

1.2.3 Aircraft Shock

In June 2017 a bird struck one of the engines of a passenger aircraft shortly after taking off from Chicago's O'Hare International Airport. The engine caught fire as a result of the shock and ensuing damage. Fortunately the airplane was able to return to the airport and land safely [6]. A similar and more widely publicized incident occurred when a flock of birds encountered US Airways flight 1549 in January 2009 resulting in the loss of thrust in both engines. That airplane was forced to ditch in the Hudson River where all of the passengers and crew were rescued [7]. Bird strikes are a somewhat common occurrence in the industry, but the risk to flight safety makes them a significant concern. Many times the damage to the aircraft is minor; however, the mechanical shock insult can destroy engines as in the case of the June 2017 incident. Other significant concerns are impacts with the windshield. Many cases have been documented of birds penetrating the windshield and entering the cockpit. This obviously has serious ramifications for pilot safety and the safety of the aircraft in general.

Other forms of aircraft shock are landing gear shock, either from hard landings on the wheels and wheel suspension systems or tailhook gear for naval aircraft. Commercial aircraft are designed to absorb energy during a crash to reduce the shock transmitted to passengers. NASA and the FAA perform research to understand the crash-worthiness of aircraft to improve design standards. The NASA Langley Research Center has a very large gantry frame shown in Fig. 1.3 from which they drop full-size private airplanes [8]. Figure 1.4 is before and after pictures of a commercial passenger aircraft fuselage drop test at the NASA Langley Crash Test Facility. The fuselage was dropped from 14 ft (4.3 m) and hit the ground at 30 ft/s (9.1 m/s). Notice the damage to the fuselage and the floor.

In addition to these obvious sources of mechanical shock, the aircraft is also a pressure vessel. As the aircraft transitions through atmospheric layers, the pressure on the aircraft increases or decreases. This change in pressure can cause joints to slip and pop sending shock pulses through the aircraft structure. These can also be quite disconcerting to the passengers even if they are not a significant concern for the structure.

Military aircraft also have similar conditions as naval vessels. Machine gun firing, missile launches, and bomb ejections all induce shock loads into the air frame. These shock loads are not large and damaging in and of themselves, but they are repetitive and can contribute to low-cycle or high-cycle fatigue damage.

1.2.4 Space Vehicle Shock

Space vehicles experience a series of minor and substantial shock events during a typical flight. The launch phase is the most severe environment that a launch vehicle and payload will experience. During this phase, shocks have the potential to cause

Fig. 1.3 A Cessna 172 suspended in the NASA Langley Research Center Crash Test Facility (Photo Credit: NASA Langley Research Center)

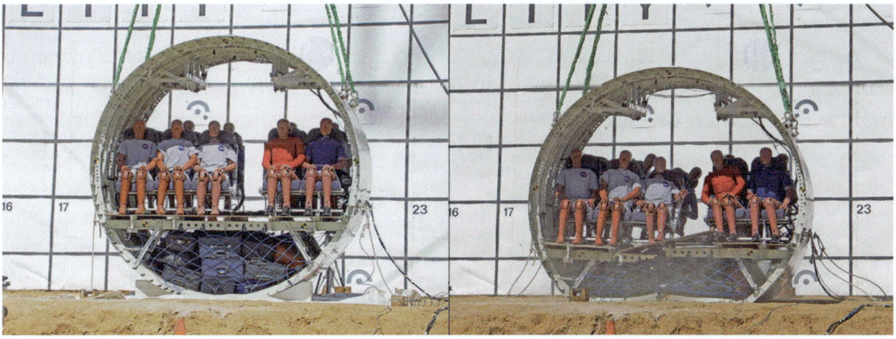

Fig. 1.4 Regional jet fuselage tests at the NASA Langley Research Center Crash Test Facility (Photo Credit: NASA Langley Research Center) [9]

mission threatening damage. Launch shock is an obvious concern with the sudden ignition of the first-stage rocket motor. In addition to launch shock, stage separations are typically conducted with pyrotechnics that impart substantial high-frequency energy to the area around the pyrotechnic device. The pyroshocks do not cause failure of the primary structure, but they can cause functional failures of shock-sensitive electronic equipment. Stage separations impart low-frequency shocks due to the momentum change from ejecting the spent motor stage and igniting the next state motor. There are also aerodynamic shocks resulting from the transition from sub-sonic to super-sonic velocities. Once the system is exoatmospheric, steering is typically performed by small rocket boosters which are activated in a pulsing fashion. These steering control motors can generate hundreds or more small shocks in rapid succession.

In February 2003, the space shuttle *Columbia* disintegrated on reentry. The accident was the result of a mechanical shock caused by the impact of foam insulation from the shuttle's external fuel tank. A piece of the fuel tank's foam insulation broke off during launch and struck the left wing of the space shuttle. During reentry, the damaged wing allowed hot atmospheric gases to enter the wing, destroying the wing's structure. Failure of the wing led to rapid failure of the entire vehicle and the loss of all aboard.

The space shuttle *Columbia* tragedy is an example of shock induced severe structural damage, which is very rare. More common is shock induced functional failure. Moening [10] compiled data on 85 flight failures from the 1960s to the 1980s. Approximately 63 of the 85 flight failures were shock induced. The most tragic of these occurred in 1971 on the Soviet Soyuz-11 capsule. The separation of the reentry capsule from the orbiting vehicle was accomplished by firing 12 explosive bolts. These shocks inadvertently opened a valve that allowed all the oxygen to escape, killing the three cosmonauts. The aerospace industry has learned a lot since then and pyroshock failures occur much less frequently. Increasing knowledge of the potentially severe consequences of pryoshock events has led to increased safety.

After a satellite reaches orbit, its mechanical shock environment is very benign compared to the launch environment; however, the sensitivity to shocks may be higher. The source of on-orbit shocks is usually temperature changes. These can cause functional failures or affect satellite operations. For example, the Hubble Space Telescope was subjected to thermally induced shocks whenever it entered or exited the eclipse phase of its orbit and the sudden temperature change caused a mechanical snap of the solar arrays. This snap was severe enough to affect the systems pointing control system [11]. Even though the disturbance seemed small (in the range of 10 mN m) the Hubble space Telescope is a sensitive and precisely pointed optical instrument and these shocks affected operations. Various mitigations were implemented and in December 1993 the solar arrays were replaced as part of the first Hubble servicing mission (STS-61).

1.3 Effects of Shock on Systems

The effects of shocks on systems can be as varied as the systems themselves. The resulting damage can also be equally varied. The most obvious damage is a structural failure. Structural failures present as yielding, cracking, rupture, or other forms of rapid system disassembly. These failures are easy to see but not always easy to predict or correct. In contrast, functional failures are often more prevalent in complex systems. A functional failure is often just as damaging, especially if the system cannot readily be reset. There is a another failure type, sometimes referred to as a secondary failure, that is caused when an unrelated piece of equipment or a component comes loose during the shock event and damages a critical component or piece of equipment. Designers can often be focused only on their component and ensuring their system functions as intended only to have their system damaged or destroyed by some unrelated item that was unexpectedly turned into a missile as a result of a shock event.

1.3.1 Structural Failure

Structural failures from shock are usually easy to see. The screen shatters on a cell phone when the device is accidentally dropped. Circuit boards can be cracked by rough handling of electronics during shipment. A car fender is dented after an impact with another vehicle. Structural components rupture after ballistic impact. All of these scenarios are common in that they result from material over-strain.

Structural failures can be classified into one of two failure types: first passage failures or fatigue failures. First passage failures are the result of a single shock event straining a structural member beyond its strain limit resulting in immediate rupture. This type of shock failure is quite common given the traditionally high shock loads to which equipment can be subjected.

Shock fatigue failures are often thought to be less common but the truth depends heavily on the application. Under a shock load, a system experiences some strain usually resulting from the initial impulse followed by a decaying oscillation where each successive oscillation generates less strain than the previous oscillation. As a result, numerous large amplitude loading and unloading cycles can be accumulated in a relatively short time period although the focus is almost always on the initial transient. It is often repeated, but not necessarily true, that if a structure survives one shock, then it will survive a few more shocks of the same magnitude. While this is frequently true, if damage is initiated with the first shock, subsequent shocks can propagate that damage with alarming rapidity.

1.3.2 Functional Failure

On November 21, 1939, a magnetic mine detonated approximately 100 m from the bow of the British Light Cruiser HMS *Belfast* steaming in the Firth of Forth. While little direct damage was done to the hull, significant structural damage was done due to hogging of the ship. The resulting whipping caused the upper decks to fracture and the ship's back was broken. The resulting damage took several years to repair with the ship recommissioned in November 1942, However, more interesting than the significant structural damage was the list of functional failures experienced as a result of the shock. Circuit breakers opened on two generators causing all the lights to go out and leaving the ship in complete darkness. All of the main circuit breakers opened either from the shock or from lack of voltage. All low-voltage power was lost due to the opening of a generator motor starter. Most interesting was that about 70% of all inter-ship telephones jumped off their hooks resulting in significant confusion and the inability to relay information and orders. While the structural damage to the *Belfast* was substantial, the functional failures would have prevented any kind of meaningful response had further threats been imminent [12]. Regardless, the functional damage significantly impacted the ship's operation and recovery.

As discussed previously, failures of a building's mechanical systems can cause numerous functional failures. Circuit breaker trips are a very common form of functional failure that have been repeatedly demonstrated through shock testing and shock events. Lighting is another related failure in naval vessels. Most modern naval vessels have little to no natural light within the ship. If the lights or power to the lights fails, then crew operations are significantly hindered. Electrical connectors are another common source of functional failures. Recent testing on a fiber-optic interface cabinet showed that after a shock test the entire cabinet lost power. When the cabinet door was opened the reason for the total failure of the system was that the power chord relied on friction to hold the connector into its socket. Friction interfaces are notorious for slipping loose under shock loads. A simple cable tie fixed that particular functional failure.

There is a very simple experiment that can be performed anywhere to demonstrate the danger of a functional failure. Figure 1.5 shows a photograph of a common household circuit breaker. If a shock is applied perpendicular to the switch mechanism such as shown on the left side of Fig. 1.5 the breaker will remain in the on position. However, if a light shock is applied in a direction parallel to the switching mechanism as shown on the right side of Fig. 1.5, then the switch will readily trip to the off position. A very small shock load is required since this experiment can be conducted by hand as shown. Nevertheless, the consequences are catastrophic as a system without power is generally useless until someone repairs the system.

While functional failures are often demonstrated in the laboratory, equipment designers are often slow to learn this repeatedly taught lesson. Part of the reason for this difficulty is the lack of an intuitive understanding of mechanical shock

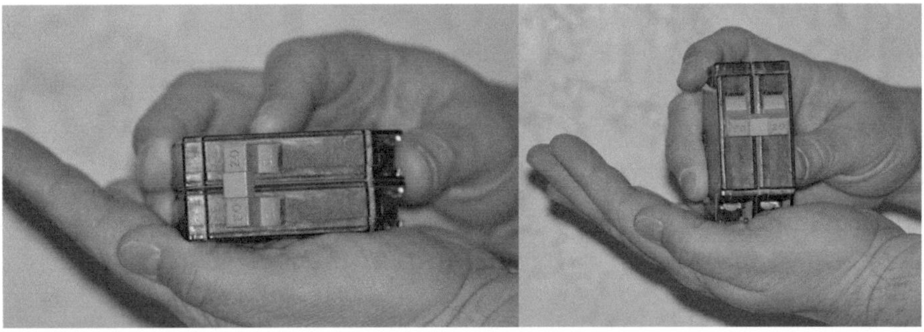

Fig. 1.5 Circuit breakers with shock applied perpendicular to the mechanism and shock applied parallel to the mechanism

phenomena. When a part is designed and fabricated, structural members and joints often appear strong, robust, and tight. However, it is very difficult to intuitively speculate what may go wrong when the component is suddenly subjected to 100 g or more shock loading. It is difficult to imagine what a 0.5 kg circuit board may do when it suddenly appears to weigh 50 kg for a brief moment. Connections that seem perfectly mated will suddenly work themselves loose. Bolts that were properly torqued can stretch and move. Nothing is rigid under shock loads and nothing is guaranteed to remain where it was left, unless it is firmly attached with adequate restraints. Even then it is likely to slip.

1.4 The Need for Dynamic Analysis

One of the simplest approximations used for shock design is to apply a static acceleration load to the component. Unfortunately, a static analysis can lead to the wrong conclusion when approximating dynamic loads. Figure 1.6 shows a sketch of two cantilever beams, one long and one short, with the same lumped mass at the free end. These two systems will be used to demonstrate the dangers of only considering a static acceleration when designing for shock. For this calculation it is assumed that the long, thin cantilever beams are massless and the system mass is concentrated at the free end.

From Fig. 1.6, a static acceleration analysis predicts that the longer beam fails first because the bending moment at the cantilever beam base is larger with a longer beam. The applied load is given by Newton's law, $F = ma$ and the bending moment is simply $M = FL_i$. Since beam two is longer than beam one and the tip masses are equal

$$mL_1 > mL_2. \tag{1.4}$$

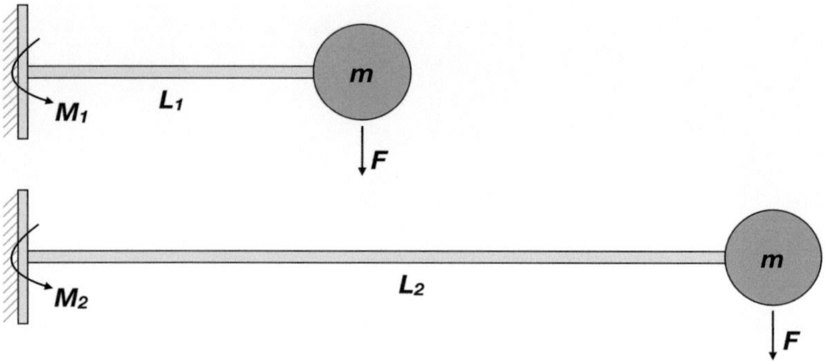

Fig. 1.6 Long and short cantilever beams subjected to a simple transverse shock load

Since the bending moment is greater, the stress is greater and the longer beam is predicted to fail first.

However, solving the same problem using the dynamic equations of motion leads to the opposite conclusion. The equation of motion for an undamped system is given by the usual equation:

$$m\ddot{y}(t) + ky(t) = 0, \tag{1.5}$$

where m is the lumped mass at the beam's tip, k is the beam's transverse bending stiffness, and $y(t)$ is the vertical displacement. The initial conditions for the solution of the differential equation are

$$\begin{aligned} \dot{y}(0) &= V_0 \\ y(0) &= 0. \end{aligned} \tag{1.6}$$

These initial conditions represent a system with an initial velocity, downward in this example, impacting a stiff surface at the origin. The transverse bending stiffness of a slender beam is given by

$$k = \frac{3EI}{L^3}, \tag{1.7}$$

where E is the modulus of elasticity, I is the cross-sectional area moment of inertia, and L is the beam length. The solution to the equation of motion in Eq. 1.5 is

$$y(t) = A\sin(\omega t) + B\cos(\omega t), \tag{1.8}$$

where $\omega = \sqrt{k/m}$ and the constants A and B are determined from the initial conditions. Substituting the initial conditions given in Eq. 1.6 into Eq. 1.8, the displacement $y(t)$ is given by

$$y(t) = \frac{V_0}{\omega}\sin(\omega t), \tag{1.9}$$

and the acceleration is given by

$$\ddot{y}(t) = -\omega V_0 \sin(\omega t). \tag{1.10}$$

The maximum acceleration occurs when the sine term is equal to -1, thus

$$\ddot{y}(t)_{max} = \omega V_0. \tag{1.11}$$

Substituting Eq. 1.11 into Newton's law gives

$$F = m\ddot{y}(t)_{max} = m\omega V_0. \tag{1.12}$$

The purpose of this derivation was to evaluate the effect of beam length on the dynamic shock loading. As such, Eq. 1.11 needs to be rearranged in terms of the beam length. The beam's length only appears in the stiffness term (recalling again the original assumption of a massless cantilever beam) which is embedded in the frequency, ω. Substituting for ω yields

$$F = m\omega V_0 = mV_0\sqrt{\frac{k}{m}} = V_0\sqrt{km} = V_0\sqrt{\frac{3EIm}{L^3}}. \tag{1.13}$$

Likewise, the bending moment at the cantilever beam base is

$$M = FL = V_0 L\sqrt{\frac{3EIm}{L^3}} = V_0\sqrt{\frac{3EIm}{L}}. \tag{1.14}$$

Thus, the bending moment is an inverse function of the beam length. The higher fundamental beam frequency of the shorter beam is a greater contributor to the shear force and bending moment than the beam length. This is directly opposed to the static solution which concludes that a longer beam has a higher bending moment for the same applied force. This somewhat counterintuitive result is actually observable in many situations. For example, it is easier to fail a short bolt than a long bolt during installation even though the dynamic loading is usually relatively slow. Short shafts almost always fail before long shafts when equally shocked. This simple example shows the need for dynamic analysis when designing and analyzing for shock. The preceding example was derived from a similar example published by Scavuzzo and Pusey [13].

1.5 Summary

Mechanical shock has been a concern for as long as people have found themselves dropping things only to see them break when hitting the ground. However, the science of mechanical shock is a relatively recent field of study with its formal beginnings only about 100 years ago. Research in mechanical shock was originally derived from the shock loads generated by the firing of a ship's own guns. As systems became more complex, operational environments more severe, and reliability needs increased, the study of mechanical shock has also increased.

The goal of this book is to present the fundamentals of mechanical shock science and engineering. The science of mechanical shock involves understanding of the basic physics of the event: how the dynamics of the system interplay with the high-speed transient nature of the excitation and the numerical methods for quantifying the shock event and its damage potential to various structures, systems, and components. The engineering aspect of mechanical shock involves understanding the various shock test methods and the practical aspects of designing systems to survive shock loads.

Unfortunately mechanical shock is a frequently misunderstood field of engineering. It is frequently treated as a minor branch of vibration engineering and often not given significant detail in its treatment. Most structural dynamics texts focus on the vibration aspect of dynamics with only a cursory nod to the shock problem. The problem is carried through into the practical world with many system designers designing their components to pass vibration testing and simply hoping that those same designs will then pass shock testing. This is clearly not an ideal approach to the problem of what is frequently a design limiting load condition.

This book intends to provide a history of mechanical shock, including the origins of the field, the development of test techniques, the justification for testing, and discussion of the many different types of shock testing performed today. The book also presents an in-depth discussion of structural dynamics as it relates to the field of mechanical shock and the mathematics of analyzing shock loads on systems. Finally, details of how to develop and specify mechanical shock tests for various shock environments are presented.

Problems

1.1 Designing a system against structural failures is understood, although not always performed correctly. Designing against functional failures is a more difficult challenge. List some ways that a system can be designed to prevent or mitigate functional failures.

1.2 Another way to guard against catastrophic structural failures is to allow some level of permanent deformation to occur during the shock event. Why is this an efficient methodology for preventing catastrophic failures? What material property is required for this method to succeed?

1.3 The need for dynamic analysis was derived using a short and long cantilever beam in bending. Derive the relationships for a short and long shaft in torsion. Are the results similar?

1.4 Derive the dynamic relationships for a short and long axially loaded beam. Are the results again similar to the beams in bending?

1.5 The dynamic analysis derivation presented here also identifies a fundamental precept in designing for mechanical shock survivability—larger components tend to tolerate shock better than smaller components. Why?

References

1. Roosevelt, T. (1882). *The naval war of 1812*. New York: G. P. Putnam's Sons.
2. Rayleigh, J. W. S. (1877). *The theory of sound*. London: Macmillan.
3. U.S. Department of Defense. (2017). *Detail Specification, Requirements for Shock Tests, H.I. (High-Impact) Shipboard Machinery, Equipment, and Systems*, MIL-DTL-901E, Washington, DC, 20 June 2017.
4. Bolt, B. A. (1990). Earthquakes: Effects of earthquakes. In *The new encyclopedia Britannica*, Vol. 17, pp. 612–614. Chicago: Encyclopædia Britannica, Inc.
5. Fleming, R.P. (1998). *Analysis of fire sprinkler system performance in the Northridge earthquake*, NIST-GCR-98-736, January 1998.
6. Owen, J. (2017). *United flight returns safely to O'Hare after bird strike*, 1 June 2017. Chicago: Chicago Sun-Times.
7. National Transportation Safety Board. (2010). *Aircraft accident report, loss of thrust in both engines after encountering a flock of birds and subsequent ditching on the Hudson River US airways flight 1549 airbus A320-214*, N106US, Weehawken, New Jersey, January 15, 2009, NTSB/AAR-10/03, Adopted May 4, 2010.
8. Odds, E. (2005). *Watch NASA Crash a Perfectly Good Plane in the Name of Science*, July 30, 2005.
9. Barnstorff, K. (2017). *NASA and FAA Put Dummies and Baggage to the Test for Airplane Safety*, March 2017.
10. Moening, C. J. (1984). Pyrotechnic shock flight failures. In *Proceedings of the Aerospace Testing Seminar* (Vol. 8, pp. 95–109).
11. Foster, C. L., Tinker, M. L., Nurre, G. S., & Till, W. A. (1995). *The solar array-induced disturbance of the Hubble Space Telescope pointing system*. NASA Technical Paper 3556, NASA Marshall Space Flight Center, May 1995.
12. Bort, R. L., Morris, J. A., Pusey, H. C., & Scavuzzo, R. J. (2001). *Practical shock analysis & design*. Course Notes. Dresher, PA: Society for Machinery Failure Prevention Technology.
13. Scavuzzo, R. J., & Pusey, H. C. (2000). *Naval shock analysis and design*. In *SVM-17*. Falls Church, VA: The Shock and Vibration Information Analysis Center, Booz-Allen and Hamilton, Inc.

Chapter 2
Common Mechanical Shock Environments

Mechanical shock events are as varied as the systems they influence. Shocks can occur in many forms, with many different amplitudes, durations, and other dynamic parameters. While mechanical shocks can be varied in their sources and characteristics, many similarities do exist. For this reason, the analysis techniques for one type of mechanical shock are applicable to many types of mechanical shock events. Additionally, mechanical shock events can be divided into just a few general types of shocks.

2.1 Impact Shock

Everyone has experienced the pain of an impact shock event, either physically from taking an unfortunate tumble or personally from dropping a valuable item only to have it shattered on impact with the ground. Impact shock is one of the most common forms of mechanical shock insults, because it happens every day. Handling shocks occur frequently too. Boxes are dropped in shipping. They are knocked off tables or countertops. Roads are not smooth; they have potholes, expansion joints, and other obstacles. Parts are dropped or tossed into storage bins. Other parts can be accidentally or carelessly dropped onto sensitive parts.

When an object is dropped from a particular height, it starts with a given potential energy proportional to its height above the impacting surface. At the moment of impact, all of the object's potential energy has been converted to kinetic energy and the object has obtained a certain impact velocity given by the usual conservation of energy relations.

$$mgh = \frac{1}{2}mv^2 \tag{2.1}$$

C. Sisemore, V. Babuška, *The Science and Engineering of Mechanical Shock*,
https://doi.org/10.1007/978-3-030-12103-7_2

Here the variables have the usual meanings: m is the object's mass, g is the gravitational acceleration, h is the drop height, and v is the impact velocity. It is also apparent from the conservation of energy that during an elastic impact the object must decelerate and come to rest momentarily before the velocity vector direction reverses and the object rebounds. The depth of impact can be determined by a second conservation of energy relation.

$$\frac{1}{2}mv^2 = \frac{1}{2}kx^2 \tag{2.2}$$

In this equation, x is the compression distance and k is the combined stiffness of the impacting component and the impacting surface. The combined stiffness is used here because the collision is effectively a pair of springs in series. The first spring is the stiffness of the falling part and the second spring is the stiffness of the impacting surface. The equivalent stiffness of two springs in series is given by:

$$\frac{1}{k_{equivalent}} = \frac{1}{k_{part}} + \frac{1}{k_{surface}}. \tag{2.3}$$

In the above equation, it is obvious that if one of the two springs is significantly softer than the other, then that stiffness dominates the equation and the combined stiffness is soft. If both springs have a similar stiffness, then the combined stiffness is approximately half of the individual values.

The equivalent stiffness of the impacting bodies defines the peak acceleration level developed during the collision. Per the energy equation above, a high stiffness will result in a small displacement. Thus, the impacting object is brought to an abrupt stop resulting in high accelerations. Likewise a low stiffness comes at the expense of larger displacements and yields correspondingly lower accelerations. However, the equations show that the relationship is not linear, but rather is a squared function of displacement.

Figure 2.1 shows a photograph of a set of glass goblets in a box. The box contains no special padding or other shock mitigation except for a set of cardboard dividers between the glasses. The purpose here is clearly not to resist all forms of shipping damage as the box could be easily crushed. Rather the dividers are present to prevent a glass-on-glass impact from occurring. The reason is given by Eq. 2.2 and the work–energy relationships given in Chap. 1. Glass is extremely stiff and also extremely brittle. A glass-on-glass impact has high stiffness and low displacement resulting in very high accelerations. In this case, the high acceleration results in cracks and brittle failure of the pieces. The cardboard, while not very substantial, does offer a significant reduction in the impact stiffness and a corresponding reduction in the acceleration loads.

Another option for preventing the glass-on-glass impact is shown in Fig. 2.2. This figure shows a set of dessert dishes packed in foam dividers. The foam provides additional cushion on the top and the bottom dish surfaces in addition to the cardboard box. The foam also prevents the glass dishes from touching each other

Fig. 2.1 Glass goblets shipped in a cardboard box with dividers to prevent impacts between glasses

Fig. 2.2 Dessert bowls shipped in a box with foam dividers to separate bowls

during shipment. In this way, the very stiff glass-on-glass impact is avoided, helping the dishes to arrive intact to the customer.

The primary goal of shipping containers is to provide protection for the contents during transportation and handling. Transportation environments can be quite harsh. There are two primary goals of shipping containers: first to protect the contents from outside elements and second to protect multiple internal items from each other. The outer shipping container can be quite simple, as is often the case with corrugated cardboard boxes or it can be extremely robust depending on the nature of the

Fig. 2.3 Musical instruments are usually transported in padded cases for protection

contents and their value. Internal packaging can also run the spectrum from simple paper or wrapping material up to highly specialized internal isolation systems, again in proportion to the value. However, in all cases the goal is the same, to make the shock trade between displacement and acceleration by choosing the stiffness accordingly.

Figure 2.3 shows a photograph of a musical instrument in its case. The instrument case is padded on all sides. Here again, the stiffness is selected to limit the acceleration imparted to the instrument. In this example, the case padding is also a tight fit against the instrument. The purpose is not to limit displacement but rather to control the displacement. If the instrument were a loose fit within the case, then displacement would be larger overall but the sudden stop when the sliding instrument struck the case side would be more damaging and result in higher accelerations than limiting displacements with a tight fit against a padded case. It is for this reason that shock isolation systems typically require tighter tolerances. The abrupt stop experienced when a slipping joint reaches its tolerance limit can sometimes do more damage than if the joint had not slipped at all.

2.1.1 *Transportation Shock*

Transportation is another very common source of impact shock. Every manufactured component ends up being transported at some point in its life cycle.

Parts are transported from their place of manufacture to the place where they are assembled into systems. Systems are transported to their point of use. Many military and aircraft systems are in a near constant state of transportation. Over-the-road transportation has pot-holes, expansion joints, debris, and other normal road hazards that can all produce impact shocks of varying severities. In addition, many systems are designed to meet the demands of accident scenarios.

One of the most common ways of mitigating transportation shocks is through the use of vehicle suspension systems. Every roadworthy vehicle has some form of suspension system to absorb and dissipate shocks caused by discontinuities in the road's surface. Here again, the primary purpose of vehicle suspension systems is to trade acceleration for displacement.

Figure 2.4 shows two common types of vehicle suspension systems. The photograph on the left side of Fig. 2.4 shows a leaf spring system with a separate shock absorber. The long leaf springs, made of spring steel in this case, are connected to the vehicle frame at the front and rear with the axle attached at the spring's center. The stiffness of a leaf spring is defined by the equation of a beam in bending. The leaf spring shown in Fig. 2.4 has two active leaves on the top and one overload spring on the bottom. The overload spring is used to increase the stiffness of the leaf spring system when the vehicle is carrying heavy loads. As the vehicle load increases, the suspension system deforms and will eventually bring the lower overload leaf into contact with the two active leaves, considerably increasing the system stiffness. This is a non-linear hardening spring system.

The shock absorber shown in Fig. 2.4 is a pneumatic damper that dissipates shock energy by converting it to heat. The shock absorber in this suspension style is set outboard of the leaf springs in order to increase the displacement in the absorber in response to road shocks. Many systems are built with the shock absorber inboard of the leaf springs. If the axle translates vertically, caused by both the left and right wheels hitting the discontinuity simultaneously, the result is the same whether the shock absorber is inboard or outboard. However, if only one wheel encounters a road discontinuity, then there is a rotation of the axle relative to the frame and an outboard shock absorber will experience more relative displacement than an inboard style suspension. Higher damping is achieved with greater relative displacement between the shock absorber and vehicle frame.

The photograph on the right side of Fig. 2.4 shows a coil over shock absorber suspension system. In this design, the shock absorber also acts as an integral structural member in the vehicle suspension system. The spring in this case is a coil spring that presses against the vehicle frame at the top and an integral flange on the shock absorber at the bottom. Both suspension systems serve the same purpose, supporting the vehicle and mitigating vertical acceleration to the vehicle from road conditions. In each case the goal is the same, to trade increased displacement for decreased acceleration and to increase the damping so that oscillations decay to zero faster resulting in a better ride for the vehicle's occupants.

While shock mounting the entire vehicle is a very elegant solution and the best solution for passenger vehicles, it is not always a practical approach for every vehicle. Many agricultural vehicles have their drive axles rigidly connected to the

Fig. 2.4 Common vehicle suspension types: leaf springs with shock absorber on left, coil over shock absorber on right

frame with no suspension system. While there are advantages to this approach for a low-speed agricultural vehicle such as a tractor, it means that every bump in the field is transmitted directly to the tractor operator. One easily implemented approach to mitigating shock loads imparted to the operator is to add a small suspension system to the operator seat as shown in Fig. 2.5. The suspension system shown in Fig. 2.5 is difficult to see but one of the two springs is circled. The seat is hinged on the front and supported on short coil springs at the rear. The padding on the seat also acts as a second spring supporting the operator. While this design can be reasonably effective, it is severely limited by the range of likely operator weights and input excitations. The spring stiffness is fixed but the operator weight can have a significant variability range. In terms of mechanical shock, the range of possible design natural frequencies is very large. Likewise, the input excitation can vary significantly with the terrain.

This is unlike a typical automotive suspension system where the weight of the vehicle is substantial compared to the weight of the occupants, significantly narrowing the relative mass range over which the spring-mass-damper system must be designed in proportion to the spring stiffness. As a result, the operational natural frequency range of a typical automobile is much smaller. This is also helped by the fact that most roads are relatively smooth (as compared to not driving on a road) keeping the range of input excitations to a more manageable design range.

Fig. 2.5 Simple shock mounted tractor operator's seat using a pair of coil springs

Another, often severe transportation environment is the result of forklift movement. Forklift shocks occur when driving across uneven surfaces and can be quite severe since most service forklifts do not have appreciable suspension systems. Additionally, the forks can often bounce and impact against the forklift lifting mechanism resulting in steel-on-steel impacts directly adjacent to the cargo being transported.

2.1.2 Rail Shock

Another relatively severe source of transportation impact shock is railcar coupling. While many items are now transported solely by truck, freight rail still carries a considerable quantity of goods and materials worldwide. In addition, intermodal shipping containers often go by ship, rail, and truck. It is common to see trailers transported on railcars. Figure 2.6 shows a photograph of the first few cars in a freight train traveling across the western USA. The photograph shows a pair of stacked intermodal shipping containers in the first car behind the engine and a third container further back in the line. Two flat cars are carrying truck trailers and a third car is carrying a 16.15 m (53 ft) intermodal shipping container configured for truck transport. This photograph is a perfect example of how truck, ship, and rail transportation are heavily intertwined.

Railcar weights in the USA can be as high as 143 t (315,000 lb) although many areas are restricted to 121.5 t (268,000 lb). At these weights, even a low velocity

Fig. 2.6 Freight train pulling cars loaded with intermodal shipping containers and truck trailers

impact between two railcars carries a significant momentum change. Although railcar impacts are usually relatively low velocity impacts, MIL-STD-810G does define a worst case impact velocity of 12.9 kph (8 mph) [1]. A fully loaded railcar weighing 121.5 t traveling at 12.9 kph (3.58 m/s) has a momentum of 435 kN-s which must be transferred or absorbed in the system during impact. While much of the impact is undoubtedly carried by the railcars, substantial load can still be transmitted to the cargo.

2.2 Drop Shock

Drop shock is a special case of impact shock; however, it is such a significant source of shock induced damage that it is frequently treated as a separate field. Drop shock test specifications were some of the first shock specifications developed for testing equipment and most of those original specifications were based on common sense everyday environments. For example, if a component was light enough to be carried by a single person, then a 0.76 m drop onto a hard surface was specified. The reason for this is that the average man's pocket is about 0.76 m from the floor [2]. If the items are too heavy for a single person to carry, then drop heights were specified according to the likely height attainable from an appropriate overhead crane.

The damage done by a drop shock is directly related to the surface that the component impacts. Dropping a glass on a hard concrete floor generally fairs very poorly for the glass. On the other hand, something as fragile as an egg can survive a considerable fall if it lands in the center of a soft, thick foam pad. The difference is easy to see—soft foam versus hard concrete—but how is it quantified? It is quantified by the stiffness of the interface, the velocity change from the impact, and the fragility of the item.

Figure 2.7 shows plots of the acceleration, velocity, and displacement from a theoretical 100 g haversine shock with a 10 ms duration. The equation for the haversine acceleration is given by:

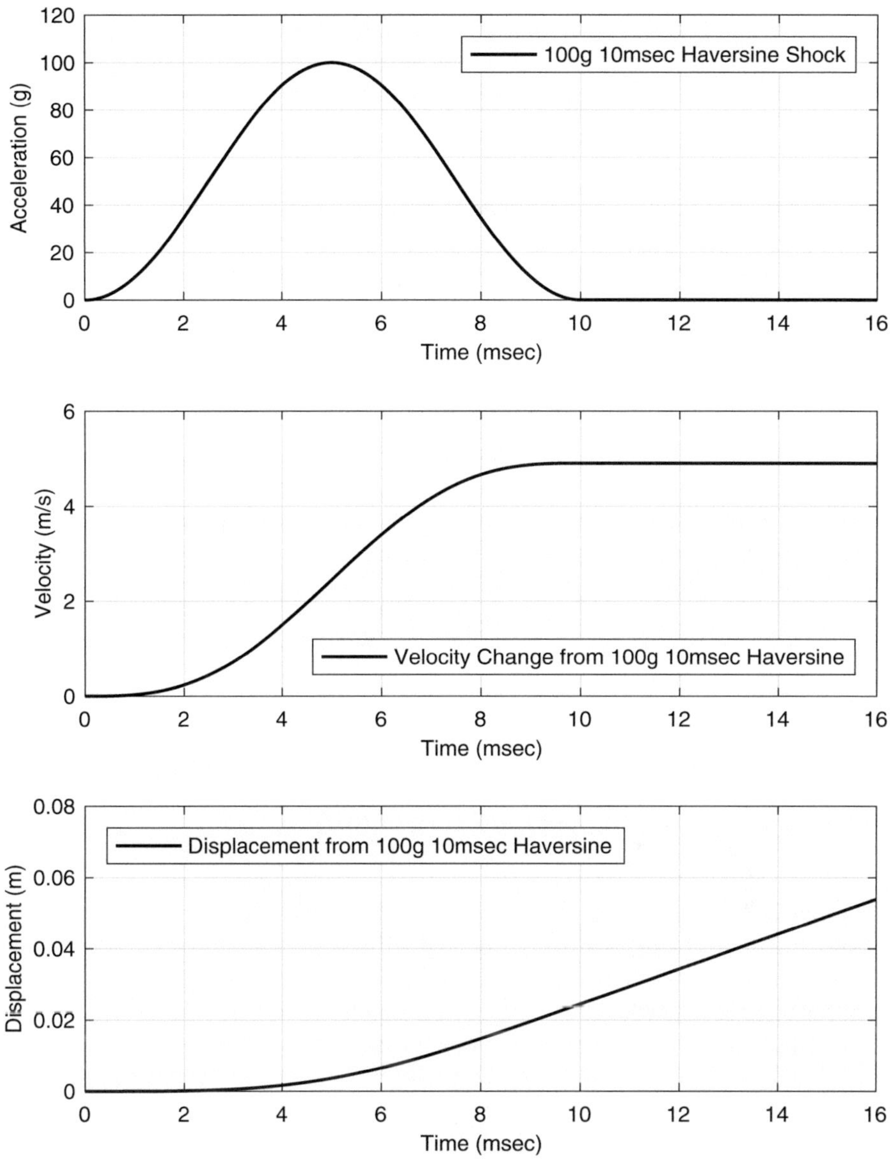

Fig. 2.7 Acceleration, velocity, and displacement from a theoretical 100 g 10 ms duration haversine drop shock with zero initial conditions

$$\ddot{w}(t) = \frac{A}{2}\left[1 - \cos\left(\frac{2\pi t}{T}\right)\right], \tag{2.4}$$

where A is the haversine amplitude and T is the haversine duration. The haversine function is simply an offset cosine function and is an impact shock idealization. However, it is also a good representation of many types of shocks and can be replicated in the laboratory using commercially available drop tables. The haversine function starts at zero with a zero slope, smoothly rises to a maximum acceleration, then smoothly returns to zero with a zero slope at the end of the shock pulse. Equation 2.4 can be integrated to obtain the velocity as:

$$\dot{w}(t) = \frac{A}{2}\left[t - \frac{T}{2\pi}\sin\left(\frac{2\pi t}{T}\right)\right] + v_0, \tag{2.5}$$

where v_0 is the initial velocity prior to the shock. From this expression, the maximum velocity change from a haversine shock can be readily deduced. The maximum velocity change must occur at the end of the shock pulse when $t = T$ and the sine term goes to zero. Thus, the maximum velocity change is given by:

$$\Delta\dot{w}(t)|_{max} = \frac{AT}{2}, \tag{2.6}$$

where A is expressed in units of length and time as opposed to multiples of g. From this expression, the 100 g, 10 ms haversine shock has a maximum velocity change of 4.9 m/s as shown.

$$\Delta\dot{w}(t)|_{max} = \frac{AT}{2} = \frac{(100\,\mathrm{g})(9.81\,\mathrm{m/s^2})(0.010\,\mathrm{s})}{2} = 4.9\,\mathrm{m/s} \tag{2.7}$$

From the energy expressions given in Eq. 2.1 this impact velocity corresponds to about a 1.2 m drop height. If the combined stiffness of the impacting surface and the dropped part are such that the component decelerates in 10 ms, then the part will experience a 100 g shock.

Integrating Eq. 2.5 yields an expression for the displacement:

$$w(t) = \frac{A}{4}\left[t^2 + \frac{T^2}{2\pi^2}\cos\left(\frac{2\pi t}{T}\right)\right]. \tag{2.8}$$

This displacement is the displacement occurring during the shock pulse only. This expression does not represent the free-falling object prior to impact, nor does it represent the linear displacement after the shock pulse has passed. The displacement predicted from the hypothetical haversine shock pulse is 2.45 cm at the end of the shock but since the velocity is now a non-zero constant, the displacement will continue to increase linearly after the shock—the results of the "push" received by the shock.

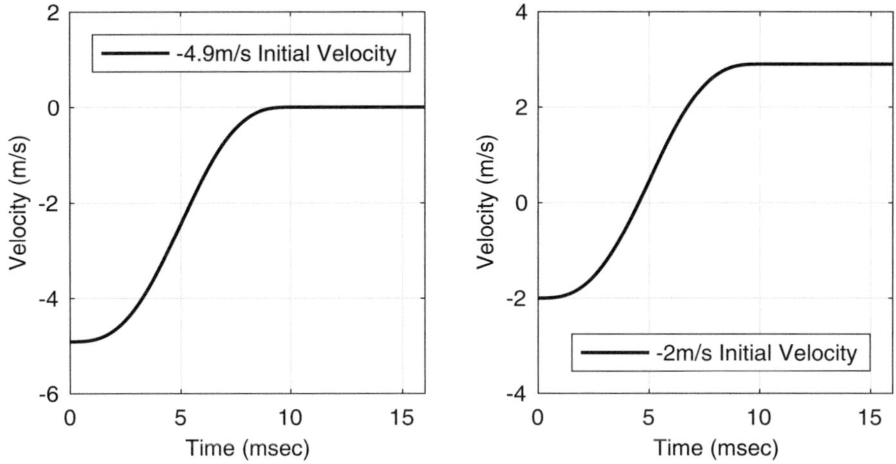

Fig. 2.8 Velocity plots from a theoretical 100 g 10 ms duration haversine drop shock with different initial velocities

The velocity plot in the center of Fig. 2.7 shows that the velocity is a smooth, continuous function but does not return to zero. As was just described, the haversine shock, like most classical shocks, results in a velocity change and a corresponding change in the object's position. In the example shown here, the initial conditions were assumed to be zero displacement and velocity so the resulting shock changed the velocity from zero to some maximum. In a true drop shock the part is falling with some velocity immediately prior to impact and has its velocity changed from some maximum value to zero. Another option that frequently occurs in the drop shock scenario is that the velocity changes from some negative value to a positive value. This occurs when the item rebounds off of the impacting surface. The component had a downward velocity prior to impact and has an upward velocity post impact. Both of these scenarios are shown in the velocity plots in Fig. 2.8. The plot on the left hand side of Fig. 2.8 shows the velocity change from the same 100 g, 10 ms shock with an initial velocity of -4.9 m/s. In this scenario, the haversine shock is just sufficient to bring the component to rest. The plot on the right hand side of Fig. 2.8 shows the velocity change from the same shock with a -2 m/s initial velocity. In this case, the final velocity is $+2.9$ m/s meaning that the part had an initial downward velocity and a final rebound velocity back in the direction from which it came.

2.3 Pyroshock

Pyroshock is the induced shock caused by the sudden release of energy. In general, a pyroshock is an intense, localized mechanical transient from a pyrotechnic device detonated on a nearby surface. It is a very high-frequency transient that decays quickly—within 5–20 ms. Pyroshock events are unique in that they typically exhibit no gross momentum transfer between the two connected bodies and no velocity change. In this respect, the pyroshock is fundamentally different from a drop shock or impact shock, which results in a net velocity change for the component.

Pyrotechnic devices are typically used to release launch locks and clamping mechanisms for the deployment of spacecraft antennas and solar arrays. They can also be used for stage separations or other types of deployment operations. Devices that generate pyroshock loads include explosive bolts, separation nuts, bolt cutters, cable cutters, pin pullers, cutting cords, and other explosively activated hardware. This type of explosively activated hardware is frequently used in military and space vehicle applications because of its simplicity and high reliability. Typically, the shock loading from a pyroshock includes not only the loading from the nearby detonation but also a mechanical shock from the localized release of material strain energy. For example, an explosive bolt is frequently used to secure a component with a V-band clamp (sometimes referred to as a Marmon clamp). In order to secure the component, the V-band clamp is installed with a sizable hoop stress due to the tightening of the explosive bolt. When the explosive bolt is cut, not only does the restrained system experience the pyrotechnic loading, but the V-band clamp pre-stress is released nearly instantaneously releasing a transient stress wave to propagate through the system.

Figure 2.9 shows a rocket body stage separation test at Northrop Grumman Innovation Systems. The stage separation is performed using a linear shape charge to separate the upper and lower rocket sections. The photograph shows the cutting operation by the fully developed pyrotechnic explosion. While this is certainly a harsh event, it is also a very short duration event.

Figure 2.10 shows a photograph from a pyroshock test of an electronic component. In this test, the electronic component is mounted to a fixture secured to a large flat plate. The plate is sized such that its fundamental resonant modes correspond to the natural frequencies of the larger system where the component is to be installed. The system is tested by detonating a small explosive charge on the underside of the plate. This in turn sends a pressure wave propagating through the test fixture assembly and into the unit under test.

Because of the frequency content and intensity of the pyroshock, the structural response is frequently not as repeatable as it is for drop shocks that excite the structure in a lower frequency range. Electronic components are particularly susceptible to pyroshock because of the high-frequency excitation. Small brittle electronic parts such as capacitors, integrated circuits, cables, and connectors may be damaged in a pyroshock event. The pyroshock may damage the part or the die bond wires that attach the parts to a printed circuit board. Even though displacements are small in high-frequency events, electronic parts are small and closely spaced so small displacements can cause damage. For example, bond wires may vibrate and create intermittent short circuits with adjacent wires.

Pyroshock does have one significant advantage over the other types of shock analyzed—systems are usually designed to withstand pyroshocks. Under most circumstances, a pyroshock is the result of an internally installed pyrotechnic device

Fig. 2.9 Photograph of a stage separation test (Photo Credit: Northrup Grumman Innovation Systems)

Fig. 2.10 Photograph of a pyrotechnic test of an electronic component (Photo Credit: NASA Johnson Space Center) [3]

being initiated. Unlike a drop shock which can occur from any height and at any orientation, often inadvertently, the system designer knows where the pyrotechnic device is located relative to shock critical components and can relocate or shock isolate components as appropriate. Thus, despite its severity, pyroshock is actually one of the more predictable shock events a system designer is likely to encounter.

2.3.1 Explosive Bolt Theory of Operation

There are many types of pyrotechnic devices in common use. One of the more common devices is the explosive bolt, or pyrotechnic bolt. Explosive bolts are used to hold two components together and then release the components when the bolt is initiated. There is a common misconception about pyrotechnic devices and explosive bolts in particular that is useful to consider in the discussion of mechanical shock. A true pyrotechnic device initiates a controlled event. This is in contrast to the free detonation of a bomb or other explosive. Explosive bolts are designed to fracture cleanly with no shrapnel at a predetermined cutting plane. Lack of shrapnel is a critical component of explosive bolt design. After all, explosive bolts are typically used on space vehicles which are decidedly opposed to shrapnel flying in their vicinity.

An explosive bolt is a bolt with a specially designed notch turned in the bolt at a specified location. A hollow cavity is bored down the bolt centerline ending near the base of the cutting notch. The cavity is then packed with explosive and one or more detonators. When the bolt is installed it is torqued to a specified value, developing a specified tensile load in the bolt body. When the pyrotechnic charge is initiated, a compressive stress wave propagates down the bolt length beginning at the base of the explosive cavity. When the stress wave reaches the end of the bolt, the impedance mismatch between the bolt and the air causes the stress wave to be reflected back toward the bolt head as a tensile wave. The reflective tensile wave now comes into interaction with the stress-concentration notch. At the notch boundary, the reflected tensile wave amplitude is nominally doubled and is added to the already present static tensile stress from the installation. The combined tensile stress from the pre-tension and the pyrotechnic shock wave results in a stress significantly above the material ultimate stress and the bolt fractures. A well-designed and manufactured explosive bolt will fracture completely through with a very high reliability and into exactly two pieces with neither of the pieces having any significant velocity. Two pieces with no velocity translates to no shrapnel.

2.4 Ballistic Shock

Ballistic shock could refer either to the recoil of a weapon system as a result of the weapon firing or to the resulting projectile impact on target. Typically, the projectile on target impact does not fall within the category of mechanical shock since the target is often significantly altered by the impact event—frequently including a material phase change.

Fig. 2.11 Photograph of a disassembled pistol with the recoiling mass on top, the recoil spring in the center, and the frame beneath

On the other hand, the weapon firing the projectile also undergoes significant mechanical ballistic shock. The conservation of momentum and Newton's third law require that the projectile momentum downrange be equivalent to the recoil force against the gun mount or weapon operator. Recoil force is generated by the reaction of the movable components to the gas pressure impulse. The recoil force builds while the projectile is in the bore and continues while the propellant gases are being exhausted after the projectile exits. The forces on the weapon are the gas force, the projectile resistance force, and the rifling force if a rifled barrel is used. For smooth bore weapons the projectile resistance force is the frictional force between the projectile and the barrel. For rifled bore weapons, the projectile resistance force is a function of the frictional coefficient and engraving ring properties, the angle of the rifling, and the projectile's radius of gyration.

Gun jump or muzzle flip is caused by the fact that the gun bore axis is typically not collinear with the center of mass of the recoiling parts. The propellant gas forces are applied along the gun bore axis creating a moment couple due to the offset between the applied load and the center of mass. The moment couple, sometimes referred to as a powder couple, results in a gun rotation typically resulting in muzzle rise [4, 5].

Figure 2.11 shows a photograph of a partially disassembled pistol. The recoiling mass is shown on top consisting of the barrel and slide assembly. The recoil spring is shown in the center of the picture along with the two bearing end caps for the spring. The frame of the weapon is shown at the bottom and contains the trigger mechanism, the hammer, and the magazine. The pistol shown here is a true spring-mass system with the barrel and slide assembly as the mass, the frame as ground, and the recoil spring as the SDOF spring. There is no damping in this system other than Coulomb damping, but the system is not intended to oscillate either. The recoiling mass makes one trip rearward when a round is fired and is locked in-battery when the recoiling mass returns to its initial position.

While most gun systems are single-degree-of-freedom systems, they do not have an oscillating breech. The gun recoils with each shot and returns to battery where it is locked in place. Some guns are a spring-mass system such as the pistol shown here, while larger caliber weapons typically utilize a full spring-mass-damper system.

2.4.1 Recoil Theory

From the conservation of momentum, the recoil of a gun system is given by:

$$M_w v_w = m_p v_p + m_c v_c, \tag{2.9}$$

where M_w and v_w are the mass and velocity of the weapon, m_p and v_p are the projectile's mass and velocity, and m_c and v_c are the same terms for the propellant charge. The charge mass is constant throughout, whether burned or unburned. From the theory of interior ballistics, the propellant charge velocity is assumed to be half the projectile velocity, which corresponds to the velocity of the mass center of the propellant combustion products. The propellant and propellant combustion products are also being accelerated down the barrel during the process of discharging the weapon; however, the volume of space available for propellant combustion products is continually increasing as the projectile travels through the barrel. Thus the center of mass of propellant and combustion products lags behind the projectile by approximately half. Rearranging Eq. 2.9 gives:

$$v_w = \frac{m_p + \frac{1}{2} m_c}{M_w} v_p. \tag{2.10}$$

It should be noted that the maximum recoil velocity occurs sometime after the projectile leaves the barrel. This is due to the continued effects of the propellant gases exiting the barrel. Experimental evidence has found that a good approximation for the maximum recoil velocity is found by assuming the propellant gas center of mass velocity is equal to 1433 m/s [4, 5]. Thus, Eq. 2.10 is approximated as:

$$v_w = \frac{m_p v_p + m_c (1433\,\text{m/s})}{M_w}. \tag{2.11}$$

Table 2.1 Recoil energies of various military weapons

Weapon system	Projectile weight	Propellant weight	Weapon weight	Projectile velocity	Recoil energy
5.56 NATO	55 grain	25 grain	3.63 kg	994 m/s	4.74 J
7.62 NATO	147 grain	45 grain	4.54 kg	833 m/s	16.2 J
0.50 BMG	647 grain	248 grain	13.6 kg	928 m/s	141 J
16 in Naval gun	860 kg	270 kg	121.5 t	820 m/s	4.95 MJ

As an example, the standard 5.56 NATO round uses a 55 grain bullet over nominally 25 grains of powder fired from an approximately 3.63 kg rifle. The muzzle velocity of the 5.56 NATO round is approximately 994 m/s meaning that the bullet spends about 6.13 ms in the rifle assuming a linear acceleration and a 0.51 m barrel. Using Eq. 2.11, the peak rifle recoil velocity is approximately 1.62 m/s, the peak recoil momentum is 5.86 N-s, and the recoil energy is 4.74 J. Most weapon system recoil calculations are given in terms of energy, as done here. The reason for this is that the felt recoil—the impulse transmitted to the operator, in the case of small arms, or the gun mount in the case of fixed or towed weapons—depends on the recoil system employed. Here again, the system designer is making the trade between acceleration and displacement. Increasing the displacement increases the time for the momentum transfer to the user or weapon mount and thereby decreases the acceleration and hence the applied force. In the case of shoulder fired small arms, such as the 5.56 NATO round described here, the user's body can flex and recoil with the rifle over a much longer time than the 6.13 ms that the bullet is in the barrel. Table 2.1 provides recoil energies for a four different military weapon systems for comparison.

2.4.2 *Shot-to-Shot Dispersion*

An interesting application of mechanical shock theory to ballistics is focused on limiting dispersion. Shot-to-shot dispersion is a measure of the difference in bullet paths from multiple, rapidly fired rounds from the same weapon. When a gun of any size is fired, the components of the weapon are excited by the shock induced by launching the projectile. If a large, mounted gun is fired in rapid succession, the intention is that all rounds will travel the same nominal flight path with some small accepted variations due to slight changes in atmospheric conditions and the variability of the projectile and propellant. However, if the gun barrel or gun mount is still vibrating from the previous shot, the motion of the gun barrel and mount can substantially influence the flight path of the next round. This occurs because the barrel is in a slightly different position when the projectile exits and a small error at the start of the flight path can easily grow to a substantial error at the flight's end.

One of the primary goals of gun mount design is to ensure that the weapon returns to rest before the next round is fired. This ensures that the gun barrel starts from the same place each time it is fired, thereby eliminating one variable in the accuracy equation. To ensure that this happens, the gun mount designer should ensure that the fundamental natural frequency of the gun mount is a least a factor of ten higher than the gun system's maximum possible fully automatic firing rate. For example, a large caliber mounted weapon might have a fully automatic firing rate of one round per second, thus the first natural frequency of the gun mount should be in the 10–20 Hz range. Likewise, a high-speed smaller caliber weapon might have a firing rate of ten rounds per second necessitating a gun mount fundamental natural frequency in the 100–200 Hz range. Damping is also necessary to bring down the oscillatory motion within the required number of cycles and many large gun recoil systems are hydraulic or pneumatic, both of which naturally have high damping coefficients.

While the previous discussion has focused on large, mounted gun systems, a similar problem exists for small arms. The mechanical shock from a round firing sets the system in motion with relative motion between the barrel and frame. If the barrel and frame do not return to their same relative position after a shock, the next round will have a different starting point than the previous round. This lowers the overall system accuracy. There are generally two common, and diametrically opposed approaches to addressing this problem. The first is to bring the frame and barrel into tight contact to ensure that there is no relative motion between them during a shot. The second approach is to separate or float the barrel relative to the frame or stock such that they vibrate completely independently and return to rest with minimal interaction between the two.

2.5 Seismic Shock

Seismic shock is the result of earthquakes. Earthquakes are actually extremely common occurrences. Hundreds of thousands of earthquakes occur each year. About 100,000 earthquakes can be felt by people around the globe each year, but only about 100 or so cause any kind of damage [6]. The earthquakes that most people think of are the big earthquakes that result in substantial destruction but those are quite rare. However, when they do occur in populated areas the consequences are severe. A major earthquake and the fire it caused destroyed San Francisco in 1906. A large earthquake in 1964 caused widespread damage in Anchorage Alaska.

The Earth's crust is constantly in motion, albeit usually very slowly. As the crust moves, it deforms, storing strain energy just like compressing a spring. The deformation and energy storage is a direct result of friction between the Earth's tectonic plates. At some point the spring force from the deformed earth is large enough to overcome the frictional force between the tectonic plates. At this point the energy is usually released quite suddenly as the frictional forces break and the two tectonic plates move relative to one another. In reality, earthquakes are very complicated events and numerous types of motion are possible from the event. Different types of earthquakes produce different shock signatures depending on how the tectonic plates move relative to each other, how far they move, how far one is located from the source of the shock, and numerous other factors. However, for the

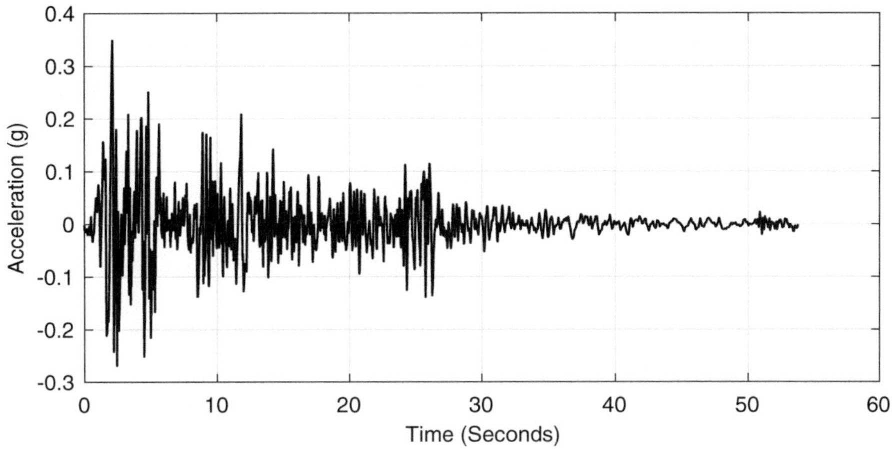

Fig. 2.12 El Centro, California 18 May 1940 north-south earthquake ground response

purposes of mechanical shock, the exact mechanism associated with the earthquake is not particularly relevant.

Figure 2.12 shows a time history plot of the El Centro, California 18 May 1940 earthquake acceleration in the north-south direction. A few things are immediately apparent from this time history. The first is that the acceleration is actually relatively low compared to many other mechanical shock insults, with a peak acceleration about 0.35 g. Second, the transient occurs over a relatively long time, the record length here is 53.7 s. The long window for earthquake shocks can allow for significant motion to develop in a structure, which in turn can induce significant strain and mechanical damage.

A full study of seismic design for structures is defined by local and international building codes and is well beyond the scope of this text. In general, the specific seismic design requirements are defined by the maximum considered earthquake (MCE) for the specific geographical region. Maps and tables of likely earthquake ground accelerations have been compiled by various international agencies including the United States Geological Survey (USGS). Most United States organizations have incorporated USGS data into their seismic design standards by reference [7, 8]. The Federal Emergency Management Agency (FEMA) sponsors a process by which seismic requirements in US building codes and standards are periodically updated [6].

Seismic design is based on shock spectra defined in terms of the short period seismic acceleration, S_S and the seismic acceleration for a 1 s period, S_1. The short period seismic accelerations are defined as the 0.2 s period ground motion acceleration. The definition here highlights the differences in spectral representations in the seismic community, given in terms of period, and the spectral representations in the mechanical engineering community, defined in terms of frequency. The differences are largely driven by the frequency, or period, ranges of interest, and

Table 2.2 Seismic occupancy categories

Category	Type of occupancy
I	Low hazard to life—agricultural or minor storage
II	General construction
III	Structures with substantial hazard to life
IV	Structures designated as essential facilities
V	National strategic military assets

the size of the structures. Buildings, bridges, and other large structures have natural frequencies that are very low, typically less than 1 Hz, making working in terms of period more convenient. Seismic design is also based on occupancy and importance categories. Table 2.2 provides a brief summary of the major seismic occupancy design categories as defined by the United States Department of Defense [7]. With each of these occupancy categories comes an increase in the design safety factor against seismic loads.

2.6 Underwater Shock

One type of mechanical shock that has already been mentioned but is likely not familiar to those outside of the naval community is underwater shock. The underwater shock phenomena is caused by an underwater explosion, typically abbreviated as UNDEX. Most of the time this is associated with a sea mine or a torpedo detonation. In the field of naval engineering this is usually defined as a non-contact detonation. The reason for this distinction is that a direct hit on a ship's hull is likely to rupture the hull which results in a different set of dynamics and frequently a different set of problems.

At the spring meeting of the Institution of Naval Architects in 1948, Commander Bonny of the Royal Navy reported on underwater explosion tests against the British destroyers HMS *Cameron*, HMS *Ambuscade*, and the submarine HMS *Proteus*. He summarized the test results by reducing all the complex phenomena of an underwater explosion to a very simple equation.

$$\mathrm{KSF} = \frac{\sqrt{W}}{R} \tag{2.12}$$

Where W is the explosive weight in pounds of TNT, R is the distance from the ship's hull to the explosive in feet, and KSF is known as the *keel shock factor*. It is rumored that his conclusions were not well received by the audience whose general option was that the relationship was too simple and too much physics had been ignored. Despite its initially poor reception, the relationship is still in common use today [9]. A sketch of the underwater shock scenario is shown in Fig. 2.13 which shows the diagonal stand-off distance R between the explosive and the hull.

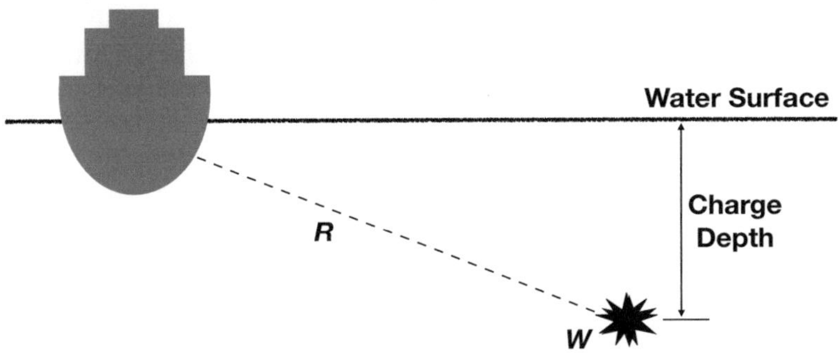

Fig. 2.13 Sketch of a typical underwater explosion configuration showing the diagonal stand-off distance

Commander Bonny's simplified approach started with the very reasonable assumption that the explosive energy should be uniformly distributed around a sphere of radius R. A plate inserted into the field would then receive energy proportional to the plate's presented area. Assuming that explosive energy is completely converted to kinetic energy, the plate's velocity can be readily determined. The velocity together with the assorted constants from the combination of equations is rolled up into a factor which he called the shock factor or keel shock factor. Bonny's equations were done assuming an explosive weight in TNT. If other explosives are used, an additional conversion to equivalent TNT weight must be added.

Using the above expression a few simple relationships can be calculated. If a charge of 4536 kg (10,000 pounds) of TNT is detonated 30.5 m (100 ft) below the ship, the keel shock factor is

$$\text{KSF} = \frac{\sqrt{10{,}000\,\text{lbf}}}{100\,\text{ft}} = 1.0. \tag{2.13}$$

Intuition readily indicates that this a very harsh shock event. On the other hand, if the same charge was detonated at a distance of 305 m (1000 ft), the resulting keel shock factor is 0.1. Still a very large charge but detonated more than a sixth of a nautical mile away naturally seems significantly more mild.

The keel shock factor is used to derive a measure of relative severity between different explosive loadings. It is used to determine if a large charge at a great distance is more or less severe than a small charge nearby. However, the dynamics of an underwater explosion are much more complex than a simple relationship between charge weight and stand-off distance.

When an explosion occurs underwater, there is an initial mass of explosive. The explosion rapidly converts the explosive material to a very hot mass of gas under tremendous pressure through a chemical process. The resulting gas pressure is far higher than the water's hydrostatic pressure at the explosion depth. Since water is a compressible medium, a pressure applied at a local region will transmit an acoustic wave to other points within the medium with a large but finite velocity. In seawater the propagation velocity is about 1500 M/S (4900 FT/S). If the acoustic waves are one dimensional, then they travel without appreciable change in magnitude or shape. However, in the underwater explosion scenario, the waves are radiated outward in a spherical fashion from a point source. As the surface area of the sphere increases, the amplitude of the shock wave decreases in a process known as spherical divergence.

The underwater shock results in two very distinct and separate events. The first is the arrival of the direct pressure wave from the explosion. For TNT, the pressure wave starts on the order of 14 GPa (2 million psi). This pressure wave front travels at the speed of sound through the water and impacts everything within the expanding sphere from the point source detonation. The second event is caused by the extremely dense mass of gas left behind by the burnt explosive. The gas then begins to expand forcing the water out and creating a gas bubble. Inertia naturally causes the gas bubble to expand beyond the pressure equalization diameter and when the expansion velocity goes to zero, the force of the water then recompresses the gas bubble. Again, inertia results in overshoot and the cycle begins again with the gas bubble expanding. With each successive cycle the gas bubble rises toward the water surface due to buoyant forces and the collapse and expansion cycle continues until the gas bubble vents when it breaks the water's surface. This sequence is sketched in Fig. 2.14 as a progression in time moving across the figure.

During an underwater shock event, the initial impact of the acoustic pressure wave results in a phenomenal amount of energy transferred to the vessel. This transfer is extremely efficient in steel ships as the speed of sound in steel is almost equal to the speed of sound in water so there is very little impedance mismatch between the two materials to reflect the incoming pressure wave. As a result, extremely high accelerations can be imparted to a ship and the ship's systems within a few milliseconds after the detonation. However, the expansion and contraction of the gas bubble also creates shock waves that emanate from the collapsing bubble like a secondary or pulsating point source. In addition to the pressure wave, the expansion and contraction of the gas bubble results in significant fluid displacement which can result in significant damage in its own right. A gas bubble rising up underneath a ship can actually lift the ship and create a void beneath the ship large

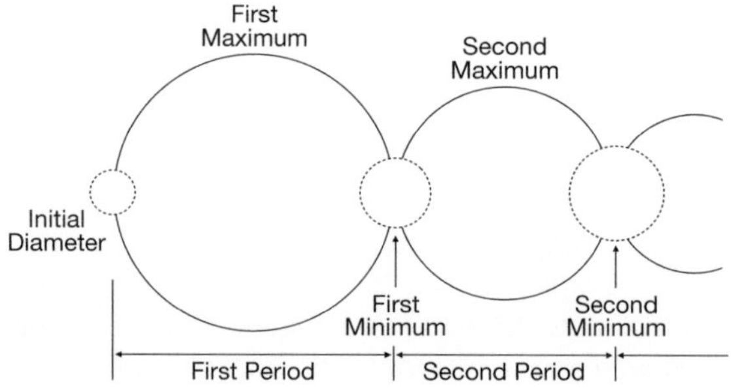

Fig. 2.14 Gas bubble pulsation from an underwater explosion

enough that the ship will break under its own weight. This is known as breaking a ship's back and it usually results in rapid sinking of the vessel. The gas bubble effects typically occur somewhat later in time, as much as a half second after the detonation. For this reason, the events are often treated as two separate insults although they are the result of a single event.

A full study of the field of underwater explosions is beyond the scope of this text. The reader is referred to the book by the late Robert Cole for a thorough treatise on the subject [10].

2.7 Summary

This chapter introduced some of the many different sources of mechanical shock in different fields. Clearly mechanical shock is a rich and diverse field and its study is of practical value. Every discipline has developed engineering methods for its specific applications. However, shock events are fundamentally similar and underlying those discipline specific methods is fundamental engineering mechanics applicable to all fields. In the subsequent chapters, we will discuss the fundamental engineering theory and application of shock.

Problems

2.1 Designing to survive impact shock requires the engineer to make the trade between acceleration and displacement. How can the felt acceleration from a shock be driven to zero? The answer explains why there are no ideal shock absorbers.

2.2 The acceleration versus displacement trade can also lead to other problems caused by allowing a system or component too much room to move. Notably, collisions between adjacent units or toppling of taller components. Discuss some options for preventing or mitigating these situations given the usual space constraints that exist in most systems.

2.3 MIL-STD-810G calls for a shock test sequence consisting of 66 shocks for every 5000 km of on-highway transportation. Based on this frequency of occurrence, how many shock events is the average automobile likely to encounter during its life? Does this number of shock events qualify as low- or high-cycle fatigue?

2.4 Rail shock is typically associated not with the actual transportation but with the railcar coupling and de-coupling events. Since railcar coupling occurs at a relatively low speed, what is the cause of the severe rail shock requirement?

2.5 Based on the discussion of gun recoil loads, there are three methods of managing recoil. What are the three methods? Which method is most practical?

2.6 Gunfire dispersion is a problem largely limited to mounted gun batteries. Dispersion is normally managed by controlling the natural frequency of the gun mount with respect to the firing frequency. What is the most effective method for limiting shot-to-shot dispersion?

2.7 Keel shock factor is a method for quantifying the relative severity of underwater shock events. The equation is somewhat simplistic but effective. The equation indicates that the shock load diminishes as the explosion is moved further away. While this is intuitive, what is the physical reason?

References

1. United States Department of Defense. (2014). Department of Defense Test Method Standard; Environmental Engineering Considerations and Laboratory Tests, MIL-STD-810G (w/Change 1), 15 April 2014.
2. Pusey, H. C. (2014). *What a fun business: A look back on the professional life of Henry C. Pusey*. Arvonia, VA: HI-TEST Laboratories, Inc.
3. Varela, G. (2016, March 10). *Pyroshock testing at ESTA is a blast*. NASA Johnson Space Center Roundup Reads.
4. Hayes, T. J. (1938). *Elements of ordnance*. New York: Wiley.
5. Carlucci, D. E., & Jacobson, S. S. (2008). *Ballistics, theory and design of guns and ammunition*. Boca Raton, FL: CRC Press.
6. National Institute of Building Sciences Building Seismic Safety Council. (2010). *Earthquake-Resistant Design Concepts: An Introduction to the NEHRP Recommended Seismic Provisions for New Buildings and Other Structures*, FEMA P-749, December 2010.
7. United States Department of Defense. (2010). *Unified Facilities Criteria (UFC), Structural Engineering*, UFC 3-301-01, 20 July 2010.
8. United States Department of Defense. (2010). *Unified Facilities Criteria (UFC), Seismic Design for Buildings*, UFC 3-310-04, 27 January 2010.
9. Bort, R. L., Morris, J. A., Pusey, H. C., & Scavuzzo, R. J. (2001). *Practical shock analysis & design*. Course Notes. Dresher, PA: Society for Machinery Failure Prevention Technology.
10. Cole, R. H. (2007). *Underwater explosions, third printing*. In *SVM-18*. Arvonia, VA: The Shock and Vibration Information Analysis Center, HI-TEST Laboratories, Inc.

Chapter 3
Single Degree-of-Freedom Systems

This chapter introduces the response of single degree-of-freedom (SDOF) systems to shock loading. An SDOF system is one whose motion is governed by a single, second-order differential equation. Only two variables, position and velocity are needed to describe the trajectory of the system. Many structures can be idealized as single degree-of-freedom systems. Figure 3.1 shows four examples of single degree-of-freedom systems. Figure 3.1a shows a photograph of a vehicle suspension system with an obvious spring and damper in parallel. Figure 3.1b is a sketch of a common traffic sign. Here the sign is the mass and the post is essentially a beam in bending with stiffness and internal damping. Figure 3.1c is the classic diagram of a theoretical spring-mass-damper system described by a mass allowed to translate on frictionless rollers. Finally, Fig. 3.1d shows a sketch of a disk hanging from a rod. The disk has mass and the rod acts as a torsional spring. Many other examples of SDOF systems can be seen in the world around us everyday.

Understanding how a single degree-of-freedom system responds to shock provides the engineer with important insights into the fundamental behavior of general systems. As you will see in Chap. 8, a linear multi-degree-of-freedom model may be transformed into a collection of parallel single degree-of-freedom models.

In this chapter, the transient response of a single degree-of-freedom system to a shock is described. Later chapters will use the concepts developed in this chapter. In this chapter, we explain

1. the equation of motion of the single degree-of-freedom oscillator;
2. the output quantities of interest for shock analyses;
3. free vibration and forced response of the SDOF oscillator;
4. the characteristics of the response to basic shock excitations—the impulse, the step input, and the ramp input;
5. the response of the SDOF system to general shock excitation;
6. digital filters that are the basis for efficiently computing the SDOF response.

C. Sisemore, V. Babuška, *The Science and Engineering of Mechanical Shock*,
https://doi.org/10.1007/978-3-030-12103-7_3

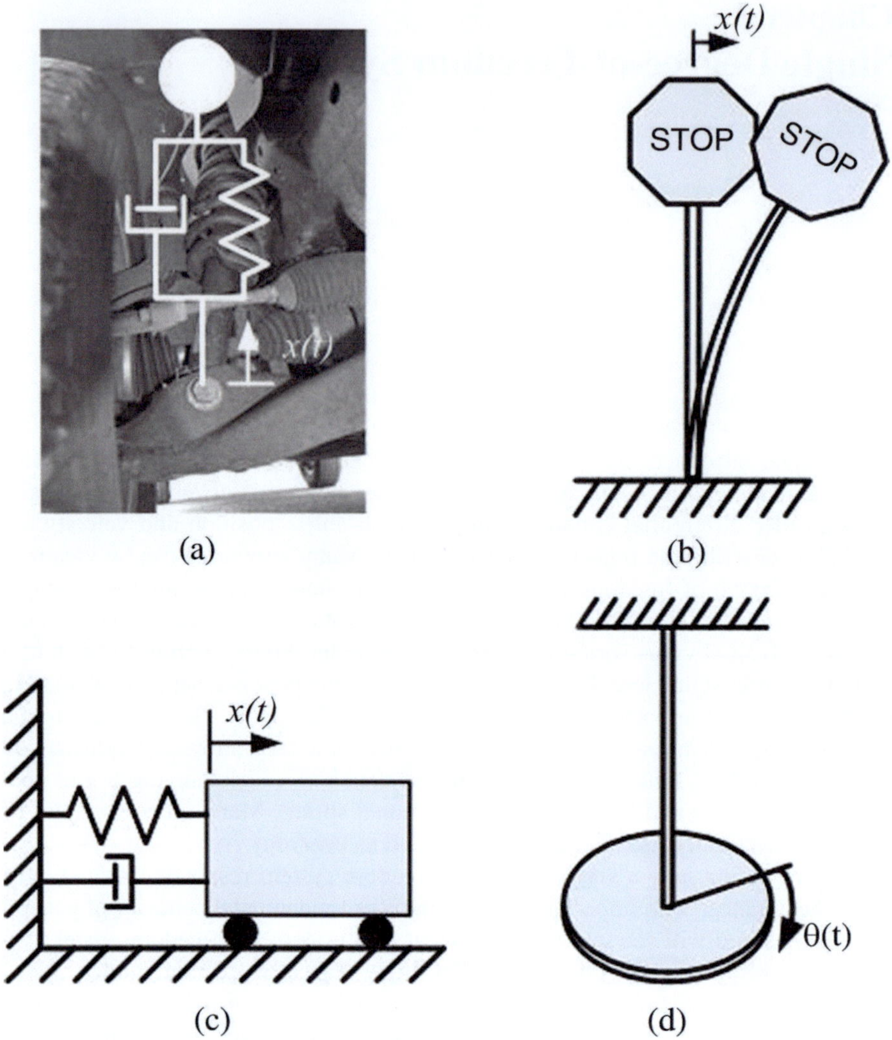

Fig. 3.1 Examples of single degree-of-freedom systems. (**a**) Vehicle suspension system represented with a spring and damper, (**b**) Stop sign represented as a beam in bending, (**c**) Spring-mass-damper system, (**d**) Disk on a shaft in torsion

3.1 SDOF Governing Equations

The linear spring-mass damper is the prototypical single degree-of-freedom system. It consists of a mass connected to ground by a spring and viscous damper as shown in Fig. 3.2. Three things are required to understand the response of a dynamic system: (1) the equation of motion, (2) initial conditions, and (3) an output equation. The equation of motion represents the dynamics of the system. The initial conditions are the state of the system from which the response evolves. The output equation relates the state variables to response quantities. The output equation need not be

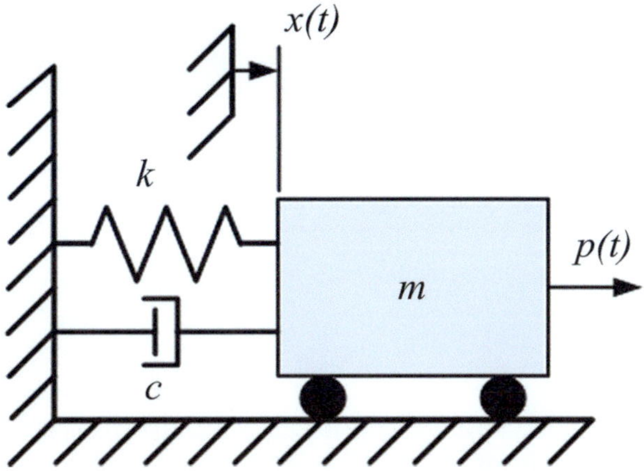

Fig. 3.2 Prototypical single degree-of-freedom model

linear. A response quantity may be something that is measured, such as acceleration at a specific location, or a global characteristic of the system such as kinetic energy. Kinetic energy is an example of a nonlinear output quantity since it is a quadratic function of the velocity.

The equation of motion of the SDOF oscillator is the second-order differential equation:

$$m\ddot{x}(t) + c\dot{x}(t) + kx(t) = p(t) \tag{3.1}$$

with initial conditions:

$$\dot{x}(0) = \dot{x}_0; \quad x(0) = x_0 \tag{3.2}$$

where m is the mass, c is the coefficient of viscous damping, k is the spring stiffness, and $p(t)$ is the force applied to the mass. The units on both sides of the equation are force, so the units of c are force/velocity and the units of k are force/distance. Of course, a consistent set of units must be used. If the units of mass are kilograms, then the units of c are N s/m and the units of k are N/m. Table 3.1 shows some common consistent unit systems. A further discussion of unit systems is provided in Appendix A.

Equation 3.1 is usually written as:

$$\ddot{x}(t) + 2\zeta\omega_n\dot{x}(t) + \omega_n^2 x(t) = \frac{p(t)}{m} \tag{3.3}$$

by dividing by the mass. The terms ω_n and ζ are fundamental parameters of vibrating SDOF systems. ω_n is defined as:

$$\omega_n = \sqrt{\frac{k}{m}} \tag{3.4}$$

Table 3.1 Consistent systems of units

Quantity	US (ft)	US (in)	SI (m)
Length	ft	in	m
Mass	slug	snail	kg
Density	$\frac{\text{slug}}{\text{ft}^3}$	$\frac{\text{snail}}{\text{in}^3}$	$\frac{\text{kg}}{\text{m}^3}$
Time	s	s	s
Force	lb_f	lb_f	N
Stress	psf $\left(\frac{\text{lb}_\text{f}}{\text{ft}^2}\right)$	psi $\left(\frac{\text{lb}_\text{f}}{\text{in}^2}\right)$	J $\frac{\text{N}}{\text{m}^2}$
Energy	ft lb_f	in lb_f	N m
Power	hp $\frac{\text{ft lb}_\text{f}}{\text{s}}$	$\frac{\text{in lb}_\text{f}}{\text{s}}$	W $\frac{\text{N m}}{\text{s}}$

and is called the undamped natural circular frequency. Its units are rad/s. We often prefer units of cycles per second or Hertz rather than rad/s. Hertz was named in honor of the German physicist Heinrich Hertz. The natural frequency or resonant frequency is defined as:

$$f_n = \frac{\omega_n}{2\pi}. \tag{3.5}$$

The reciprocal of the natural frequency is the fundamental period of oscillation:

$$T_n = \frac{2\pi}{\omega_n} = \frac{1}{f_n}. \tag{3.6}$$

The parameter ζ is called the viscous damping factor or damping ratio. It is dimensionless and is defined as:

$$\zeta = \frac{c}{c_{cr}} = \frac{c}{2\sqrt{km}}. \tag{3.7}$$

The typical range of values for the viscous damping factor of vibrating systems is $0.005 < \zeta < 0.05$, half a percent to five percent. Another commonly used parameter is the quality factor, Q, which is defined as:

$$Q = \frac{1}{2\zeta}. \tag{3.8}$$

The quality factor is the factor by which the peak response of a system vibrating at its natural frequency exceeds its static response. For example, for $\zeta = 0.05$, $Q = 10$, so the resonant response will be ten times the static response.

Equations 3.1 and 3.3 are two forms of the equation of motion of an SDOF system. In these equations, the base is fixed and the force is applied to the mass as shown in Fig. 3.2. In shock problems, often we are interested in the motion of the SDOF oscillator relative to a moving base. Base motion occurs during an earthquake when the ground moves and transmits forces into a building. When equipment is dropped, it suffers a shock through the sudden change of movement that occurs on impact. This is called base excitation and is illustrated in Fig. 3.3.

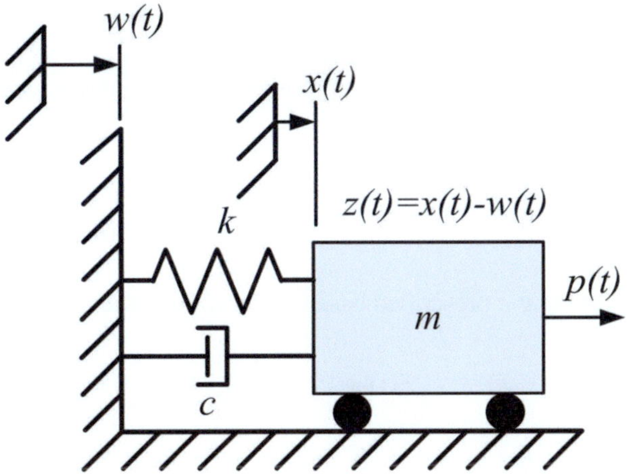

Fig. 3.3 Base excitation of a single degree-of-freedom oscillator shown with a moving base

The equation of motion of an SDOF system subject to base excitation and a force applied to the mass can be written as:

$$m\ddot{x}(t) + c(\dot{x}(t) - \dot{w}(t)) + k(x(t) - w(t)) = p(t) \tag{3.9}$$

with initial conditions:

$$\dot{x}(0) = \dot{x}_0; \quad x(0) = x_0; \tag{3.10}$$

The equation of motion is typically written in a mass normalized form by dividing all the terms in Eq. 3.9 by m:

$$\ddot{x}(t) + 2\zeta\omega_n(\dot{x}(t) - \dot{w}(t)) + \omega_n^2(x(t) - w(t)) = \frac{p(t)}{m}. \tag{3.11}$$

This mass normalized form is preferred because it uses the parameters of vibrating systems, ζ and ω_n.

In Eq. 3.11, the base motion is specified by $w(t)$, $\dot{w}(t)$, $t \geq 0$, so it can be written with all the known terms on the right-hand side:

$$\ddot{x}(t) + 2\zeta\omega_n\dot{x}(t) + \omega_n^2 x(t) = 2\zeta\omega_n\dot{w}(t) + \omega_n^2 w(t) + \frac{p(t)}{m}. \tag{3.12}$$

This form of the equation of motion is in terms of absolute displacement, velocity, and acceleration. However, the damping and elastic forces are functions of the relative motion between the mass and the base. Furthermore, shock problems involve situations where the base motion is defined by the acceleration of the base, $\ddot{w}(t)$, not the displacement and velocity, so we write the equation of motion in terms of relative coordinates. Letting $z(t)$ be the displacement of the mass relative to the base, $z(t) = x(t) - w(t)$, the equation of motion is

$$\ddot{x}(t) + 2\zeta\omega_n\dot{z}(t) + \omega_n^2 z(t) = \frac{p(t)}{m} \tag{3.13}$$

with initial conditions:

$$z(0) = z_0; \quad \dot{z}(0) = \dot{z}_0. \tag{3.14}$$

Moving the prescribed base acceleration to the right-hand side gives

$$\ddot{z}(t) + 2\zeta\omega_n\dot{z}(t) + \omega_n^2 z(t) = -\ddot{w}(t) + \frac{p(t)}{m}. \tag{3.15}$$

This is the equation of motion of a fixed base SDOF system where the base acceleration acts as an external load, as shown in Fig. 3.4. Usually, there is no force applied to the mass in Eq. 3.15 (i.e., $p(t) = 0$) and the SDOF system is excited only by the base motion given by $\ddot{w}(t)$. When $p(t) = 0$, Eq. 3.13 says that the specific inertial force is equilibrated by the damping and elastic forces.

Equation 3.15 is the fundamental equation of motion for shock analysis. Its form is identical to Eq. 3.3 when $p(t)/m = -\ddot{w}(t)$ in Eq. 3.3. Structural dynamics textbooks generally focus on Eqs. 3.1 and 3.3. In this book, we focus on Eq. 3.15 with $p(t) = 0$, and its initial conditions given in Eq. 3.14. In Chap. 12, we also will use Eq. 3.15.

Both $z(t)$ and $\dot{z}(t)$ are needed to completely define the trajectory of the moving mass in the SDOF oscillator with Eq. 3.15. The pair $z(t)$ and $\dot{z}(t)$ are called *state variables*. Similarly, the state variables in Eqs. 3.1 and 3.3 are $\{x(t),\ \dot{x}(t)\}$. The term state variables is used commonly in control systems and less so in the structural dynamics and shock fields, but it is the correct way to describe the fundamental motion variables of a mechanical system.

An output equation is needed to accompany the equation of motion and define quantities of interest. For SDOF systems, the output equation is often overlooked because there are only two states and there is no geometric complexity; however in

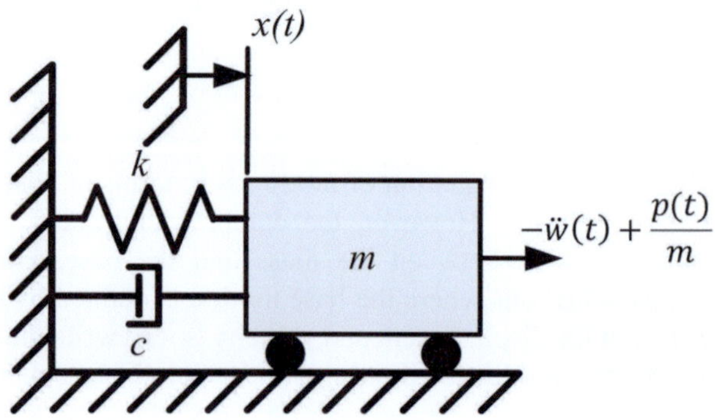

Fig. 3.4 Base excitation of an SDOF oscillator shown as a fixed base system with an inertial force applied to the mass

multi-degree-of-freedom systems that we discuss in a later chapter, the equation of motion may have a very large number of state variables and only a few may be of specific interest.

The output equation is a function of the state variables and the input in the equation of motion. The output equation for Eq. 3.15 has the general form:

$$\mathbf{y}(t) = \mathbf{g}(z(t), \dot{z}(t), \ddot{w}(t)) \tag{3.16}$$

for $p(t) = 0$ where $\mathbf{y}(t)$ is a real-valued vector of output variables. Most often, the output equation is linear and the output variables are linear combinations of the state variables and input. For example, an accelerometer senses absolute acceleration. Equation 3.15 describes the motion of the mass relative to the moving base, but we need the output equation for the absolute acceleration that an accelerometer on the mass would sense. The absolute acceleration is function of the relative acceleration and the base acceleration (i.e., input), so the output equation is

$$y(t) = \ddot{x}(t) = \ddot{z}(t) + \ddot{w}(t). \tag{3.17}$$

The more common output equation of absolute acceleration, which does not explicitly use the prescribed base motion, is

$$y(t) = \ddot{x}(t) = -\omega_n^2 z(t) - 2\zeta\omega_n \dot{z}(t), \tag{3.18}$$

which is obtained by rearranging Eq. 3.13, for $p(t) = 0$.

The output quantities that are of most interest in shock analysis are pseudo-velocity, relative velocity, and absolute acceleration. These are the three output quantities we consider.

Pseudo-velocity is ubiquitous in the naval and seismic communities. The pseudo-velocity is the relative displacement scaled by the circular natural frequency of the SDOF oscillator [1]:

$$PV(t) = \omega_n z(t) = 2\pi f_n z(t). \tag{3.19}$$

Pseudo-velocity has units of velocity but it is not velocity. It is proportional to relative displacement, hence the "pseudo" in the name. Pseudo-velocity is in-phase with relative displacement and 90° out-of-phase with relative velocity.

Equation 3.18 reveals a useful relationship between pseudo-velocity and absolute acceleration. For an undamped system, $\zeta = 0$, so Eq. 3.18 becomes

$$\ddot{x}(t) = -\omega_n^2 z(t), \tag{3.20}$$

which can be rearranged as:

$$-\frac{\ddot{x}(t)}{\omega_n} = \omega_n z(t) = PV(t). \tag{3.21}$$

In the absence of damping, pseudo-velocity is the negative absolute acceleration divided by the circular natural frequency.

Relative velocity is important because it is one of the state variables and because kinetic energy is a quadratic function of relative velocity. Absolute acceleration is an output quantity of interest because it is something that is frequently measured and it is the basis for one of the most common shock response spectrum (SRS). Response spectra are discussed in Chap. 4.

3.2 Solution of the Differential Equation

The general solution to a differential equation, such as Eq. 3.15 consists of two parts—the forced or steady state response, $z_f(t)$ and the transient response, $z_c(t)$:

$$z(t) = z_f(t) + z_c(t) \tag{3.22}$$

In mathematical parlance, $z_f(t)$ is called the particular solution and $z_c(t)$ is called the homogeneous or complimentary solution. Both parts are needed to completely represent the motion for arbitrary initial conditions and loading. When the right-hand side of Eq. 3.15 is zero, i.e., $\ddot{w}(t) = 0$ and $p(t) = 0$, $z_f(t)$ is identically zero and $z(t) = z_c(t)$. This situation is called free vibration because all motion of the system is a result of the initial conditions. When the system is initially at rest, and the right-hand side is not zero (i.e., $\ddot{w}(t) \neq 0$), then all of the motion is due to the applied load. This is called the forced response and $z(t) = z_f(t)$. We will discuss both cases in this chapter because both are necessary to understand the response of an SDOF system to shock.

The general solution of a second-order ordinary differential equation with constant coefficients depends on the magnitude of the damping ratio. The underdamped case, where $0 < \zeta < 1$, is the most important case for shock problems. The motion is oscillatory and decaying. Usually $\zeta \ll 1$, typically in the range $0.005 \leq \zeta \leq 0.05$. The other cases are the critically damped case where $\zeta = 1$, and the overdamped case, in which $\zeta > 1$. In both of these cases, there are no oscillations and the response magnitude decays slowly. These cases are not covered here, and the interested reader may find them in any text or online references on dynamics or structural dynamics.

Shock loading is, by definition, a short duration transient input. Let t_{shock} be the duration of the shock excitation. The shock response of a dynamic system has two eras: the forced vibration era, $0 < t \leq t_{shock}$, where $\ddot{w}(t) \neq 0$, and the residual vibration era, $t > t_{shock}$, where $\ddot{w}(t) = 0$. In the shock community, the system response during the forced vibration era is called the *primary response* and during the free vibration era it is referred to as the *residual response*.

3.3 Free Vibration

Linear structural dynamics systems have no memory, which means that the response of the system at $t > t_0$ depends only on the state of the system at time t_0 and the loading $\ddot{w}(t)$, $p(t)$, $t \geq t_0$. Everything that happened before time t_0 is contained in the state variables at time t_0. That means that all of the information needed for computing the response in the residual vibration era is contained in the state variables at $t = t_{shock}$. This is exactly the free vibration case because in free vibration there is no applied load and all motion is due to nonzero initial conditions.

3.3.1 Free Vibration of Undamped SDOF Systems

A special case of the SDOF oscillator is a system that has no damping, i.e., $\zeta = 0$. This is called the undamped SDOF oscillator and its equation of motion is

$$\ddot{z}(t) + \omega_n^2 z(t) = -\ddot{w}(t) + \frac{p(t)}{m}. \tag{3.23}$$

This is Eq. 3.15 without the middle term. The initial conditions are those given in Eq. 3.14, repeated here:

$$z(0) = z_0; \quad \dot{z}(0) = \dot{z}_0. \tag{3.24}$$

In free vibration, the right-hand side is zero, by definition. The complimentary solution of Eq. 3.23 is the sum of trigonometric functions:

$$z(t) = A_1 \cos(\omega_n t) + A_2 \sin(\omega_n t). \tag{3.25}$$

The constants A_1 and A_2 depend on the initial conditions:

$$A_1 = z_0 \tag{3.26}$$

$$A_2 = \frac{\dot{z}_0}{\omega_n}. \tag{3.27}$$

Equation 3.25 can also be written in terms of magnitude and phase as:

$$z(t) = Z \cos(\omega_n t - \alpha), \tag{3.28}$$

where

$$Z = \sqrt{A_1^2 + A_2^2} = \sqrt{z_0^2 + \left(\frac{\dot{z}_0}{\omega_n}\right)^2} \tag{3.29}$$

and

$$\tan\alpha = \frac{A_2}{A_1} = \frac{\dot{z}_0}{z_0 \omega_n}. \tag{3.30}$$

3.3.2 Free Vibration of Damped SDOF Systems

For an underdamped SDOF system (i.e., $0 < \zeta < 1$), the form of the complimentary solution is very similar to the form of the complimentary solution of an undamped system. The complimentary solution of Eq. 3.15 is

$$z(t) = e^{-\zeta\omega_n t}[A_1\cos(\omega_d t) + A_2\sin(\omega_d t)], \tag{3.31}$$

where

$$\omega_d = \omega_n\sqrt{1-\zeta^2} \tag{3.32}$$

is called the damped circular natural frequency. Like the undamped circular natural frequency, the units are rad/s. Constants A_1 and A_2 depend on the initial conditions, z_0 and $\dot{z}_0$. For arbitrary initial conditions $z(0) = z_0$ and $\dot{z}(0) = \dot{z}_0$,

$$A_1 = z_0 \tag{3.33}$$

and

$$A_2 = \frac{\dot{z}_0 + \zeta\omega_n z_0}{\omega_d}. \tag{3.34}$$

Equation 3.31 can be written in magnitude-phase form as:

$$z(t) = Ze^{-\zeta\omega_n t}\cos(\omega_d t - \alpha), \tag{3.35}$$

where

$$Z = \sqrt{A_1^2 + A_2^2} = \sqrt{z_0^2 + \left(\frac{\dot{z}_0 + \zeta\omega_n z_0}{\omega_d}\right)^2} \tag{3.36}$$

and

$$\tan\alpha = \frac{A_2}{A_1} = \frac{\dot{z}_0 + \zeta\omega_n z_0}{z_0\omega_d}. \tag{3.37}$$

From Eq. 3.35 or 3.36, we obtain the relative velocity and acceleration by differentiation. The absolute acceleration free vibration, or residual response, is the same as the relative acceleration response since there is no base excitation. Table 3.2 summarizes the residual responses.

Table 3.2 Free vibration response

Output quantity	Response
Pseudo-velocity	$\omega_n e^{-\zeta\omega_n t}\left[z(T)\cos(\omega_d t) + \left(\frac{\dot{z}(T)+\zeta\omega_n z(T)}{\omega_d}\right)\sin(\omega_d t)\right]$
Relative velocity	$e^{-\zeta\omega_n t}\left[\dot{z}(T)\cos(\omega_d t) - \left(\frac{\dot{z}(T)\zeta\omega_n+\omega_n^2 z(T)}{\omega_d}\right)\sin(\omega_d t)\right]$
Absolute acceleration	$e^{-\zeta\omega_n t}\left[\omega_n^2 z(T)\cos(\omega_d t) - \left(\frac{\omega_n^2(z(T)\zeta\omega_n+\dot{z}(T))}{\omega_d}\right)\sin(\omega_d t)\right]$

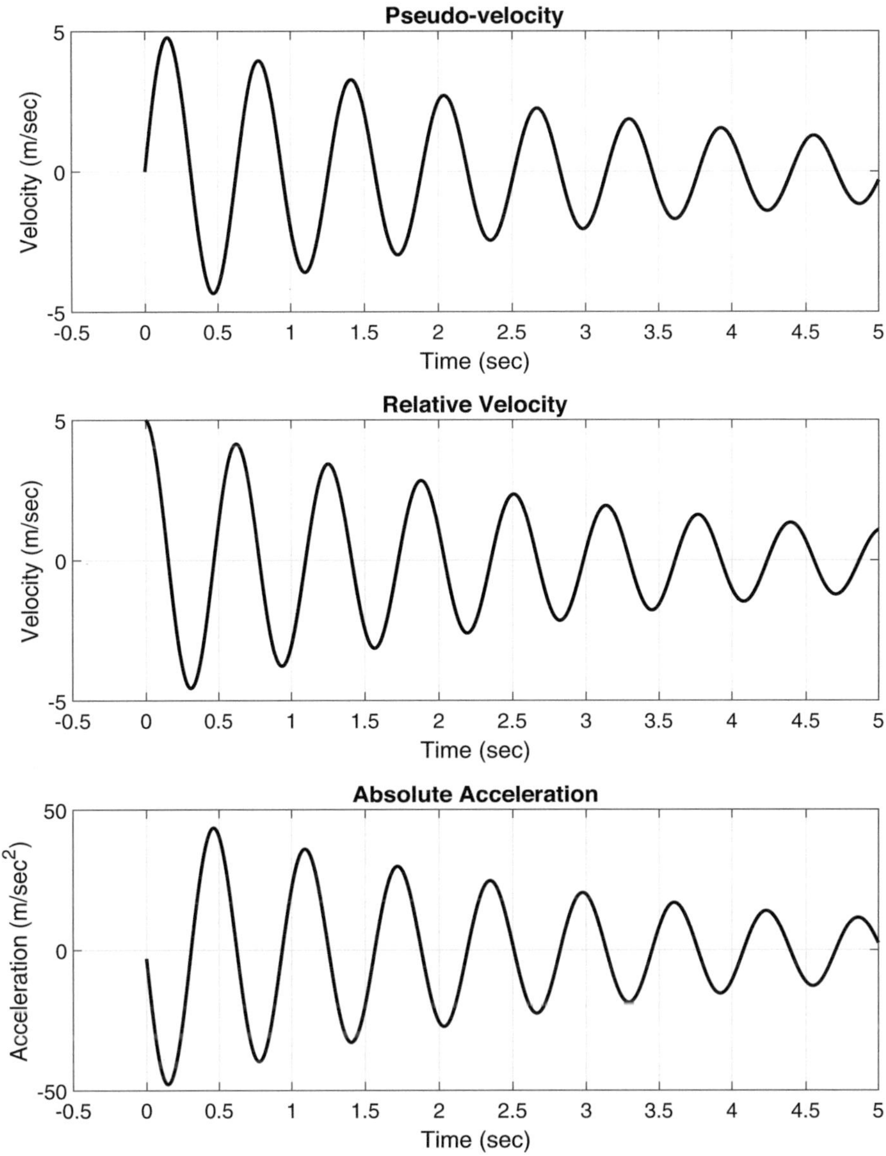

Fig. 3.5 Pseudo-velocity, relative velocity, and absolute acceleration free vibration response to an initial velocity

The response to initial conditions is essentially the response to an ideal shock. The initial velocity case is the limiting case of an impact, like a hammer blow. Figure 3.5 shows the pseudo-velocity, relative velocity, and acceleration responses of the SDOF mass to an initial velocity of 5 m/s. The circular natural frequency

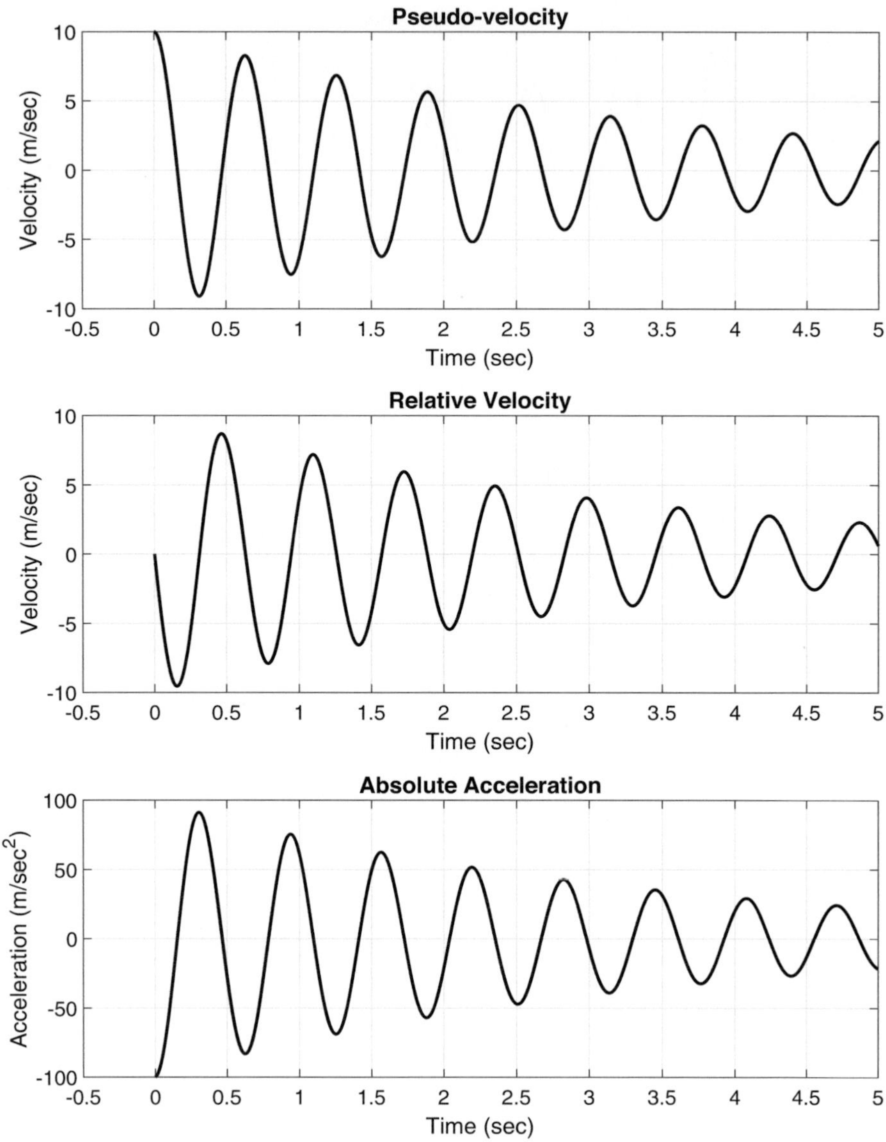

Fig. 3.6 Pseudo-velocity, relative velocity, and absolute acceleration free vibration response initial displacement

of the SDOF system is 10 rad/s. The initial displacement case is the limiting case of an impact like a drop. Figure 3.6 shows the pseudo-velocity, relative velocity, and absolute acceleration responses of the SDOF mass to an initial displacement of 1 m. Figure 3.7 shows the responses to both initial conditions; they are the sums

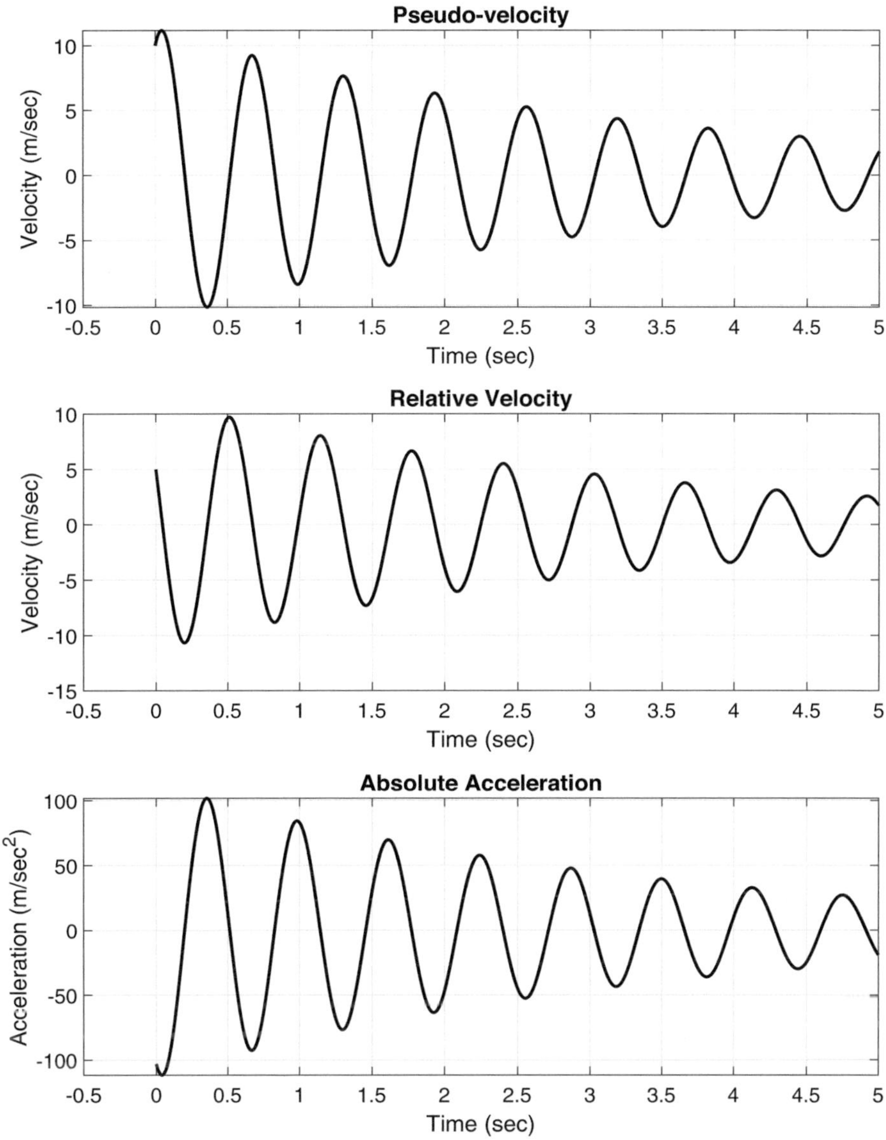

Fig. 3.7 Pseudo-velocity, relative velocity, and absolute acceleration free vibration response to both an initial velocity and initial displacement

of the responses in Figs. 3.5 and 3.6. This superposition principle is an important fundamental property of linear systems that will be relied upon when we discuss multi-degree-of-freedom systems.

3.4 Forced Response

Forced response of an SDOF system is a richer topic than free vibration because the types of forcing functions are many. Since this is a book about shock, we only consider in detail shock forcing caused by an applied force and base motion. Other types of forcing functions, such as random or sinusoidal forcing functions, are discussed in many texts on dynamics and structural dynamics.

First, we discuss the response of a damped SDOF oscillator to three basic excitations. Then, we use the results to solve the problem of response to general shock excitation.

3.4.1 Basic Shock Excitation

In this section, we introduce the response of the SDOF oscillator to three basic excitations:

1. Impulse
2. Step
3. Ramp

We will use the mass normalized form of the equation of motion, specifically Eq. 3.15 which is the equation of relative motion of a system subject to a prescribed base acceleration. The results apply equally to the case where the shock is applied as a force (Eq. 3.3). The three output quantities of interest are: pseudo-velocity, relative velocity, and absolute acceleration.

3.4.1.1 Impulse

The impulsive input is perhaps the most important basic input in shock and structural dynamics. The response of an SDOF system to an impulse with unity magnitude has a standard name; it is called the *impulse response function*. The impulse response is important for a number of reasons. Any type of excitation may be constructed as a series of impulses. This means that the response of an SDOF system is the accumulated response to the series of impulses. Also, the Laplace transform of the impulse response function is the transfer function.

Figure 3.8 shows the impulse loading as an enforced acceleration. The forcing is a unity magnitude delta function at $t = 0$. The impulse essentially gives the system an initial velocity so the impulse response is the same as the free vibration response to an initial velocity, $\dot{z}_0 = -1$, and zero initial displacement.

The governing equation of motion is

$$\ddot{z}(t) + 2\zeta\omega_n\dot{z}(t) + \omega_n^2 z(t) = -\delta(0) \tag{3.38}$$

with zero initial conditions:

$$z(0) = \dot{z}(0) = 0. \tag{3.39}$$

The term *impulse response* is usually associated with displacement as the output quantity. The impulse response of an underdamped SDOF system is

$$z(t) = \left(\frac{1}{\omega_d}\right) e^{-\zeta\omega_n t}\sin(\omega_d t). \tag{3.40}$$

The pseudo-velocity, relative velocity, and absolute acceleration impulse responses are summarized in Table 3.3 and in Fig. 3.9.

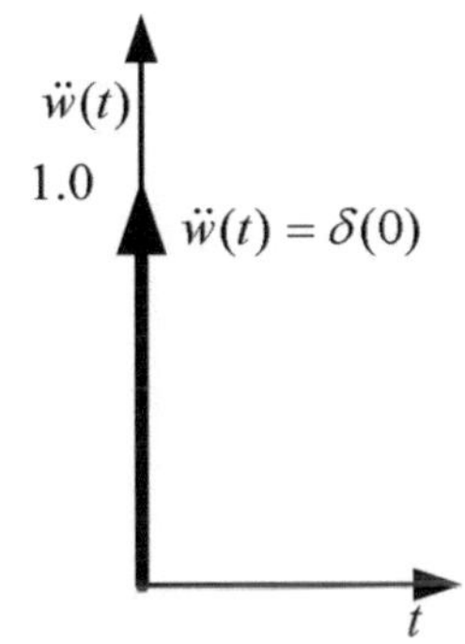

Fig. 3.8 Ideal base acceleration impulse

Table 3.3 Impulse response

Output quantity	Impulse response
Pseudo-velocity	$\left(\frac{-\omega_n}{\omega_d}\right)e^{-\zeta\omega_n t}\sin(\omega_d t)$
Relative velocity	$-e^{-\zeta\omega_n t}\left[\cos(\omega_d t) - \frac{\zeta\omega_n}{\omega_d}\sin(\omega_d t)\right]$
Absolute acceleration	$e^{-\zeta\omega_n t}\left[2\zeta\omega_n\cos(\omega_d t) + \frac{\omega_n^2(1-2\zeta^2)}{\omega_d}\sin(\omega_d t)\right]$

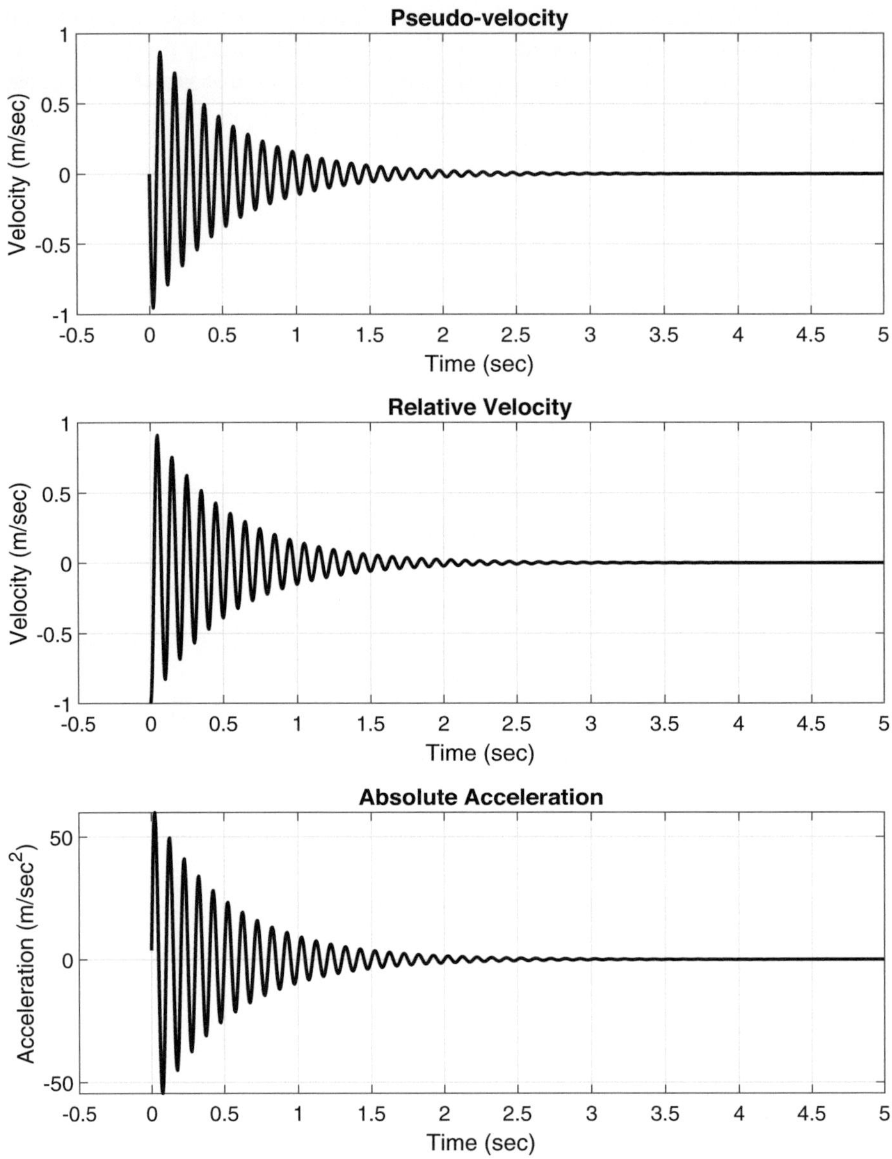

Fig. 3.9 Pseudo-velocity, relative velocity, and absolute acceleration responses to an ideal base acceleration impulse

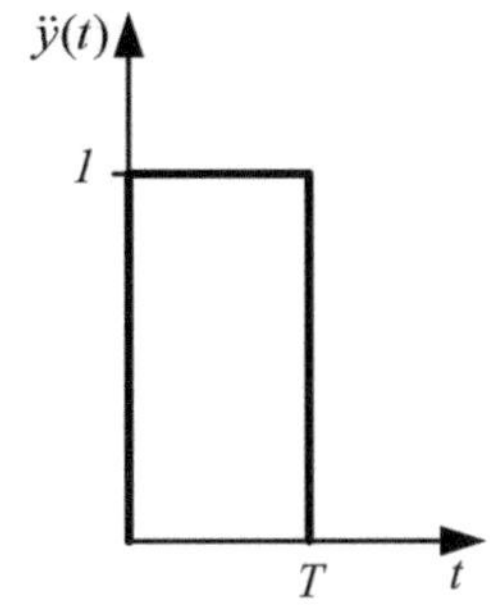

Fig. 3.10 Base acceleration rectangular pulse

3.4.1.2 Step Input

In this section, we consider the response of the SDOF oscillator to a finite step input as shown in Fig. 3.10. The response to a unit step input is called the *step response function*. We assume that the system is at rest prior to the imposition of the step.

The equation of motion is

$$\ddot{z}(t) + 2\zeta\omega_n\dot{z}(t) + \omega_n^2 z(t) = \begin{cases} -1 & 0 \le t \le T \\ 0 & t > T \end{cases} \tag{3.41}$$

with zero initial conditions:

$$z(0) = \dot{z}(0) = 0. \tag{3.42}$$

The duration of the step is T seconds. During the forced vibration era, the response consists of the particular solution and a complimentary solution. The step response in this era is the static deformation plus the free vibration response to an initial displacement of the same magnitude and opposite sign as the particular solution. The residual response is a function of the final conditions of the forced response era, $z(T)$, $\dot{z}(T)$, in Eq. 3.31.

$$z(t) = \begin{cases} \frac{1}{\omega_n^2}\left[1 - e^{-\zeta\omega_n t}\left(\cos(\omega_d t) + \frac{\zeta\omega_n}{\omega_d}\sin(\omega_d t)\right)\right] & 0 \le t \le T \\ e^{-\zeta\omega_n(t-T)}\left[z(T)\cos(\omega_d(t-T)) \right. & \\ \left. -\frac{\dot{z}(T) + \zeta\omega_n z(T)}{\omega_d}\sin(\omega_d(t-T))\right] & t > T \end{cases} \tag{3.43}$$

The *response ratio* or *dynamic load factor* is the ratio of the dynamic response to the static deformation. Figure 3.11 shows the response ratio for two values of T, for a system with circular natural frequency, $\omega_n = 10\,\text{rad/s}$ and a damping ratio, $\zeta = 0.03$. In the first case, $T < \pi/\omega_n$. The maximum response occurs during the residual vibration era. In the second case, when $T > \pi/\omega_n$, the maximum response occurs during the forced vibration era. Notice that the magnitude of the maximum

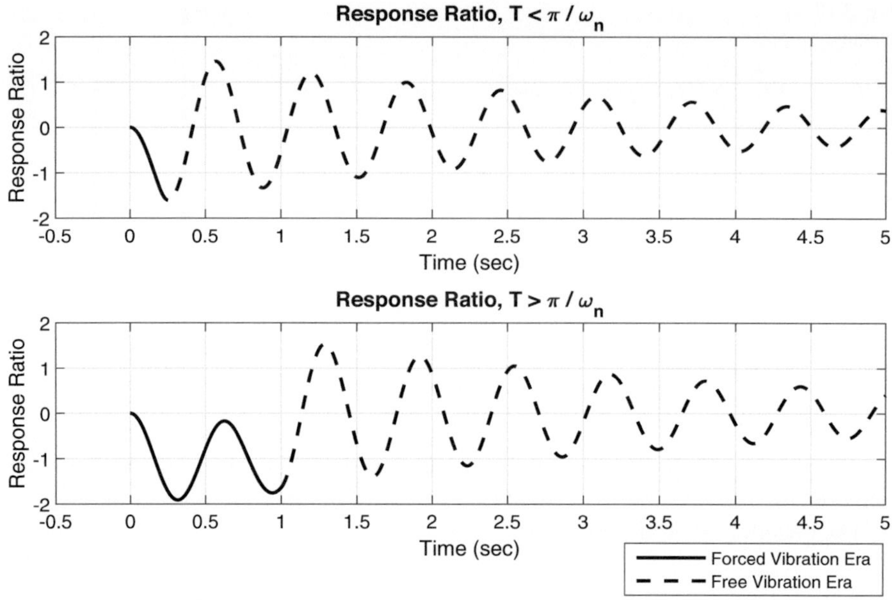

Fig. 3.11 Relative displacement response ratio for a short duration step input (top) and a long duration step input (bottom)

Table 3.4 Forced vibration era step response

Output quantity	Step response
Pseudo-velocity	$\frac{1}{\omega_n}\left[-1+e^{-\zeta\omega_n t}\left(\cos(\omega_d t)+\frac{\zeta\omega_n}{\omega_d}\sin(\omega_d t)\right)\right]$
Relative velocity	$\left(\frac{-1}{\omega_d}\right)e^{-\zeta\omega_n t}\sin(\omega_d t)$
Absolute acceleration	$1-e^{-\zeta\omega_n t}\left[\cos(\omega_d t)-\frac{\zeta\omega_n}{\omega_d}\sin(\omega_d t)\right]$

displacement response ratio in Fig. 3.11 is less than 2. The factor of two is for an undamped system. This suggests a useful way to estimate the dynamics effects of a shock; the system design should be able to tolerate a peak response that is two times the static response.

The pseudo-velocity, relative velocity, and absolute acceleration step responses during the forced vibration era are summarized in Table 3.4. The relative velocity step response is the same as the relative displacement impulse response, Eq. 3.40. The complimentary solution portion of the absolute acceleration step response in the forced vibration era is the same as the relative velocity impulse response. The absolute acceleration step response is the relative acceleration response plus the step. This is exactly the relative velocity impulse response plus the step input.

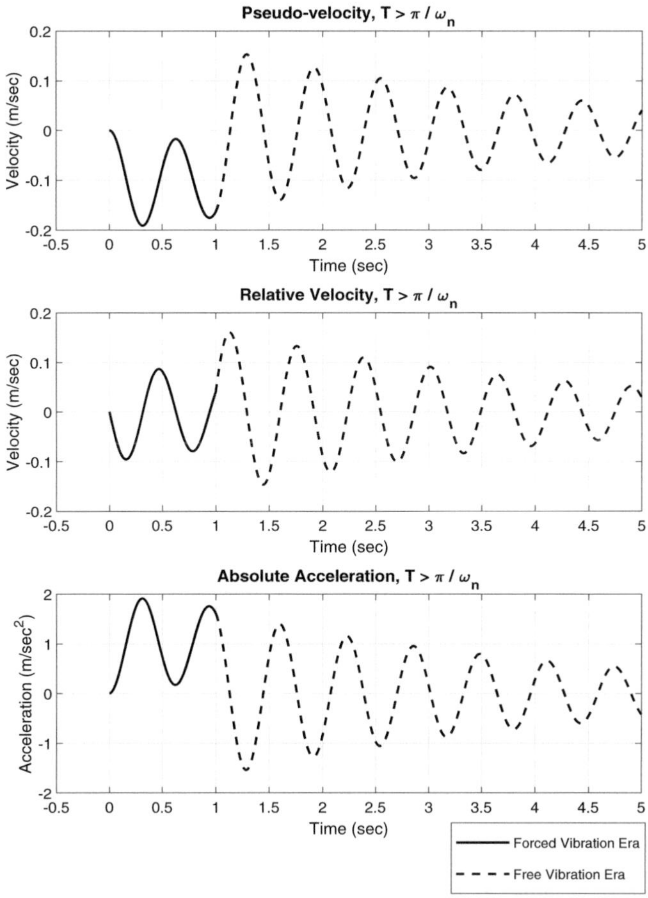

Fig. 3.12 Pseudo-velocity, relative velocity, and absolute acceleration responses to a rectangular pulse base acceleration

The pseudo-velocity, relative velocity, and absolute acceleration step responses during the free vibration era are those shown in the free vibration section, except that the initial conditions are the final conditions of the forced response era, $z(T)$, $\dot{z}(T)$.

Figure 3.12 shows the pseudo-velocity, relative velocity, and absolute acceleration responses to a unit step input of duration $T = 1$ s, which is longer than π/ω_n, so the peak responses occur in the forced vibration era.

3.4.1.3 Ramp Input

A ramp load is the third fundamental input type. This load type is often called a sawtooth input. The response to a ramp is called the *ramp response* function. Unlike the impulse or step inputs in which the excitation is applied instantaneously, the ramp input builds over a finite time. The duration of the ramp is called the rise time and it is denoted as T_R in Fig. 3.13. The positive ramp pulse in Fig. 3.13 is also called a terminal peak saw-tooth shock.

The ramp input is important because it shows the relationship between loading duration, rise time, and the dynamic response characteristics of the system. It is also used in the ramp invariant filter with which the response to general forcing is computed.

As with the other loading cases, the forced vibration response is the sum of the particular solution and the complimentary solution of the differential equation of motion:

$$\ddot{z}(t) + 2\zeta\omega_n\dot{z}(t) + \omega_n^2 z(t) = \begin{cases} -\frac{t}{T} & 0 \le t \le T \\ 0 & t > T \end{cases} \tag{3.44}$$

with zero initial conditions:

$$z(0) = \dot{z}(0) = 0. \tag{3.45}$$

The relative displacement ramp response is

$$z(t) = \begin{cases} \frac{2\zeta - \omega_n t}{\omega_n^3 T} + e^{-\zeta\omega_n t}\left[A_1 \cos(\omega_d t) + A_2 \sin(\omega_d t)\right] & 0 \le t \le T \\ e^{-\zeta\omega_n (t-T)}\left[z(T)\cos(\omega_d (t-T)) + \frac{\dot{z}(T) + \zeta\omega_n z(T)}{\omega_d}\sin(\omega_d (t-T))\right] & t > T \end{cases} \tag{3.46}$$

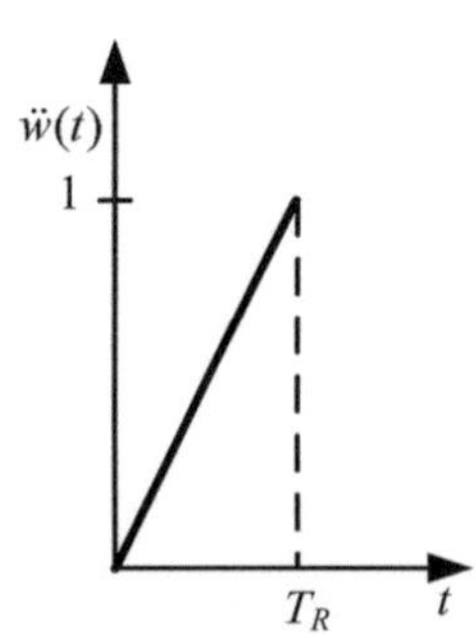

Fig. 3.13 Base acceleration positive ramp pulse

Table 3.5 Forced vibration era ramp response

Output quantity	Ramp response
Pseudo-velocity	$\frac{2\zeta-\omega_n t}{\omega_n^2 T}+e^{-\zeta\omega_n t}[A_1\cos(\omega_d t)+A_2\sin(\omega_d t)]$
Relative velocity	$\frac{1}{\omega_n^2 T}\big[-1+e^{-\zeta\omega_n t}\big(\cos(\omega_d t)+\frac{\zeta\omega_n}{\omega_d}\sin(\omega_d t)\big)\big]$
Absolute acceleration	$\frac{t}{T}-\left(\frac{1}{\omega_d}\right)e^{-\zeta\omega_n t}\sin(\omega_d t)$

where

$$A_1=\frac{-2\zeta}{\omega_n^3 T} \tag{3.47}$$

and

$$A_2=\frac{(1-2\zeta^2)}{\omega_n^2\omega_d T}. \tag{3.48}$$

In the forced response era ramp response, the first term is the particular solution and the second term is the complimentary solution.

The pseudo-velocity, relative velocity, and absolute acceleration ramp responses during the forced vibration era are summarized in Table 3.5. The absolute acceleration ramp response is interesting. It is the impulse response shown in Eq. 3.40 superimposed on the ramp acceleration. Figure 3.14 shows the pseudo-velocity, relative velocity, and absolute acceleration responses of an SDOF oscillator to the ramp input in Fig. 3.13. The oscillatory response is superimposed on the ramp input up to $t = T_R$, at which point the loading stops and the system is in free vibration, which is the same as the step response free vibration response, given in Table 3.2.

The rise time is long relative to the fundamental period of the SDOF oscillator, so the peak response occurs in the forced vibration era. Since the rise time is slow, the SDOF system response is essentially quasi-static. The terms fast and slow are defined relative to the fundamental period of the system.

This leads to a very useful "rule of thumb." If the rise time is greater than approximately $3/f_n$, then the SDOF system response is essentially quasi-static, which means the system does not experience a shock. This characteristic is true for other types of shock environments, which were covered in Chap. 2.

The ramp excitation shown in Fig. 3.13 is a ramp input increasing from zero to a peak value at T_R. We can also consider the decreasing ramp shown in Fig. 3.15, which is also called an initial peak sawtooth shock. The response to this loading is the sum of the impulse response and the ramp response for a ramp whose slope is negative. Figure 3.16 shows the pseudo-velocity, relative velocity, and absolute acceleration responses of an SDOF oscillator to the ramp input in Fig. 3.15.

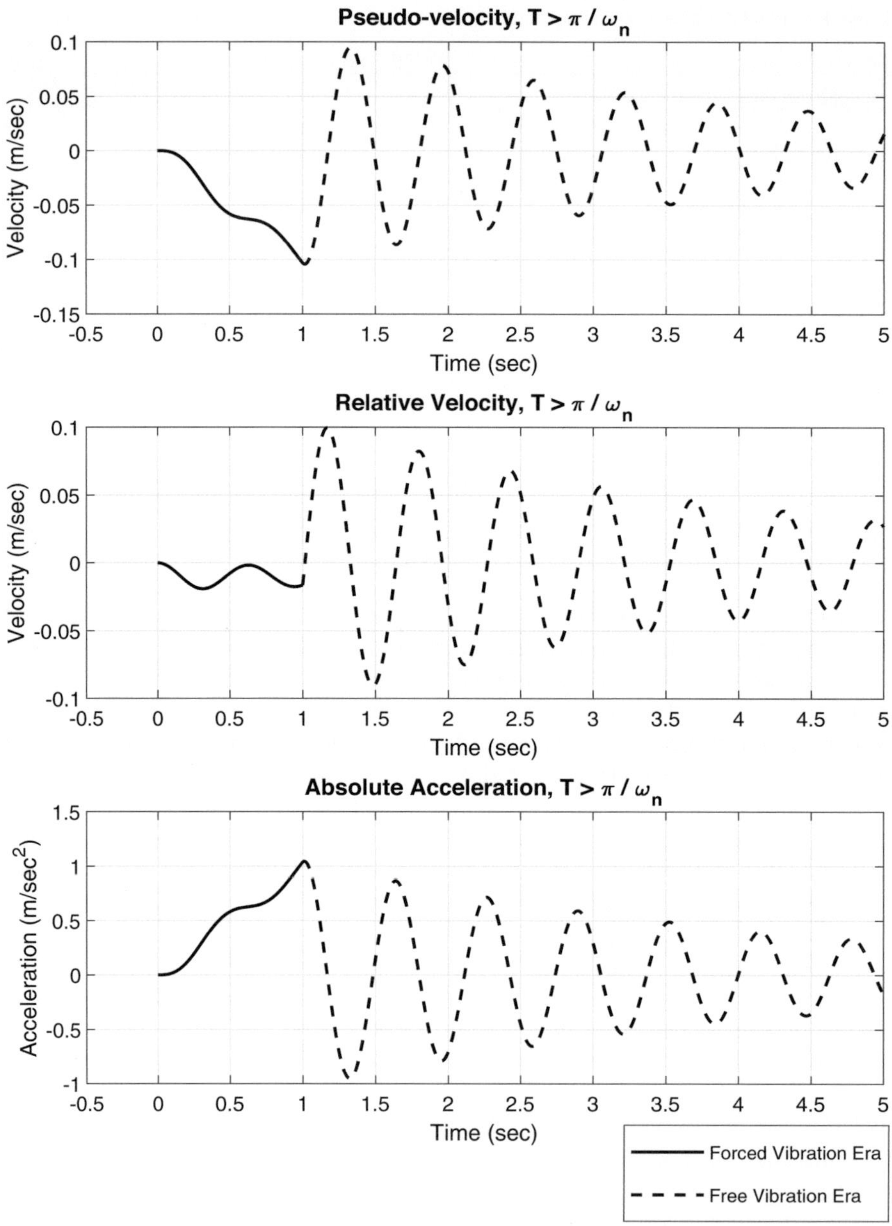

Fig. 3.14 Pseudo-velocity, relative velocity, and absolute acceleration responses to a positive ramp pulse

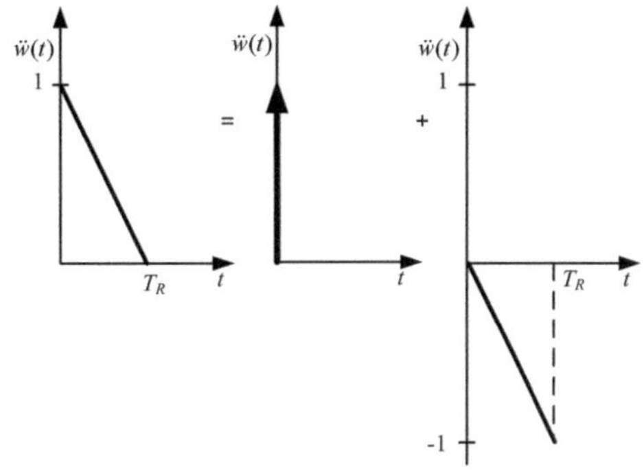

Fig. 3.15 A descending ramp base acceleration pulse made up of an impulse and a negative slope ramp pulse

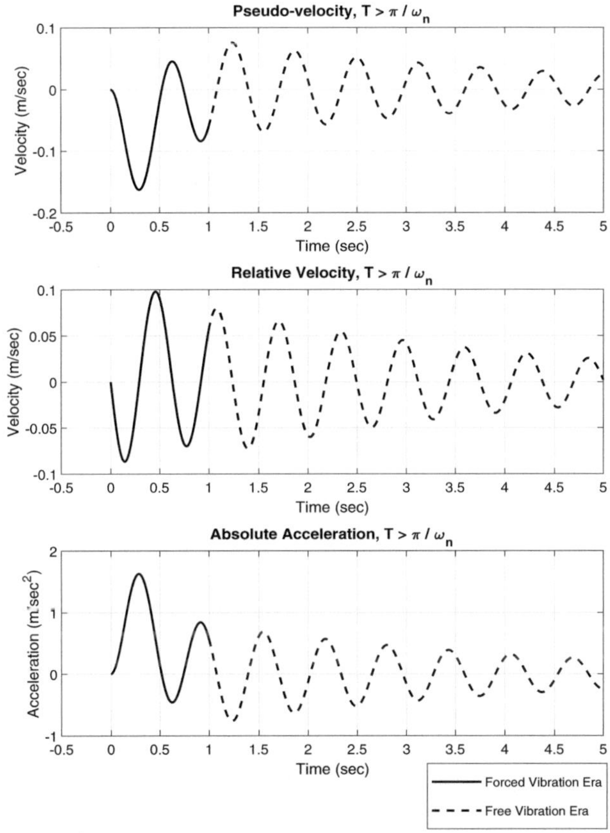

Fig. 3.16 Pseudo-velocity, relative velocity, and absolute acceleration responses to a descending ramp base acceleration pulse

3.4.2 General Shock Excitation

The displacement response to a general excitation of an underdamped SDOF system is the sum of the particular solution and the complimentary solution as described in Sect. 3.2.

$$z(t) = e^{-\zeta\omega_n t}\left[z_0\cos(\omega_d t) + \frac{z_0\zeta\omega_n + \dot{z}_0}{\omega_d}\sin(\omega_d t)\right] - \frac{1}{\omega_d}\int_0^t \ddot{w}(\tau)e^{-\zeta\omega_n(t-\tau)}\sin(\omega_d(t-\tau))d\tau, \tag{3.49}$$

where z_0 and $\dot{z}_0$ are the initial relative displacement and relative velocity, respectively. For the typical case with zero initial conditions, Eq. 3.49 reduces to

$$z(t) = -\frac{1}{\omega_d}\int_0^t \ddot{w}(\tau)e^{-\zeta\omega_n(t-\tau)}\sin(\omega_d(t-\tau))d\tau. \tag{3.50}$$

The solution is simply the convolution of $\ddot{w}(t)$ with the unit impulse response of the SDOF system. Equation 3.50 is called the Duhamel integral equation after Jean-Marie Duhamel, a nineteenth century French mathematician and physicist. The Duhamel integral is valid only for linear systems because it is based on the principle of superposition. The total response at time t is the sum of all the impulse responses up to time t.

The Duhamel integral is a special form of the *convolution integral*:

$$z_p(t) = \int_0^t \ddot{w}(\tau)h(t-\tau)d\tau \tag{3.51}$$

where $h(t-\tau)$ is the *response kernel*. In Eq. 3.50, the response kernel is the displacement impulse response function.

$$h(t-\tau) = \frac{1}{\omega_d}e^{-\zeta\omega_n\tau}\sin(\omega_d(t-\tau)) \tag{3.52}$$

For the undamped case, the response kernel is similar:

$$h(t-\tau) = \frac{1}{\omega_n}\sin(\omega_n(t-\tau)) \tag{3.53}$$

and the displacement response to general excitation of an undamped SDOF system is

$$z(t) = \left[z_0\cos(\omega_n t) + \frac{\dot{z}_0}{\omega_n}\sin(\omega_n t)\right] - \frac{1}{\omega_n}\int_0^t \ddot{w}(\tau)\sin(\omega_n(t-\tau))d\tau, \tag{3.54}$$

where z_0 and $\dot{z}_0$ are the initial relative displacement and relative velocity as before.

So, the problem of obtaining the response to a general excitation boils down to solving for the particular solution of the equation of motion for the given excitation. There are a number of ways to do that, but for the general case this integral must be evaluated numerically. Numerical methods are the only practical way to solve the general excitation problem. Solving the general response problem numerically means that the excitation can be truly general. It need not be a function; rather it can be, and usually is, a set of discrete points. The discrete points can be samples from a function, samples from a transient random process, or test measurements.

There are three basic ways to obtain the response of an SDOF system to general transient excitation. First, we can numerically integrate the Duhamel integral directly using quadrature formulas (e.g., trapezoidal rule or Simpson's rule). The second, and more general method, is integration of the equation of motion for a specific excitation, which means solving an ordinary differential equation numerically. There are many standard numerical recipes available, such as the Runge–Kutta family of algorithms (e.g., Runge–Kutta, Runge–Kutta–Fehlberg) or the Newmark-Beta algorithm. Most scientific software, like MATLAB, have robust integration functions (e.g., ODE45).

The third way is to treat the SDOF system as a filter through which the input passes to generate the output quantities of interest. This is the most common and numerically efficient method. This is essentially equivalent to solving the discrete version of the convolution integral.

Figure 3.17 shows a flow diagram of the discretized solution of the general excitation problem. First, the loading function is sampled at a specific sampling rate, T. That means we only know the values of the loading at specific times, $\ddot{w}(kT)$, $\ddot{w}((k+1)T), \ldots \ddot{w}((k+N)T)$. Of course, we can start with sampled data rather than discretizing a continuous function. In both cases, we do not know how the loading behaves in between the sampled points. This lack of knowledge implies we want to use a fast sampling rate. The influence of sample rate on the numerical solution is discussed later, in Sect. 3.4.2.4.

Next, a hold is applied. This hold essentially fills the knowledge gap of what the input does between samples. The most common hold is the zero-order hold (ZOH), in which the values in between the sampled points are assumed to remain constant, as shown in Fig. 3.18. The digital filter created using the ZOH is called a *step invariant filter*.

$$\ddot{w}(t) = \ddot{w}(kT) \quad kT \leq t < (k+1)T \tag{3.55}$$

Another common hold is the first-order hold (FOH), where the values in between the sample points are taken to be piecewise linear. Rather surprisingly, the term

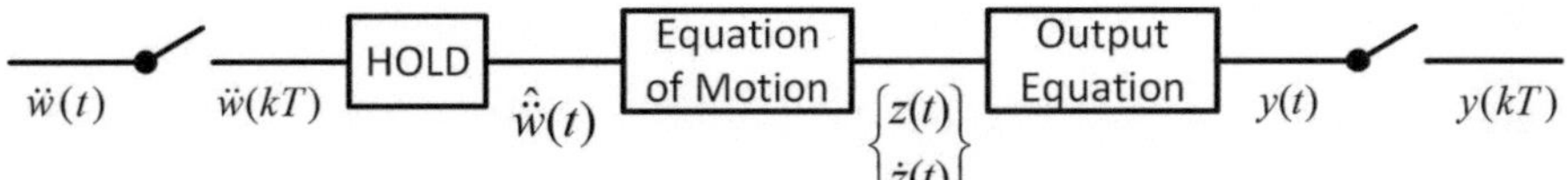

Fig. 3.17 Signal flow diagram for general excitation

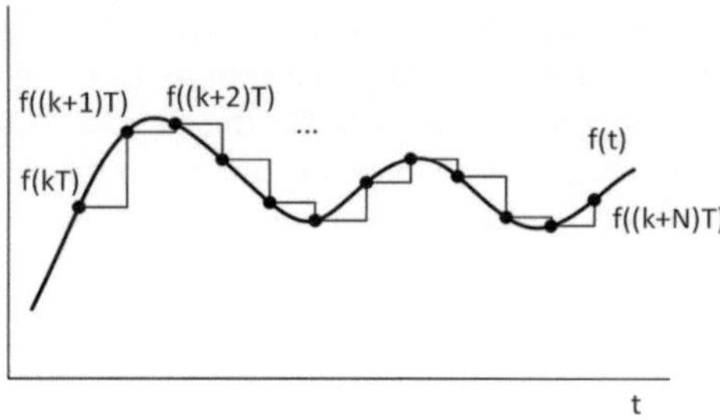

Fig. 3.18 Zero-order-hold discretization of a continuous waveform

Fig. 3.19 Causal first-order-hold discretization of a continuous waveform

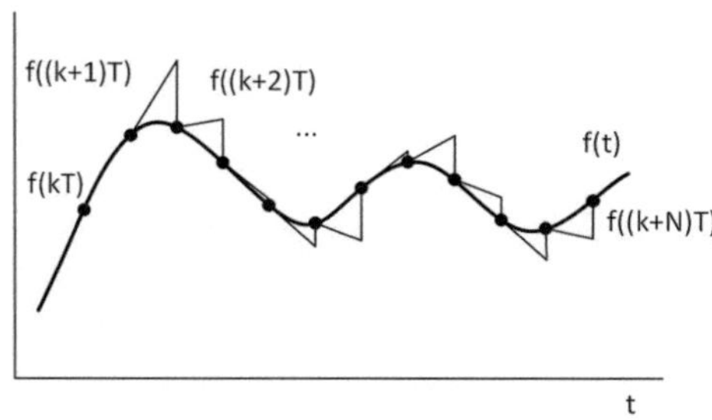

Fig. 3.20 Noncausal first-order-hold or triangle hold discretization of a continuous waveform

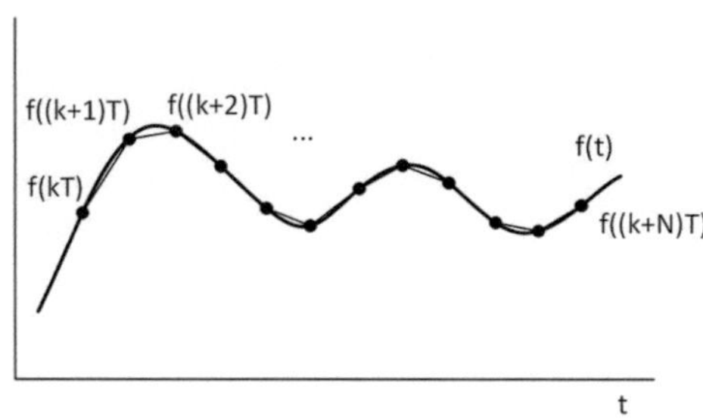

"first-order-hold" does not have an unambiguous definition. There is the causal first-order hold:

$$\ddot{w}((k+1)T) = \ddot{w}(kT) + \frac{t-kT}{T}(\ddot{w}(kT) - \ddot{w}(k(T-1))) \quad kT \le t < (k+1)T \tag{3.56}$$

where the slope is extrapolated from the previous values of the input. This is illustrated in Fig. 3.19. The causal first-order hold has some advantages over the zero-order hold, but as illustrated in Fig. 3.19, it has more amplitude distortion than the zero-order hold. It is not used for shock simulation.

There is also the noncausal first-order hold in which the value of the function depends on the current and future sampled value:

$$p(t) = p(kT) + \frac{t-kT}{T}(p(k(T+1)) - p(kT)) \quad kT \le t < (k+1)T \tag{3.57}$$

This hold is shown in Fig. 3.20.

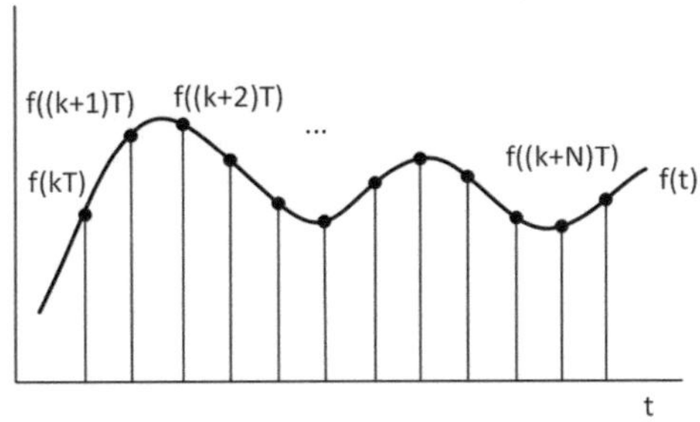

Fig. 3.21 Impulse hold discretization of a continuous waveform

The noncausal first-order hold is also called the *triangle hold* and the filter obtained using this hold is called a *ramp invariant filter*. We use the term triangle hold to avoid confusion with the causal first-order hold. While it is a noncausal discretization, the resulting discrete function is causal.

There is also the impulse hold, also called a z-transform hold, which is no hold at all:

$$\ddot{w}(t) = \begin{cases} \ddot{w}(kT) & t = kT \\ 0 & kT < t < (k+1)T \end{cases} \tag{3.58}$$

This hold is shown in Fig. 3.21. The discrete Duhamel integral is based on the impulse hold. A filter created with the impulse hold is called an *impulse invariant filter*. The main disadvantage of the impulse invariant filter is that its DC gain error is nonzero.

Next, we apply the held input to the equation of motion and solve for the state variables. The state variables are then passed to the output equation to get the output quantities of interest, which are sampled to produce the discrete output.

The input to the equation of motion, $\hat{\ddot{w}}(t)$, in Fig. 3.17 depends on the hold that is used. If the hold is the ZOH, the input is a step input. When the hold is the triangle hold or the first-order hold, the input is a ramp input. When no hold is used, the input is an impulse. These are the basic excitations discussed in Sect. 3.4. The forced vibration era (i.e., the primary response) starts at $t = kT$ and ends at $t = (k+1)T$. This means that the discretized general solution is the particular solution at $t = (k+1)T$ plus the complimentary solution since the $\{z(kT), \dot{z}(kT)\}$ are not zero at $t = kT$.

It is not efficient to compute the general response literally with the process outlined. In addition to computational inefficiencies, it is not efficient because we must do a lot of up front work to derive the specific ramp or step or impulse responses. We would prefer a solution procedure that is applicable to a wide range of linear systems with constant coefficients. Figure 3.22 shows the solution flow diagram where the equation of motion and output equation have been combined into a transfer function. Furthermore, we group the hold, the transfer function, and the output sampler into a discrete transfer function. This discrete transfer function

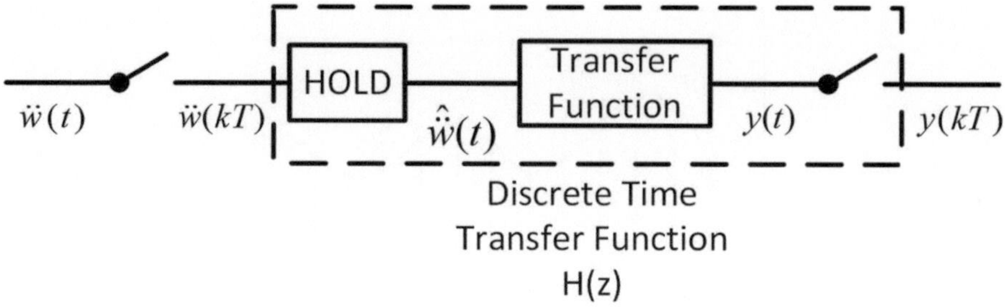

Fig. 3.22 Signal flow diagram for general excitation with transfer function

is the digital filter that transforms the sampled input into the sampled output. There are significant computational speed benefits that come with the use of digital filters versus solving the equations of motion directly.

3.4.2.1 Transfer Function of the SDOF Oscillator

The equations of motion and their associated output equations fundamentally represent the dynamics of the SDOF oscillator. As illustrated in Fig. 3.22, we can create an *input–output* model by combining the equation of motion and output equation into a single function called a transfer function.

A transfer function is a linear model that describes the input–output characteristics of a dynamic system in the Laplace domain. If $F(s)$ is the Laplace transform of an input function $f(t)$, and $Y(s)$ is the Laplace transform of the output, $y(t)$, and the system is initially at rest (i.e., zero initial conditions), the transfer function is the ratio of the output to the input in the Laplace domain:

$$H(s) = \frac{Y(s)}{F(s)}. \tag{3.59}$$

If the initial conditions are not zero, the transfer function is

$$H(s) = \frac{Y(s)}{P(s)} + z_0 + s\dot{z}_0. \tag{3.60}$$

However, when we use transfer functions to compute the forced response, the initial conditions are almost always zero. Equation 3.60 is included for completeness.

The denominator polynomial in the transfer function is called the *characteristic equation*. It is the Laplace transform of the equation of motion and is independent of the input or output. The numerator polynomial is entirely a function of the input and output types and locations. Let $Z(s)$ be the Laplace transform of $z(t)$ and let $\ddot{W}(s)$ be the Laplace transform of the base acceleration, $\ddot{w}(t)$. Note that the double-dot on

Table 3.6 Transfer functions of typical shock output quantities

Output	Transfer function
Pseudo-velocity	$\frac{-\omega_n}{s^2+2\zeta\omega_n s+\omega_n^2}$
Relative velocity	$\frac{-s}{s^2+2\zeta\omega_n s+\omega_n^2}$
Absolute acceleration	$\frac{2\zeta\omega_n+\omega_n^2}{s^2+2\zeta\omega_n s+\omega_n^2}$

$\ddot{W}(s)$ is notational only, indicating the Laplace transform of the acceleration. The characteristic equation of the relative response model, Eq. 3.15 (with $p(t) = 0$) is

$$s^2 + 2\zeta\omega_n s + \omega_n^2. \tag{3.61}$$

Let the output quantity of interest be the absolute acceleration, given by Eq. 3.18. The Laplace transform of the output equation is

$$Y(s) = \left(-\omega_n^2 - 2s\zeta\omega_n\right) Z(s). \tag{3.62}$$

So, the transfer function from prescribed base acceleration to absolute acceleration of the SDOF mass is

$$H(s) = \frac{-\omega_n^2 - 2s\zeta\omega_n}{s^2 + 2\zeta\omega_n s + \omega_n^2}. \tag{3.63}$$

Table 3.6 shows the SDOF oscillator transfer functions for the output quantities normally used in shock analysis. When we discuss multi-degree-of-freedom systems in Chap. 8, you will see how location affects the coefficients of the transfer function numerator.

The term *frequency response function* or FRF is often used synonymously with *transfer function*. Strictly speaking this is not true. A transfer function is an input–output *model*, while a frequency response function is a sequence of complex numbers representing the ratio of the output to the input at specific frequencies. An easy way to distinguish the two terms is that a transfer function is a mathematical expression, while a frequency response function is a plot of the transfer function.

Figures 3.23, 3.24, and 3.25 show the frequency response functions of the transfer functions in Table 3.6 for an SDOF oscillator with $f_n = 10\,\text{Hz}$ and $\zeta = 0.03$. The pseudo-velocity FRF is flat at low frequencies, below the SDOF resonant frequency. The magnitude is $1/2\pi f_n$ and the phase is $-180°$. The mass is moving in the opposite direction of the base at low frequencies. This is a consequence of the negative sign on the base acceleration term in Eq. 3.15. At frequencies above the SDOF resonant frequency, the phase angle is zero and the magnitude rolls off at 12 db/octave. Relative velocity is the derivative of relative displacement (i.e., the derivative of the pseudo-velocity divided by ω_n). At frequencies below the SDOF natural frequency, the magnitude increases at 6 dB/octave. Above the resonant frequency, the magnitude rolls off at 6 dB/octave. The phase angle starts

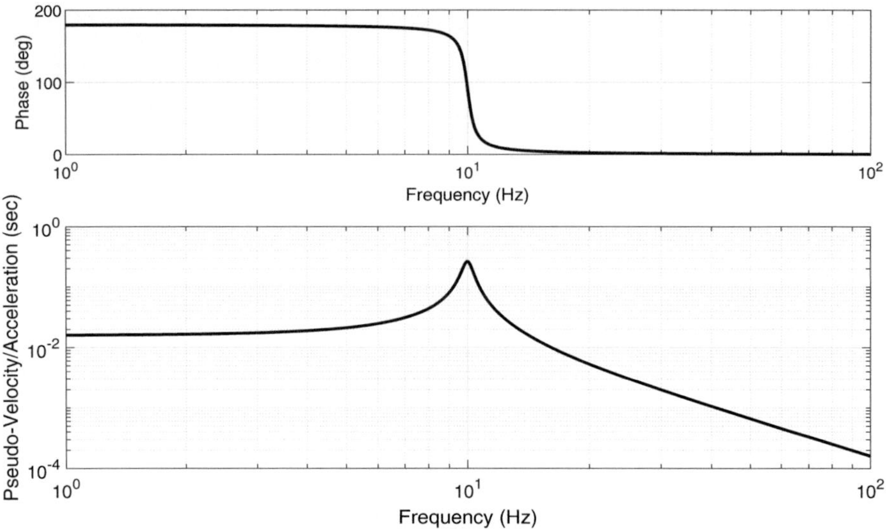

Fig. 3.23 Pseudo-velocity frequency response function, $PV/\ddot{z}$; $\omega_n = 10\,\text{Hz}$, $\zeta = 0.3$

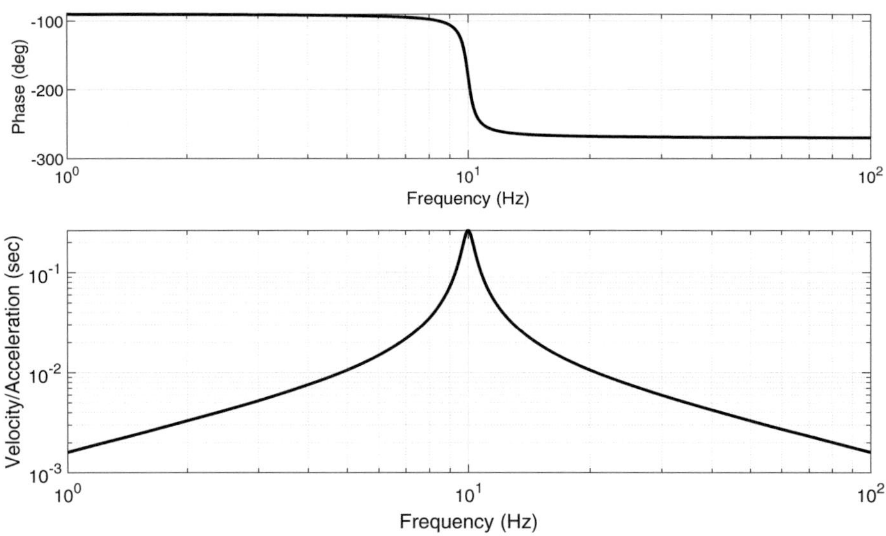

Fig. 3.24 Relative velocity frequency response function, $RV/\ddot{z}$; $\omega_n = 10\,\text{Hz}$, $\zeta = 0.3$

at $-90°$ and ends at $-270°$. The absolute acceleration FRF magnitude is similar to the pseudo-velocity FRF magnitude, differing by a factor of ω_n at low frequency. The low-frequency magnitude of the absolute acceleration is one and the phase angle is zero. The acceleration of the mass is in phase with the acceleration of the base at low frequencies. The magnitude decays at 12 db/octave above the resonant

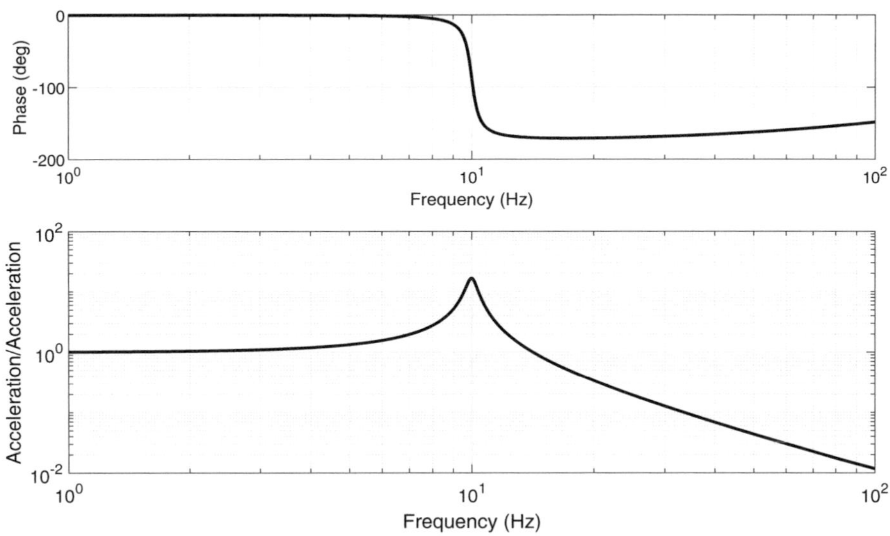

Fig. 3.25 Absolute acceleration frequency response function, $AA/\ddot{z}$; $\omega_n = 10\,\text{Hz}$, $\zeta = 0.3$

frequency, but the decay rate changes to 6 dB/octave at higher frequencies. The phase angle of the absolute acceleration FRF goes from 0° to −180° around the SDOF resonant frequency but it increases from −180° at higher frequencies, asymptotically approaching −90°.

3.4.2.2 Discrete Transfer Function

Figure 3.22 shows that a discrete time transfer function is the product of the hold function, the held input, and the continuous time transfer function followed by the output sampler. The discrete time transfer function is realized with the z-transform. An in-depth description of the z-transform and its properties and uses are beyond the scope of this book. The reader should consult texts such as Ref. [2] or online references on discrete time systems.

The discrete transfer function is

$$H(z) = \frac{Y(z)}{P(z)}, \tag{3.64}$$

which is a ratio of polynomials in z. Note that z is not the relative displacement of the SDOF oscillator; it is the inverse of the delay operator. The transform of the sequence $\{y(k)\}$ given by:

$$Y(z) = \sum_{k=-\infty}^{k=\infty} y(k) z^{-k} \tag{3.65}$$

and is called the z-transform of the sequence. One of the properties of the z-transform the we exploit is

$$\sum_{k=-\infty}^{k=\infty} y(k-1)z^{-k} = \sum_{j=-\infty}^{j=\infty} y(j)z^{-(j+1)} = z^{-1}Y(z). \tag{3.66}$$

The term z^{-1} is the delay operator and it is the discrete time analog of the Laplace operator, s. In the Laplace domain, the derivative of a function is the product of the Laplace operator and the Laplace transform of the function. If $Y(s)$ is the Laplace transform of $y(t)$, then $sY(s)$ is the Laplace transform of $\dot{y}(t)$ and so on. The Laplace transform converts an ordinary differential equation into an algebraic equation. This is the property that we used to create the transfer function given in Eq. 3.63.

In the z-domain, the sampled point $y(kT)$ is related to its predecessor $y((k-1)T)$ by the delay operator z^{-1}, $y((k-1)T) = z^{-1}y(kT)$. This is a recursive relationship which means that $y((k-2)T) = z^{-2}y(kT)$, and so on. The z-transform converts a difference equation into an algebraic equation just like the Laplace transform makes an algebraic equation from an ordinary differential equation. Using this property, we can compute the z-transform of the hold operator in Fig. 3.22.

The z-transform of the zero-order hold is

$$\mathrm{ZOH}(z) = 1 - z^{-1} = \frac{z-1}{z} \tag{3.67}$$

and the z-transform of the triangle hold (i.e., the noncausal first-order hold) is

$$\mathrm{FOH}(z) = \frac{(z-1)^2}{Tz}. \tag{3.68}$$

The transforms are given without proofs because their derivations can be found in texts or online references on discrete time systems.

Again referring to Fig. 3.22, the discrete transfer function, $H(z)$ is

$$H(z) = \mathscr{Z}(\mathrm{Hold})\mathscr{Z}(\hat{\ddot{w}}(t))\mathscr{Z}(H(s)), \tag{3.69}$$

where $\mathscr{Z}$ is the z-transform. The discrete transfer function depends on the hold we select. Table 3.7 shows the z-transform equations for the four hold functions described in Sect. 3.4.2.

The discrete transfer function is a ratio of polynomials in z as defined in Eq. 3.64. The order of the denominator polynomial is same as the order of the equation of motion. For the SDOF oscillator the order is two. The order of the numerator polynomial will always be one less than the order of the denominator polynomial. The only exception to this rule is a transfer function created with the triangle hold because it is noncausal. In this case, the orders of the numerator and denominator

Table 3.7 z-Transform functions of transfer functions with holds

Hold	$H(z)$
ZOH	$\frac{z-1}{z}\mathscr{Z}\left(\frac{H(s)}{s}\right)$
FOH	$\frac{(z-1)^2}{Tz^2}\mathscr{Z}\left(\frac{H(s)(Ts+1)}{s^2}\right)$
Triangle	$\frac{(z-1)^2}{Tz^2}\mathscr{Z}\left(\frac{H(s)}{s^2}\right)$
Impulse	$\mathscr{Z}(H(s))$

polynomials are the same. Therefore, the discrete transfer function of the SDOF oscillator is of the form:

$$H(z) = \frac{\beta_0 + \beta_1 z^{-1} + \beta_2 z^{-2}}{1 + \alpha_1 z^{-1} + \alpha_2 z^{-2}}. \tag{3.70}$$

Note that β_0 is nonzero only when we use the triangle hold.

The discrete transfer function in Eq. 3.70 can be written as a difference equation using the fact that z^{-1} is the delay operator:

$$\begin{aligned} y(kT) = \beta_0 \ddot{w}(kT) + \beta_1 \ddot{w}((k-1)T) + \beta_2 \ddot{w}((k-2)T) \\ -\alpha_1 y((k-1)T) - \alpha_2 y((k-2)T). \end{aligned} \tag{3.71}$$

Equation 3.71 is called an IIR (Infinite Impulse Response) filter. With it, we can compute the output, $y(kT)$, using only the previous two output values and the current and two previous values of the input $\ddot{w}$. This difference equation is the digital filter we sought. Since it is recursive, it is computationally efficient. Each output computation requires only five multiplications and four additions, and only five data points so the memory requirements are trivial.

The triangle hold and the ramp invariant filter were introduced to the shock community by Smallwood in 1981 [3] and in the intervening years the ramp invariant filter has become the de facto standard way to compute shock response. The ISO Standard 18431-4 [4] defines the digital filter and the filter coefficients for calculating the most common output quantities used in shock analyses based on the ramp invariant filter. The filter coefficients in the ISO standard are derived from Smallwood's work [3, 5].

The denominator coefficients are based on the equation of motion and they are independent of the hold used. For the SDOF oscillator equations of motion, Eq. 3.3 or 3.15:

$$\begin{aligned} \alpha_1 &= 2e^{-A}\cos(B), \\ \alpha_2 &= e^{-2A}. \end{aligned} \tag{3.72}$$

The constants A and B in Eq. 3.72 along with an additional constant C that will be used later are given by:

$$
\begin{aligned}
A &= \frac{\omega_n T}{2Q}, \\
B &= \omega_n T\sqrt{1-\frac{1}{4Q^2}}, \\
C &= \frac{\frac{1}{2Q^2}-1}{\sqrt{1-\frac{1}{4Q^2}}}.
\end{aligned}
\tag{3.73}
$$

where Q is the quality factor, defined in Eq. 3.8. Rewriting Eq. 3.73 in terms of ζ yields

$$
\begin{aligned}
A &= \zeta\omega_n T, \\
B &= \omega_n T\sqrt{1-\zeta^2}, \\
C &= \frac{2\zeta^2-1}{\sqrt{1-\zeta^2}}.
\end{aligned}
\tag{3.74}
$$

The α_1 and α_2 coefficients given in Eq. 3.72 along with the constants A and B are the same for all transfer function variants with C being used for only some of the filter variants. Only the numerator coefficients β_1, β_2, and β_3 depend on the output quantity. The ISO standard provides filter coefficients for several types of shock spectra and since the coefficients are codified, most calculations are performed using these algorithms. For example, the numerator coefficients for the absolute acceleration output are given in the ISO standard as:

$$
\begin{aligned}
\beta_0 &= 1-e^{-A}\left[\frac{\sin(B)}{B}\right] \\
\beta_1 &= 2e^{-A}\left[\frac{\sin(B)}{B}-\cos(B)\right] \\
\beta_2 &= e^{-2A}-e^{-A}\left[\frac{\sin(B)}{B}\right]
\end{aligned}
\tag{3.75}
$$

If one wants to create a digital filter other than those in the ISO standard, the algebra to determine the analytical expressions for the coefficients is very tedious, but that can be overcome using symbolic mathematical tools like MathCAD or Mathematica (i.e., Wolfram Alpha). However, the analytical work is unnecessary because the discretization of a transfer function can be easily done numerically.

Any linear system can be expressed in state-space form:

$$\begin{aligned} \dot{x}(t) &= Fx(t) + Gp(t), \\ y(t) &= Cx(t) + Dp(t). \end{aligned} \tag{3.76}$$

For example, the state-space model matrices of the SDOF oscillator represented by the equation of motion in Eq. 3.9 and pseudo-velocity output (Eq. 3.19) are

$$\begin{aligned} F &= \begin{bmatrix} 0 & 1 \\ -\omega^2 & -2\zeta\omega \end{bmatrix}, \\ G &= \begin{bmatrix} 0 \\ -1 \end{bmatrix}, \\ C &= [\omega_n \quad 0], \\ D &= 0. \end{aligned} \tag{3.77}$$

The discrete state-space model of a linear system, discretized with the triangle hold with sampling period T, is (e.g., [2, 6])

$$\begin{aligned} x(k+1) &= F_d x(k) + G_d p(k) \\ y(k) &= C_d x(k) + D_d p(k), \end{aligned} \tag{3.78}$$

where

$$\begin{aligned} F_d &= e^{TF}, \\ G_d &= \frac{1}{T} F^{-2} \left(e^{TF} - I\right)^2 G_d, \\ C_d &= C, \\ D_d &- D + \frac{1}{T} C F^{-2} (e^{TF} - I) G - F^{-1} G. \end{aligned} \tag{3.79}$$

The matrices F_d, G_d, C_d, D_d can be easily computed with the above equations.

The step invariant filter is created by discretizing the state-space model with a zero-order hold. The result is similar:

$$\begin{aligned} F_d &= e^{TF}, \\ G_d &= F^{-1} \left(e^{TF} - I\right) G_d, \\ C_d &= C, \\ D_d &= 0. \end{aligned} \tag{3.80}$$

These transformation have been implemented in many engineering computational packages. For example, the MATLAB function *c2d* does them.

The state-space model, Eq. 3.78 can be transformed into a transfer function:

$$H(z) = C_d(zI - F_d)^{-1}G_d + D_d. \tag{3.81}$$

Again, this operation can be found in many computational packages (e.g., in MATLAB the function is ss2tf).

Comparing Eqs.3.79 and 3.80, the discrete state matrices, F_d, are the same. The differences are in the input and output matrices. These differences are significant because the step invariant filter is not discrete-time positive real. The damped SDOF oscillator is a positive real system. A positive real dynamic system is one in which the real parts of the poles are greater than zero. This means that positive real systems are stable and dissipate energy. It is desirable to retain this property in the digital filter model. A discrete time system must have a nonzero feedthrough term to be positive real (i.e., $D_d \neq 0$) [6]. The feed through term of the step invariant filter, D_d, is zero whereas the feed-through term of the ramp invariant filter is not, so positive realness is preserved by the ramp invariant transformation. Admittedly positive realness of the digital filter is a subtle point. In almost all situations either the step invariant filter or the ramp invariant filter should work just fine.

3.4.2.3 Digital Filters and Discontinuous Functions

The discretization of the SDOF oscillator dynamics introduces an error in the response when the shock has discontinuities. This is because of the assumption that is required about the shape of the shock between sample points. Consider an impulsive shock. The responses to an impulse, which is the delta function at $t = 0$, are summarized in Table 3.3. The response calculated with the digital filter must be scaled by $1/dt$ because the discretization spreads the impulse over the time step. For a train of impulses, the impulse invariant filter is preferred. Similarly, for quantized shocks, the step invariant filter produces the best results. Most shocks, while short, are smooth if the sample rate is high enough, so the ramp invariant filter is the best choice.

3.4.2.4 Selection of Sample Rate

The selection of the sample rate is perhaps the most important aspect of the numerical problem. When performing simulations, usually we have the freedom to create our own discrete input sequence. If the sample rate is too low relative to the natural frequency of the SDOF system, the peak response may be significantly underestimated. If the sample rate is high, the solution will be slow. A rule of thumb is that the sampling rate should be at least ten times the natural frequency if the peak dynamic response must be estimated accurately.

The accuracy of the digital filters is a function of the ratio of the SDOF oscillator natural frequency and the sampling frequency. Figure 3.26 shows the effect of

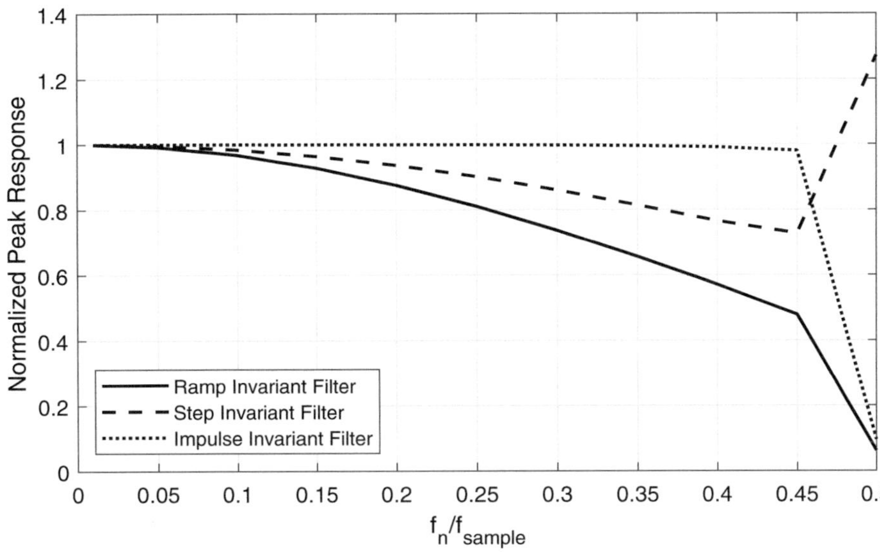

Fig. 3.26 Low-pass filter attenuation as a function of sampling frequency ratio

sample rate on the peak magnitude of the absolute acceleration frequency response of the three digital filters derived previously. The hold in the step invariant and ramp invariant filters acts as a low-pass filter, so as the SDOF natural frequency approaches the Nyquist frequency (i.e., one half of the sampling frequency) the response will be attenuated.

The ZOH in the step invariant filter multiplies the SDOF response by the magnitude of a sinc function:

$$|\text{ZOH}| = \frac{\sin(\pi f_n T)}{\pi f_n T}, \tag{3.82}$$

where f_n is the natural frequency of the SDOF oscillator and T is the sample time ($T = 1/f_{sample}$).

The low-pass filter effect of the ramp invariant filter is the square of the sinc function:

$$|TRI| = \left(\frac{\sin(\pi f_n T)}{\pi f_n T}\right)^2. \tag{3.83}$$

When the sample rate is too low, the frequency response of the discrete transfer function deviates from the frequency response of the continuous transfer function. This deviation has a potentially severe impact on the computed response. This is shown in Fig. 3.26. Note that at 45% of the sampling frequency all the filters experience numerical issues.

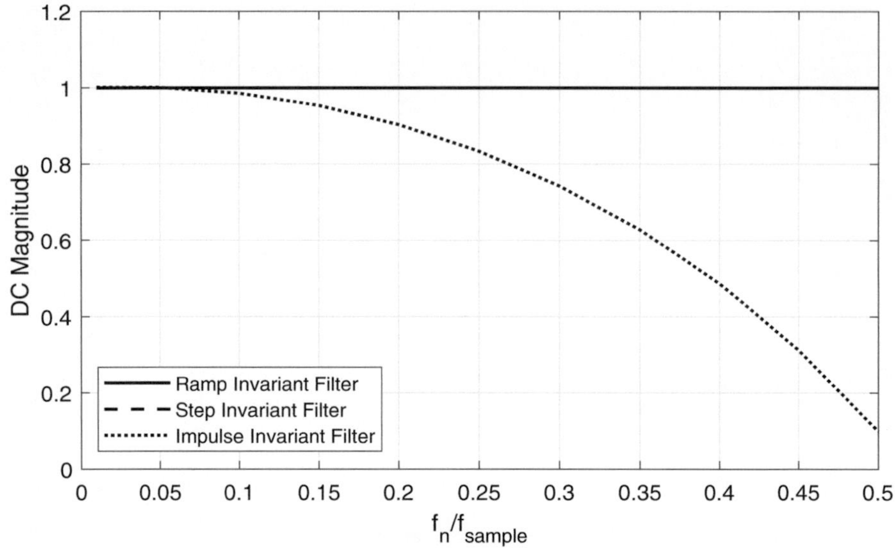

Fig. 3.27 DC gain error as a function of sampling frequency ratio

The step invariant filter shows less attenuation than the ramp invariant filter and the impulse invariant filter has no low-pass effect. This might suggest that the impulse invariant filter is the best choice. However, the impulse invariant filter has one major deficiency—its DC response is not error free. This is shown in Fig. 3.27.

As a general rule, the sampling frequency should be at least ten times the SDOF oscillator natural frequency to limit the low-pass filter attenuation. There is an additional source of error associated with the digital signal processing. Since the true analog signal is only sampled at discrete intervals, it is likely that the true maximum will occur between samples. This error can add to the filter bias error. With today's modern data acquisition systems, this is generally not a concern with MHz sampling rates becoming relatively common. Nevertheless, it has been a significant enough concern that certain SRS calculations using Duhamel's integral were performed long after the much faster digital filter methods became mainstream.

3.5 Summary

In this chapter, we introduced the equations of motion of a single degree-of-freedom oscillator and derived the equations for free and forced response. The forced response is also called the primary response and the free response is also called the residual response.

We derived the responses of an SDOF oscillator to three basic transient excitations. These excitations are used in the general shock response computations.

The general response solution involves the numerical evaluation of a convolution integral, which is efficiently done with digital filtering. We derived the equations for three digital filters: the impulse invariant filter, the step invariant filter, and the ramp invariant filter. These are differentiated by the hold function used.

The ramp invariant filter, which involves the (noncausal) triangle hold function is somewhat standard, as evidenced by its use in the ISO standard. The zero-order hold and the triangle hold act as low-pass filters. This introduces an attenuation error if the sampling frequency is not much higher than the SDOF oscillator natural frequency. The low-pass filter attenuation of the step invariant filter is smaller than that in the ramp invariant filter, which suggests that it has better performance. However, this is a bit of a moot point because today's computers are so fast, and numerical computing software is so capable and versatile that the engineer is not limited to any one method. For example, MATLAB contains functions that can easily perform any shock calculations. The main MATLAB functions are *ss*, *tf*, *ss2tf*, *lsim*, *c2d*, and *filter*.

Problems

3.1 A vehicle traveling along a road at a constant speed V can be modeled as the single degree-of-freedom system shown in Fig. 3.28. The vehicle encounters a speed bump that can be represented as a half-sine pulse $w(s) = A\sin(2\pi s/L)$. Assume the wheel is massless.

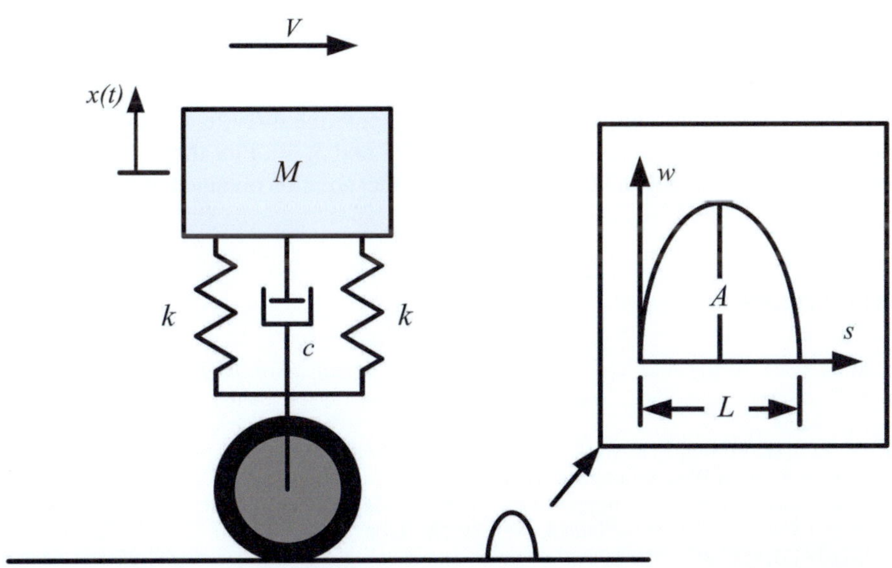

Fig. 3.28 Problem 3.1, vehicle encounters a bump

1. What is the equation of motion in terms of the absolute motion quantities?
2. What is the equation of motion in terms of the relative motion quantities?
3. What is the output equation if quantity of interest is the force in the damper?
4. What is the output equation if the quantity of interest is the force in each spring?

3.2 The vehicle in Fig. 3.28 has a mass of 1000 kg. Each spring has a stiffness of $k = 8 \times 10^6$ N/m. The damping coefficient is $c = 25 \times 10^3$ N s/m.

1. What is the undamped natural frequency of the vehicle?
2. What is the damping ratio of the vehicle?
3. Is this damping ratio appropriate for a shock absorber?

3.3 Find the free vibration (i.e., before it encounters the bump, so $w(t) = 0$) motion $x(t)$ for the vehicle in Problem 3.1 for initial conditions $x(0) = 10$ mm and $\dot{x}(0) = 0$.

3.4 For the vehicle in Problem 3.1, assume that it is traveling at $V = 22.5$ miles/h. The height of the half-sine bump is $A = 100$ mm, and the length is $L = 0.1$ m. Using the mass properties given in Problem 3.2, derive the expressions for the forced response and subsequent free vibration response of the following quantities:

1. the absolute acceleration of the mass (i.e., $\ddot{x}(t)$).
2. the total force in the springs.

Plot the vehicle response from the expressions that you derived for some suitable time.

3.5 Again using the SDOF vehicle in Problem 3.1, write the equation of motion and the output equations in state-space form (see Eq. 3.76).

3.6 Plot the frequency response functions from base acceleration to spring force and from base acceleration to absolute acceleration of the mass.

3.7 Select a suitable sampling period and compute the discrete state-space model discretized with the triangle hold (see Eqs. 3.78 and 3.79). Plot the response of the vehicle with this model and compare it to the exact solution obtained in Problem 3.4.

References

1. Kelly, R. D., & Richman, G. (1969). *Principles and techniques of shock data analysis. SVM-5*. Washington: The Shock and Vibration Information Center, United States Department of Defense.
2. Franklin, G., Powell, J., & Workman, M. (1990). *Digital control of dynamic systems* (2nd ed.). Boston: Addison-Wesley Longman Publishing.
3. Smallwood, D. (1981). An improved recursive formula for calculating shock response spectra. In *51st Shock and Vibration Bulletin* (Vol. 51, No. 2, pp. 211–217).
4. ISO 18431-4. (2007). *Mechanical vibration and shock — Signal processing — Part 4: Shock-response spectrum analysis*. Geneva: International Standard.

5. Smallwood, D. (2005). Derivation of the ramp invariant filter for shock response spectrum calculations. In *Proceedings of the 76th Shock and Vibration Symposium*.
6. Hoagg, J. B., Lacy, S. L., Erwin, R. S., & Bernstein, D. S. (2004). First-order-hold sampling of positive real systems and subspace identification of positive real models. In *Proceedings of the 2004 American Control Conference* (pp. 861–866), Paper WeM07.3.

Chapter 4
Shock Environment Characterization Using Shock Response Spectra

This chapter introduces the shock response spectrum or SRS. The SRS is one of the fundamental tools in shock analysis. Because a shock is a transient phenomenon, typical analysis techniques like the Fourier transform and power spectral density, which are very effective in studying random vibration phenomena, are not completely appropriate. The shock response spectrum provides information about the spectral content of the transient signal much the same way that the power spectral density provides information about the spectral content of long duration random vibrations.

As discussed in Chap. 1, damage created by shocks, whether it is structural or functional, is usually a function of some type of overloading of the structure. For example, straining a structural member beyond its strain limit causes fracture in brittle materials and plastic deformation in ductile materials. Large displacements can cause functional failures such as circuit breaker trips as mentioned in Chap. 1. Therefore, in assessing the damage potential of a shock, we are primarily interested in the peak response of a structure to the shock event.

The SRS is a plot of an extremal response quantity of interest from a series of SDOF oscillators subjected to a transient or shock excitation. The SDOF oscillators are parameterized by their natural frequency and damping ratios. This makes the SRS a tool to assess the potential severity of shock excitation on a structure. The response quantity is a peak quantity because that is what causes damage. Figure 4.1 illustrates the process of calculating the SRS. For each SDOF oscillator, we compute the response of interest, the maximum absolute acceleration of the SDOF mass to the base excitation, for example. From that time history, the peak value is selected and that value is plotted for that SDOF natural frequency. This is repeated for a range of SDOF oscillators. Often we will compute the response quantity for various damping ratio values as well.

The SRS has many uses in shock analysis. In addition to providing information about the spectral content of the transient excitation, it can be used when designing structures for shock loading. We usually solve vibration problems in which the

C. Sisemore, V. Babuška, *The Science and Engineering of Mechanical Shock*,
https://doi.org/10.1007/978-3-030-12103-7_4

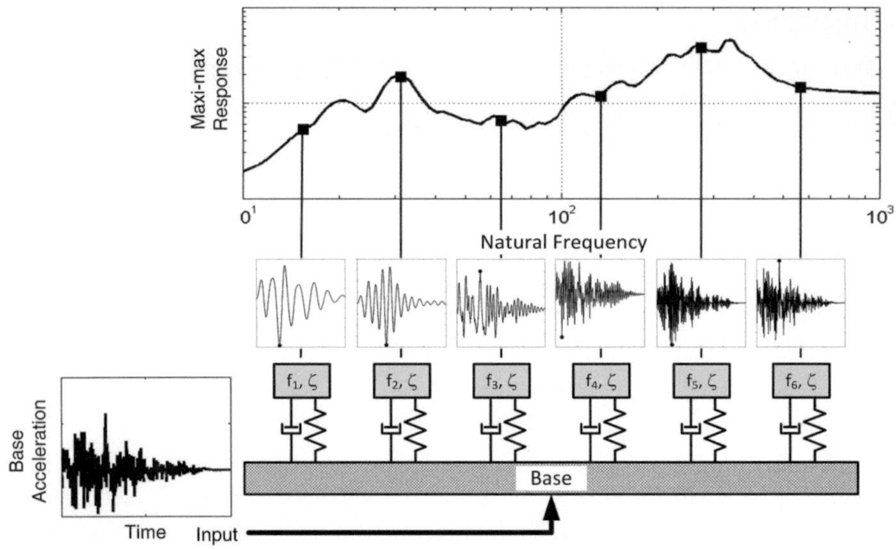

Fig. 4.1 Shock response spectrum derived from a series of SDOF oscillators

system parameters—mass, stiffness, and damping—are known a priori. However, when designing a structure we have the ability to select the system parameters so that the system response to a particular shock is within an acceptable range. The SRS is a tool for that because the response quantity of interest is plotted as a function of structural parameters, and the SRS can be used to inform the selection of those parameters. For example, if a structure can be adequately approximated by an SDOF oscillator, then the structure's natural frequency should be located in a region where the SRS is low. For the SRS shown in Fig. 4.1, this is between 60 and 100 Hz. Likewise, regions with a high SRS indicate frequency ranges where the shock has the greatest damage potential. The base excitation example shown in Fig. 4.1 has the largest damage potential between 200 and 600 Hz.

Because the response quantity in the SRS is an extremal value, any excitation that produces the same SRS should be equally damaging to a single degree-of-freedom system. This characteristic is often exploited when developing test environments. Consider the scenario where we have an aircraft component for which we measured the shock environment during flight tests, and we want to replicate this shock severity in the laboratory. If the SRS of the laboratory environment matches the SRS of the flight environment, both should be equally damaging and the laboratory environment will be a reasonable representation of the flight environment's damage potential.

In this chapter we explain:

1. the history of the SRS;
2. the main features and limitations of the SRS;
3. the primary and residual SRS;

4. the different types of shock spectra used in various industries;
5. numerical techniques to efficiently compute the SRS.

In later chapters we discuss how the SRS is used for design and environmental definition. Techniques for analyzing and understanding the response to shock excitation are covered in later chapters as well.

It is important to note that the term "response spectrum" refers generally to some quantity computed for a series of parameterized dynamic systems. For example, in an energy response spectrum, the output quantity of interest is energy. In the fatigue damage spectrum, the output quantity of interest is a fatigue damage index [1]. Almost always, the dynamic system for which a response spectrum is computed is the linear SDOF oscillator, but that is not a firm constraint. A response spectrum based on non-linear SDOF oscillators could be calculated if one were so inclined.

4.1 History of the Shock Response Spectra

Maurice Biot is generally credited with the idea of the response spectrum for analyzing transient and shock excitation. The theory was first put forth in his Ph.D. thesis on the analysis of buildings subject to earthquake excitations at the California Institute of Technology in 1932 [2, 3]. Biot did not actually publish any response spectra in his thesis; however, the concepts for the shock response spectra are contained there. Biot further developed his response spectrum theory in the 1930s and early 1940s. By the early 1940s spectra plots were beginning to appear in the common literature, specifically in other publications by Biot [4] as well as Housner [5] with both focused on earthquake analysis. George Housner actually published several earthquake response spectra in his thesis, although in earthquake engineering spectra are often plotted with the period (the inverse of frequency) on the abscissa.

The concept of the response spectrum as proposed by Biot was derived from the Fourier series. Jean-Baptiste Joseph Fourier proposed the idea that any periodic function can be expressed by a series summation of sine and cosine terms.

$$f(t) = \frac{a_0}{2} + \sum_{n=1}^{\infty} a_n \cos(n\omega_0 t) + \sum_{n=1}^{\infty} b_n \sin(n\omega_0 t) \tag{4.1}$$

Fourier first published this theory in an essay titled the *Mathematical Theory of Heat* to the French Academy of Science in 1812. Although his essay won the academy prize, he was criticized by the panel (Joseph Lagrange, Pierre Laplace, and Adrien Legendre) for a certain looseness in his reasoning [6]. The work was later published more formally as the *Analytical Theory of Heat* in 1822 [7]. Convergence of the series was not proven until 1829 by Dirichlet [6]. It can only be added here that if the theory of the Fourier series was originally considered to be loosely proven, the shock response spectrum theory is somewhat looser still.

Fundamentally, the Fourier series works because the summation of sines and cosines forms a complete basis and because the sines and cosines are orthogonal on any $2\pi/T$ interval. The completeness of the Fourier series means that any periodic waveform may be converted to an infinite Fourier series. Its completeness also implies that it can be inverted. Thus, if all the terms in the series are used, the original time history is completely recoverable. In practice of course, one never uses all the terms in an infinite series for simple reasons of practicality.

How does the Fourier series relate to the relatively modern concept of shock response spectra? The Fourier series decomposes a function $f(t)$ into a summation of pure sine and cosine signals. Each pure sine or cosine response corresponds to a single frequency given by $n\omega_0$. Recall the free vibration response of the undamped SDOF oscillator from Sect. 3.3.1. The equation of motion for the undamped single degree-of-freedom system in free vibration is

$$\ddot{z}(t) + \omega_n^2 z(t) = 0. \tag{4.2}$$

The free vibration response is given by Eq. 3.25 as

$$z(t) = A_1 \cos(\omega_n t) + A_2 \sin(\omega_n t). \tag{4.3}$$

Note the similarity to Eq. 4.1. The free vibration response of the undamped SDOF oscillator is essentially one term in the Fourier series. If a structure can be represented as a series of undamped SDOF oscillators, then the free vibration response of the structure is the sum of the free vibration responses of the SDOF oscillators.

In Biot's original work there was no accounting for damping effects in the spectra. The justification for this was that earthquake transients are relatively short and damping mechanisms do not fully develop during a short transient event. This is somewhat ironic since almost all modern SRS calculations assume some small level of damping and nearly all mechanical shock events are of a shorter duration than the earthquake excitation from which the undamped SRS was developed. The undamped response is also slightly more severe making it appropriate (i.e., conservative) for design.

There is no requirement to evaluate the Fourier series at every frequency in the infinite series, assuming that it could even be accomplished. The Fourier series naturally relates to the Fourier transform. When the Fourier transform is calculated, the frequencies for the transform are naturally derived from the algorithm with a regular spacing derived from the sample rate and length of the original time history. However, there is nothing in the algorithm preventing the analyst from choosing a set of arbitrary frequencies at which to perform the transform. The only limitation is that if an arbitrary subset of frequencies are chosen then the transform is no longer reversible.

Shock response spectra deviate from the Fourier series approach in three primary ways. First, a shock is not a periodic signal, it is a transient signal. Second, shock response spectra are calculated at arbitrary frequencies. The shock response spectrum methodology starts by picking a somewhat arbitrary set of frequencies for which to calculate the single degree-of-freedom response. By performing the calculation in this manner, a significant number of terms in the Fourier series are simply ignored. The third difference is that the entire response history of each SDOF oscillator is not used. Only a scalar, extremal response of interest from the output equation, like the absolute acceleration, is retained. This means that the temporal phasing of the SDOF responses is lost, therefore the Fourier sum cannot be computed and the SRS is not reversible.

These considerable simplifications fundamentally make the shock response spectrum calculation a non-linear and non-unique operation. This means that the original function from which the SRS was generated cannot be recovered. Furthermore, it means that more than one function can generate the same SRS. These simplifications were no doubt necessary and desirable in the 1930s with the limited computational capabilities. It should be recalled the fast Fourier transform algorithm was not yet developed, nor any type of digital computing capability.

4.1.1 Shock Spectra Calculations

The earliest shock response spectra calculations were performed either by hand or with analog computers. This was laborious and inefficient. In Biot's 1941 mechanical analyzer paper he describes a pendulum type mechanical analyzer (otherwise known as an analog computer) used to calculate the acceleration spectrum from an earthquake event [4]. The paper notes that the machine could calculate an acceleration spectrum in about 8 h on average and the spectrum plots shown contain about 40 data points. No doubt quite sophisticated at the time.

The actual calculation of the SRS is derived from the theory of the SDOF oscillator derived in Chap. 3. To calculate a shock response spectra, a series of frequencies are selected. The selection of frequencies can be completely arbitrary but it is common to select a sequence that is logarithmically spaced. The reason for this is that SRS are almost always presented as log–log plots and a logarithmic spacing of frequencies is most efficient for this type of plot. The other reason for a logarithmic distribution of frequencies is that higher resolution is typically not required as the frequency increases. Typically the engineer is most interested in the first few system modes and the response in the vicinity of those modes. In addition, while acceleration can be high at high frequencies, displacements are typically very low, consequently strain is low and damage potential is typically low for most ductile materials.

With the calculation frequencies selected, the SDOF response at the selected frequencies is calculated. This can be accomplished several ways depending on the complexity of the signal. If the shock is a very simple, classical shock, it may be possible to integrate Duhamel's integral using a closed-form technique. However, it is more likely that some form of numerical technique will be required to calculate the SDOF response. Solving the SDOF differential equations of motion is not complicated and was reviewed in Sect. 3.4.2. While solving the equation of motion directly is straightforward enough, it is also a relatively time-consuming operation. With the advent of digital computers and the increasing use of shock spectra, proposals making use of digital filters to perform the spectra calculations began to appear in the literature. Digital filters offer a significant advantage in computational speed with some limitations at high frequencies. Due to the significant computational speed advantage, nearly all SRS calculations are now accomplished with digital filters, specifically ramp invariant filters. The coefficients of the ramp invariant filters for typical quantities of interest are codified in an ISO standard [8]. The details were described in Sect. 3.4.2.2.

In order to present the basics of the SRS, it is useful to consider a classical example. Figure 4.2 shows a plot of a classical haversine shock pulse. The pulse shown here has a maximum amplitude of 1000 g and a pulse width of 5 ms. The shape of the haversine pulse is a shifted cosine waveform that starts and ends at zero with zero slope at the pulse ends. Although this is a classical pulse, it is also a pulse that is very representative of many common shock environments shown in Sect. 3.4.2.2.

To calculate the SRS, the input shock pulse is used as an input to the SDOF oscillator equation of motion. The output response quantity of interest is calculated and the time history is searched to find the minimum and maximum of the oscillatory response. Examples of this are shown in Figs. 4.3, 4.4, 4.5, and 4.6 for four frequencies of SDOF oscillator. Figure 4.3 shows the response of a 30 Hz SDOF oscillator to the example haversine shock. As can be seen here, the response is lower than the input and with a significantly longer duration. The maximum acceleration here is about 445 g compared to the 1000 g input. The 30 Hz SDOF oscillator is "slow" and the shock pulse is over before the oscillator can get moving. Figure 4.4 shows the response of a 100 Hz SDOF oscillator to the same haversine input. In this

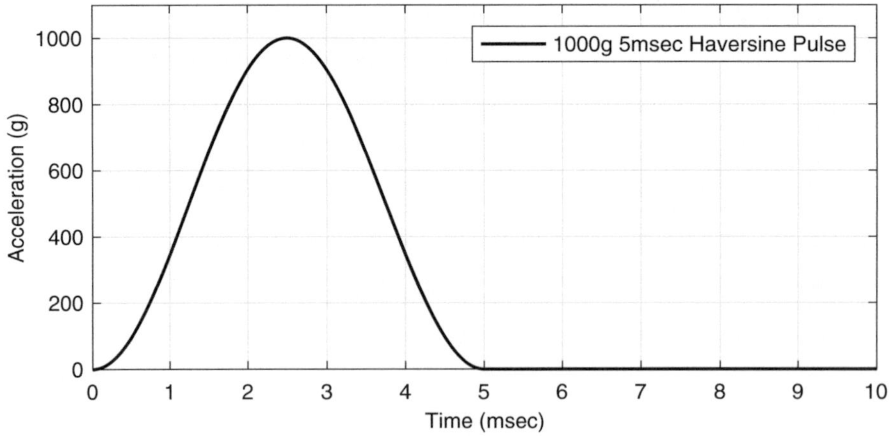

Fig. 4.2 Plot of a 1000 g 5 ms duration haversine shock pulse

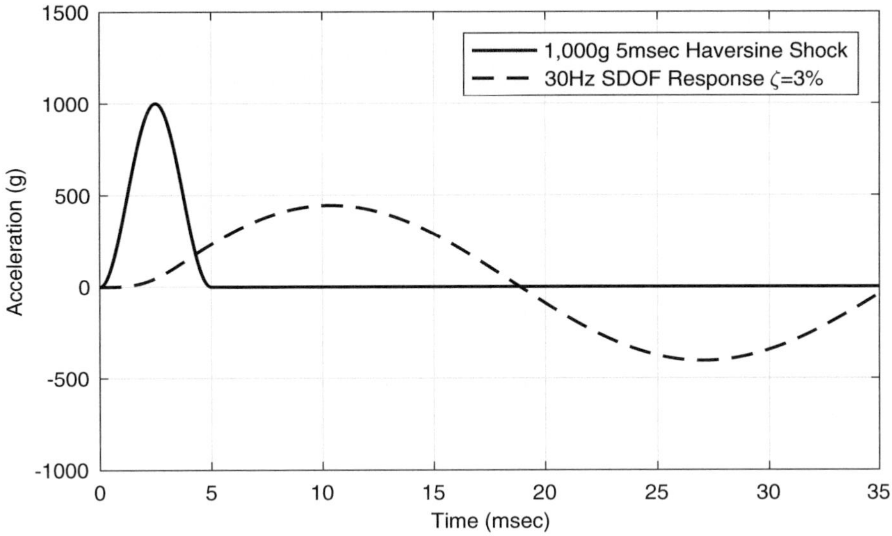

Fig. 4.3 SDOF 30 Hz response to 1000 g 5 ms duration haversine shock pulse

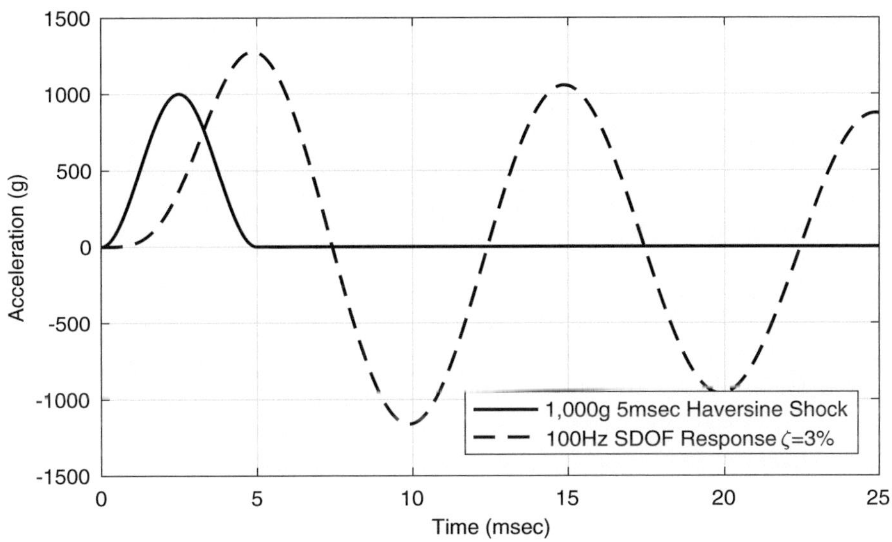

Fig. 4.4 SDOF 100 Hz response to 1000 g 5 ms duration haversine shock pulse

case, the peak response is greater than the input at 1276 g although the response period is still longer than the input shock duration. The shock pulse is exciting the resonant response of the 100 Hz oscillator. Figure 4.5 shows the 500 Hz SDOF oscillator response. In this case the oscillator frequency is now higher than the input shock frequency and the output response is starting to track much closer to the

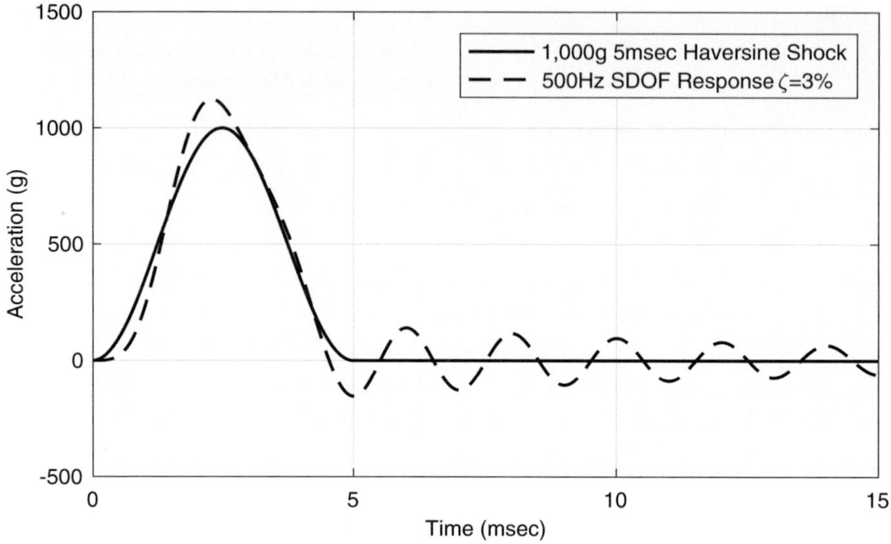

Fig. 4.5 SDOF 500 Hz response to 1000 g 5 ms duration haversine shock pulse

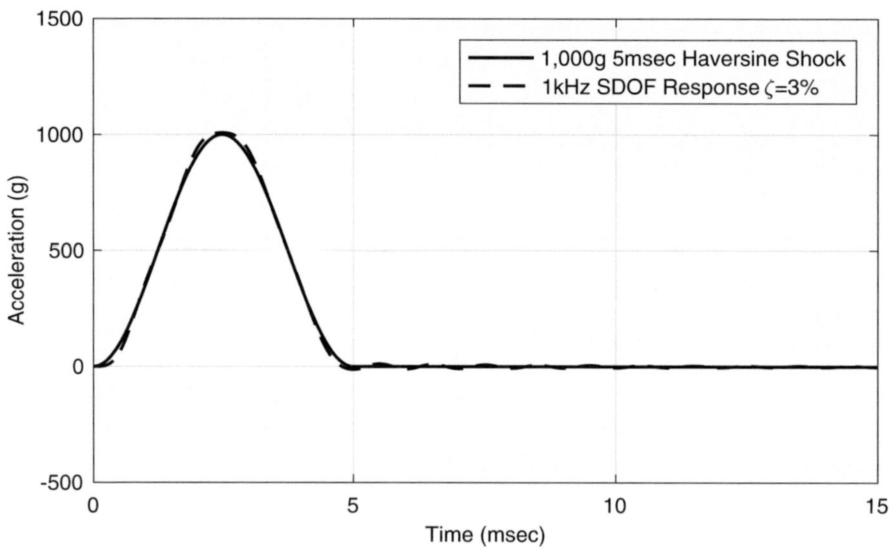

Fig. 4.6 SDOF 1 kHz response to 1000 g 5 ms duration haversine shock pulse

input excitation. The peak acceleration here is 1123 g, less than the response for the 100 Hz oscillator, but there is still some resonant amplification. Finally, Fig. 4.6 shows the response of a 1 kHz oscillator to this same haversine excitation. The response here almost exactly tracks with the input indicating that there is relatively little dynamic excitation of the SDOF oscillator. The peak response acceleration is 1008 g, differing less than 1% from the input. This is entirely reasonable since the 5 ms pulse duration corresponds to a 200 Hz input. The 1 kHz oscillator response frequency is five times higher than the input excitation and the system is essentially behaving like a rigid system for all practical purposes.

The SDOF oscillator responses shown in Figs. 4.3, 4.4, 4.5, and 4.6 were all calculated using an assumed viscous damping ratio of 3% of critical damping. When an SRS is calculated and displayed, it is important to always indicate the level of damping used as this directly impacts the appearance and magnitude of the SRS. Typically low values of damping are selected such as 3% or 5% of critical damping. Sometimes there is a benefit to assuming no damping—such was the case when the SRS was originally developed. Higher damping values can be used but it is cautioned that they may not represent physically realizable systems.

It should also be noted in Figs. 4.3, 4.4, 4.5, and 4.6 that the response was calculated in all cases for a substantial amount of time after the shock pulse had ended. Typically the response is calculated for about one full period of the oscillator period beyond the transient. This is necessary to ensure both the maximum and minimum oscillator residual response can be captured. This is shown most clearly in Fig. 4.3 where the peak response occurs in the residual vibration era, well after the shock has ended.

When SRS are calculated at each oscillator natural frequency, a portion of the response occurs while the shock event is still driving the SDOF oscillator and a portion occurs after the shock has ended. These two regions are called the forced response era and the free vibration era. The SRS during the forced response era is called the *primary spectrum* and the SRS in the free vibration era is called the *residual spectrum*. The maximum response amplitude can occur during either the positive or negative portions of the stroke. The maxi-max spectrum is found by searching the SDOF response time history for the greatest absolute value response whether it occurs before or after the end of the shock transient. This can be seen in Fig. 4.3 where the maximum response of 445 g occurs well after the shock transient has past and in Fig. 4.5 where the maximum response of 1123 g occurs near the peak of the transient.

4.1.2 *Maxi-Max SRS*

The most common presentation of SRS results is the maxi-max spectrum. The maxi-max absolute acceleration (often abbreviated as MMAA) SRS is given by

$$\mathrm{MMAA}(f) = \max_t |\ddot{x}(t)| \forall t, \ 0 < t < \infty. \tag{4.4}$$

Figure 4.7 shows a plot of the maxi-max absolute acceleration SRS of the ideal haversine time history shown in Fig. 4.2. The SDOF damping ratio was chosen

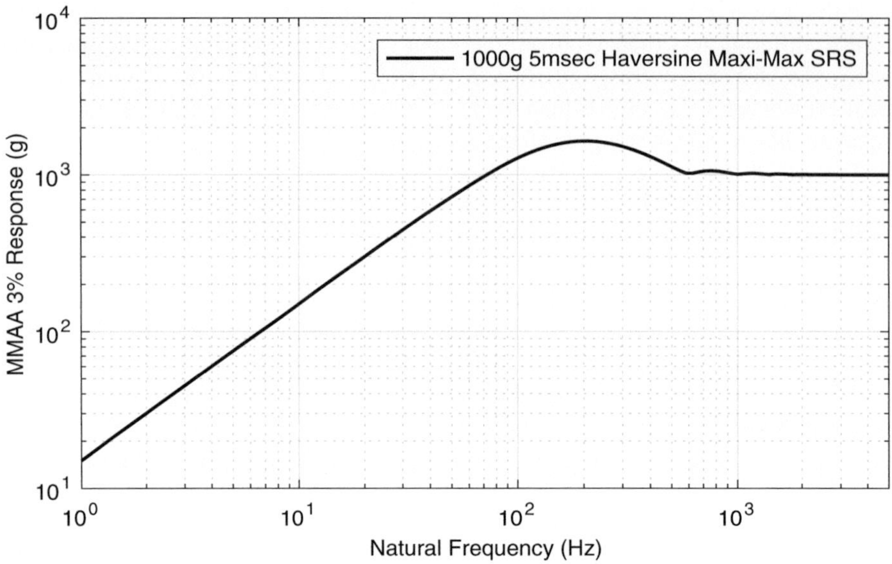

Fig. 4.7 Maxi-max absolute acceleration SRS plot of a 1000 g 5 ms duration haversine shock pulse

as 3%, which is a typical value. This SRS was calculated at 128 frequencies logarithmically spaced between 1 Hz and 10 kHz to create a smooth SRS plot. This plot shows how the data presented in the time history responses are combined to create the SRS. At frequencies well below the frequency of the transient, 200 Hz in this example, the maximum oscillator response is lower than the maximum input of the shock transient (refer back to Fig. 4.3). The oscillator response at frequencies near the shock transient frequency is greater than the input shock amplitude, 1000 g in this example (refer back to Figs. 4.4 and 4.5). Finally, at oscillator frequencies well above the shock transient frequency the oscillator response is essentially equal to the shock input and the plot levels off to the peak acceleration of the input shock (refer back to Fig. 4.6).

Figure 4.8 shows a plot of the maxi-max pseudo-velocity SRS for the same ideal haversine time history shown in Fig. 4.2. Pseudo-velocity was introduced in Chap. 3 and the pseudo-velocity SRS will be discussed further in Sect. 4.2.3. The pseudo-velocity SRS is defined as

$$\mathrm{MMPV}(f) = \max_t |2\pi f z(t)| \forall t, \ 0 < t < \infty. \tag{4.5}$$

A cursory examination of Figs. 4.7 and 4.8 indicates that very similar information is displayed in these two figures, although there is obviously some difference in the way the SRS are calculated. The pseudo-velocity is almost a rotation of the MMAA SRS curve. This is because pseudo-velocity is approximately the absolute acceleration divided by circular natural frequency (see Eq. 3.21).

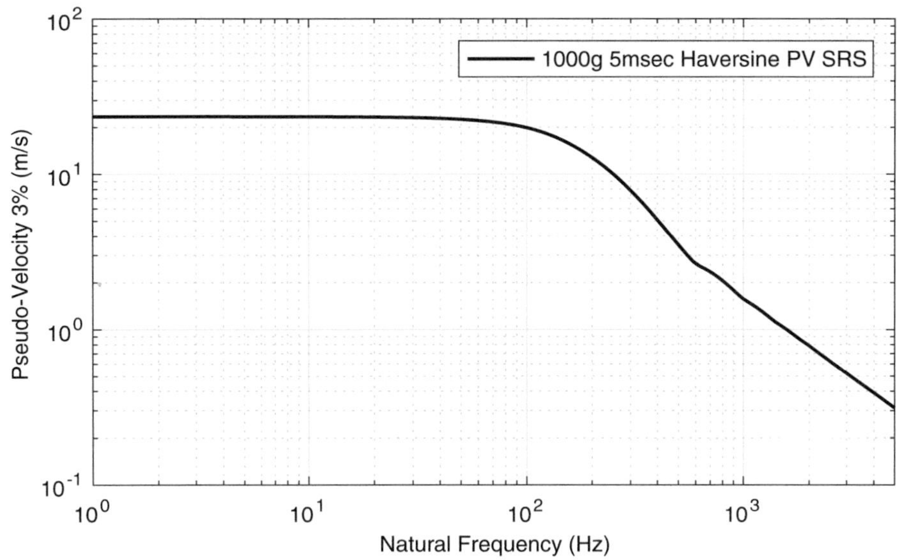

Fig. 4.8 Maxi-max pseudo-velocity SRS plot of a 1000 g 5 ms duration haversine shock pulse

The reason that this SRS is referred to as "pseudo-velocity" as opposed to simply velocity is that while it has the units of velocity, it is not exactly a true velocity; however, it does provide a good estimate of the velocity change associated with the haversine shock. Figure 4.8 shows that the change in velocity is approximately 23.4 m/s for this shock transient. This result can be obtained from the slope of the low-frequency portion of the spectrum in the MMAA SRS plot shown in Fig. 4.7 but it cannot be read directly. In contrast, the pseudo-velocity, which is almost equal to the velocity change, can be read directly from the MMPV SRS plot shown in Fig. 4.8.

Pseudo-velocity is also more closely related to the induced stress than the absolute acceleration. In this case, Fig. 4.8 indicates that the induced stress in the component is greatest up until near the primary frequency of the shock transient. While the response accelerations are lower for the low-frequency SDOF oscillators as shown in Fig. 4.3, the displacements are typically much higher, resulting in higher stresses. Likewise, the pseudo-velocity SRS begins to fall off at frequencies above the fundamental transient frequency. This effect can be seen in Fig. 4.6, which shows that the 1 kHz oscillator response tracks very closely with the input shock. As a result, there is little relative motion of the SDOF oscillator, implying that there would be equally little induced stress at that frequency or higher frequencies.

4.1.3 Primary SRS

While nearly all SRS data are presented in terms of maxi-max spectra, it is useful to examine the spectra that make up the maxi-max spectra. As was shown previously, the SRS is calculated from the SDOF system response to the transient shock excitation. At some frequencies the maximum response occurs during the forced response era and at other frequencies the maximum response may occur during the free vibration era. The portion of the time window during which the transient excitation is occurring is known as the primary. After the shock has passed, there is still motion of the SDOF oscillator—this portion of the window is known as the residual. The maxi-max spectrum is a combination of the positive primary response, the negative primary response, the positive residual response, and the negative residual response.

The positive primary SRS is composed of the maximum response of all the SDOF oscillators during the time window corresponding to the length of the shock excitation. This can often be a somewhat arbitrary designation since it is common to record the time signal from a test for some time after the event has finished. Nevertheless, for this classical example it is very easy to define the cut-off time of the shock pulse. In contrast, the negative primary SRS is composed of the minimum response of all the SDOF oscillators during the shock excitation time window.

Figure 4.9 shows a plot of the positive and negative primary SRS for the classical haversine shown in Fig. 4.2. The positive primary SRS has a steeper slope at low frequencies. This is a result of the fact that the low-frequency SDOF oscillators

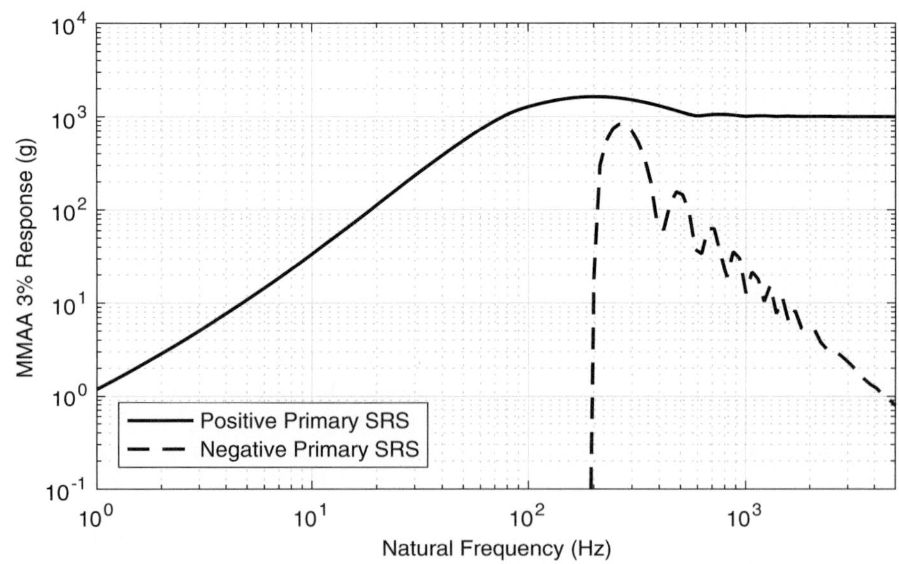

Fig. 4.9 Positive and negative primary acceleration SRS plot of a 1000 g 5 ms duration haversine shock pulse

have not reached their peak response before the shock transient passes. The positive primary SRS above about 200 Hz, the primary frequency of the 5 ms example haversine, is identical to the MMAA SRS shown in Fig. 4.7.

The negative primary SRS is a more interesting plot. Since the example haversine is a one-sided positive pulse there is no negative response at all in the time history at SDOF oscillator frequencies below the reciprocal of the shock pulse duration. This is evidenced by the sharp drop-off to zero at 200 Hz. Above the 5 ms pulse period, the one-sided nature of this pulse ensures that the minimum always occurs later in time and is always less than the positive SRS due to the fact that the SDOF oscillator tracks closely to the input shock at high frequencies. If the haversine pulse applied were inverted, then the positive and negative SRS would be reversed. If the shock pulse were a more featured pulse with a combination of positive and negative peaks, it is entirely likely that the negative primary SRS might exceed the positive primary SRS at some frequencies.

4.1.4 Residual SRS

The residual SRS is comprised of the maximum and minimum SDOF oscillator response that occurs after the shock transient has passed. As the name implies, the positive residual SRS is composed of the maximum response of all the SDOF oscillators after the end of the shock excitation. The negative residual SRS is the minimum response of the SDOF oscillators post shock. In order to calculate the residual response it is necessary to extend the SDOF response for one full oscillatory cycle at the SDOF frequency past the end of the shock transient. This can be quite a significant computational requirement if the SRS is analyzed at extremely low frequencies; however, in general it is a relatively quick calculation. There are mathematical techniques to predict the residual peaks since the equation of motion is known and dynamic conditions of the SDOF oscillator at the end of the transient are known. These can be used as initial conditions to solve for the residual response.

Figure 4.10 shows a plot of the positive and negative residual SRS for the example haversine shock. In this figure it is apparent that the two are extremely similar with the positive residual SRS being slightly higher than the negative residual SRS up until just before the 200 Hz frequency of the shock pulse. The largest difference at the low frequencies is often the result of the damping used in the SRS calculation. Even low levels of damping, such as the 3% critical damping used here, attenuate the response of the return oscillator stroke sufficiently that the magnitude of the two half cycles differs slightly. At frequencies higher than the shock pulse frequency the residual response falls off rapidly. The reason for this was shown in the time history plots in Figs. 4.5 and 4.6 that show the SDOF oscillator tracks the shock pulse more closely as the frequency increases. As such, Fig. 4.6 indicates that there is very little oscillatory motion of the SDOF system and therefore the residual response is negligible.

Finally, Fig. 4.11 shows both the primary SRS and the residual SRS. The maximax SRS is the envelope of these two SRS curves. It is easy to visually compare the

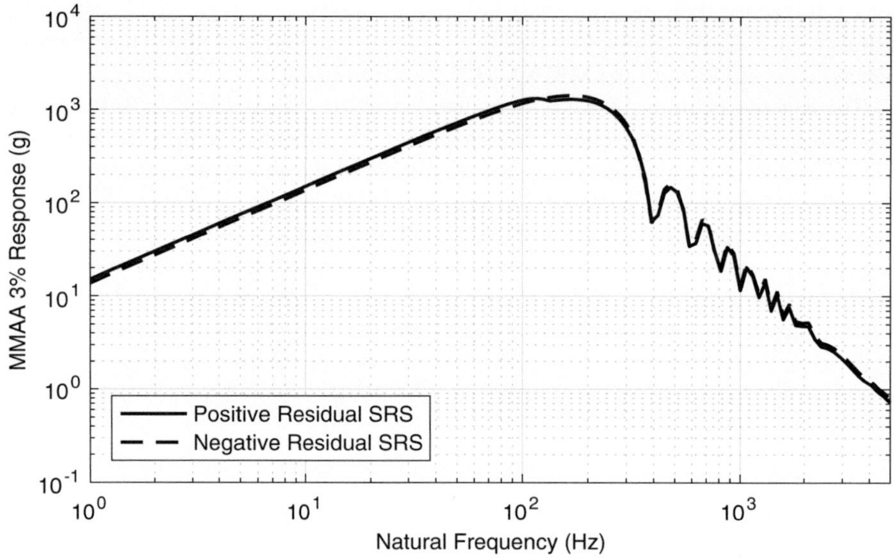

Fig. 4.10 Positive and negative residual acceleration SRS plot of a 1000 g 5 ms duration haversine shock pulse

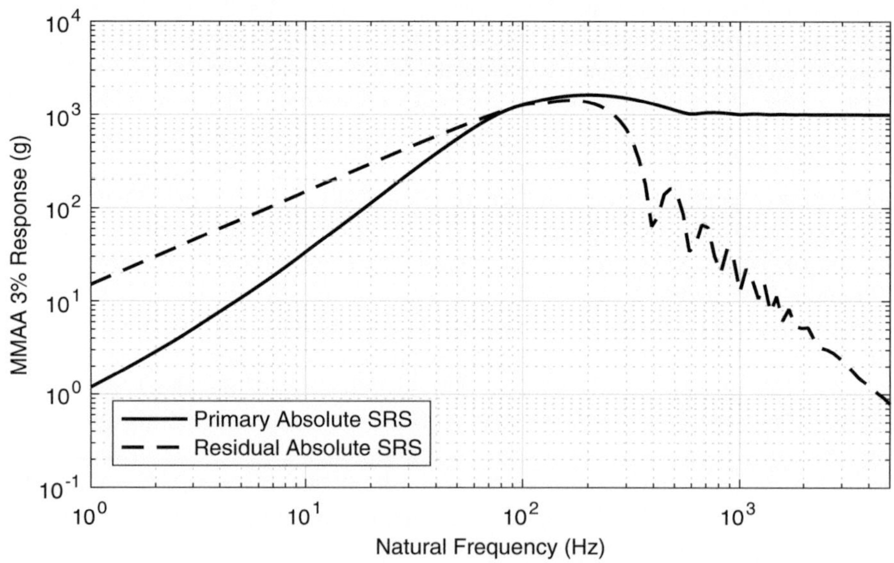

Fig. 4.11 Maximum absolute acceleration SRS along with the underlying maximum absolute primary and residual acceleration SRS of a 1000 g 5 ms duration haversine shock pulse

envelope of these two curves with the MMAA SRS shown previously in Fig. 4.7. The maximum absolute residual response defines the low-frequency portion of the spectrum while the high-frequency portion of the spectrum is defined by the maximum absolute primary response. The residual SRS dominates up to the frequency where the response first exceeds the haversine peak, which is about 80 Hz in the example shown in Fig. 4.11. At natural frequencies above that, the primary SRS dominates. Thus, both are necessary to fully define the SRS.

4.1.5 Uses of the Positive and Negative SRS

The positive and negative SRS can be useful when the loading direction is well understood. For example, in naval ship shock, the loading is always initially upward. Figure 4.12 shows a photograph of a lathe bolted down to a work stand. The photograph shows the lathe's tailstock end and it is readily apparent that the tailstock foot is quite substantial. If the shock loading were applied vertically upward, the tailstock foot would be initially in compression and it would appear to be practically impossible to fail this machine. On the other hand, if the shock loading was downward, the only thing securing the tailstock to the work stand is a pair of 13 mm bolts (one shown and a matching bolt on the back side). Significantly less shock

Fig. 4.12 Photograph of a lathe tail-stock showing the strength differences between an upward loading placing the foot in compression and a downward loading placing the bolts in tension

loading is required to rupture two 13 mm bolts in tension than would be needed to fail the machine's large iron foot in compression.

While the lathe example is interesting and applicable in many situations, in other loading scenarios the initial shock direction is unknown, or at least not guaranteed to occur in a specific direction. If the loading direction is uncertain, the maxi-max spectrum is more appropriate for design purposes.

In addition to using the positive and negative SRS to understand certain loading scenarios, the two responses are also useful for understanding the nature of the shock pulse itself. When the positive and negative SRS are very similar, it indicates a reasonably symmetric shock pulse—approximately the same positive and negative accelerations. If there is a significant disparity between the positive and negative SRS, then the shock pulse can be assumed to be substantially one-sided—appreciably more acceleration on one side of zero than the other.

4.1.6 Relationship to Fourier Spectra

The shock response spectrum is naturally related to the Fourier spectrum as the SRS theory is loosely derived from the Fourier spectrum theory. In fact, there is a commonly stated belief that the undamped residual pseudo-velocity SRS is equal to the Fourier spectrum magnitude [9, 10]. This is not exactly true although it has been stated often enough that many people believe it to be true. In a paper by Gertel and Holland [11], they show the comparison between the Fourier magnitude and the residual SRS. However, the comparison is not as direct as is often stated. Their paper contains a note stating that the "residual shock spectrum is also Fourier spectrum read on the velocity scale for an acceleration record." This is a somewhat circuitous way of stating the obvious fact that there is a mismatch of units between the two spectra. The Fourier transform of an acceleration record has units of g. In contrast, nearly all shock spectra from the early days of shock analysis were calculated in terms of velocity. Thus the residual shock spectrum has units of meters per second. To equivalence the two spectra, it is necessary to scale the Fourier transform to velocity units by the relationship:

$$\mathrm{SRS}_{\mathrm{Residual}}(y) = [\mathrm{FFT}(Y)]gT, \tag{4.6}$$

where g is the acceleration of gravity and T is the total time over which the signal was recorded.[1] This is shown in Figs. 4.13 and 4.14. Figure 4.13 shows a comparison of the FFT magnitude of the El Centro North-South time history shown previously in Fig. 2.12 and the residual pseudo-velocity SRS calculation of the same acceleration record. It is obvious that the shape of the two plots is very similar but the magnitudes are substantially different. In contrast, Fig. 4.14 shows the FFT magnitude of the

[1]The FFT in Eq. 4.6 is the magnitude of a two-sided FFT. Of course, the SRS is not defined for negative frequencies but one could imagine that it would work suitably well.

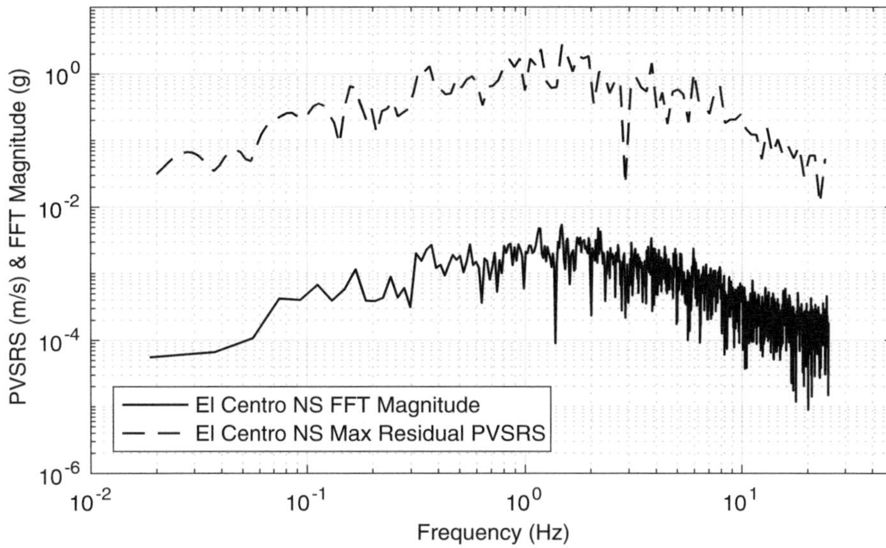

Fig. 4.13 FFT magnitude compared to the pseudo-velocity SRS of the El Centro, California 18 May 1940 North-South earthquake ground response

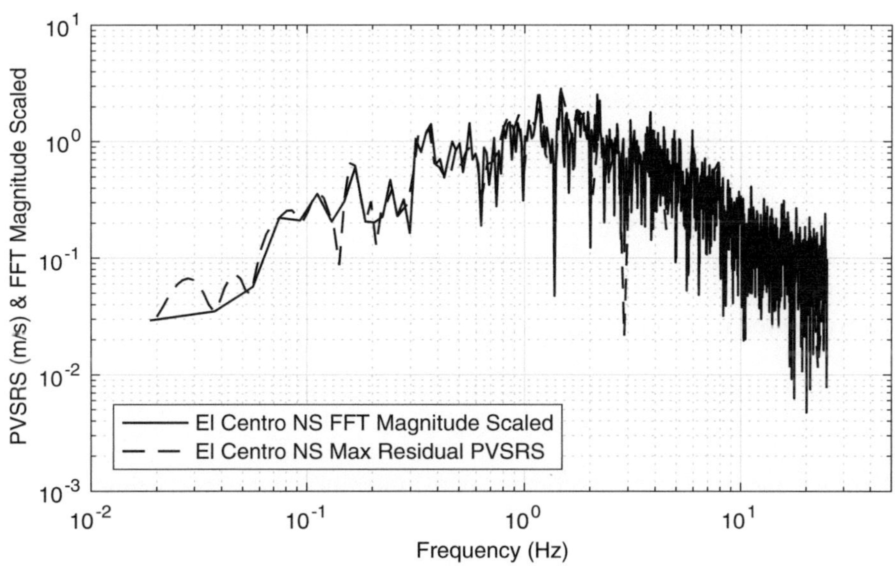

Fig. 4.14 FFT magnitude scaled compared to the pseudo-velocity SRS of the El Centro, California 18 May 1940 North-South earthquake ground response

same El Centro time history multiplied by the acceleration of gravity and the total measured time of the record. Now, the two curves lay essentially one atop the other.

It should also be noted from Fig. 4.14 that the FFT magnitude, at least up until the system noise becomes dominate, is generally equal to, or slightly less, than the SRS magnitude. There are a few frequencies where the SRS magnitude falls below the scaled FFT magnitude. If the scaled FFT magnitude had been plotted against the maxi-max SRS, then the FFT magnitude would have been universally at or below the SRS magnitude. This comparison again indicates that the FFT magnitude and the SRS are not quite as equivalent as has been proposed.

4.2 Types of Shock Spectra

Shock response spectra generally are computed with the base excited equation of motion in relative coordinates (Eq. 3.15):

$$\ddot{z}(t) + 2\zeta\omega_n\dot{z}(t) + \omega_n^2 z(t) = -\ddot{w}(t) + p(t)/m \tag{4.7}$$

with $p(t) = 0$ and zero initial conditions. This is the standard equation but a response spectrum can also be computed with an applied force or absolute coordinates or both (i.e., Eq. 3.3). The only requirement is that there is a single input.

Shock response spectra are distinguished by the output equation. The most common responses are pseudo-velocity, relative velocity, and absolute acceleration. These are summarized in Table 4.1.

Since shock response spectra require computing the responses of many SDOF oscillators, computational efficiency is a virtue even if the burden of each computation is not large. The SRS are commonly calculated using the recursive digital filter method described in Sect. 3.4.2.2 largely because of its computational efficiency. However, numerical integration was the original SRS computation method and has been used for many years. Numerical integration algorithms are also relatively simple to implement. To demonstrate the relative computational cost of calculating an SRS, the SRS was computed for the haversine input shown in Fig. 4.2 at 128 logarithmically spaced natural frequencies using three numerical methods. Table 4.2 shows the relative computational cost of three methods for two different SRS frequency ranges. The ramp invariant digital filter method for calculating the SRS

Table 4.1 Free vibration response

Output quantity	Response
Pseudo-velocity	$y(t) = PV(t) = \omega_n z(t)$
Relative velocity	$y(t) = RV(t) = \dot{z}(t)$
Absolute acceleration	$y(t) = AA(t) = -\omega_n^2 z(t) - 2\zeta\omega_n\dot{z}(t)$

Table 4.2 Relative computational time cost of different SRS calculation methods

SRS calculation method	10 Hz to 10 kHz range	1 Hz to 10 kHz range
Digital filters	1	5.5
Newmark-beta integration	110	530
MATLAB® ODE45	500	1900

over a 10 Hz to 10 kHz range is considered to be the baseline with a relative computational cost of one. The Newmark-beta integration method is approximately 100 times more computationally expensive than the digital filter method and the MATLAB® ODE45 method is about 500 times more expensive. Table 4.2 also shows the significant increase in computation time that comes with expanding the low-frequency range. Expanding the calculation range from 10 Hz down to 1 Hz increased the computational cost by about a factor of four to five for each method. The reason for this is the need to calculate the residual response for a full oscillation past the shock excitation time. This can be a considerable computational burden at very low SDOF oscillator frequencies.

Recall from Sect. 3.4.2.2 that the SDOF oscillator response is given by the difference equation:

$$\begin{aligned} y(kT) = {} & \beta_0 \ddot{w}(kT) + \beta_1 \ddot{w}((k-1)T) + \beta_2 \ddot{w}((k-2)T) \\ & -\alpha_1 y((k-1)T) - \alpha_2 y((k-2)T). \end{aligned} \tag{4.8}$$

The discrete time transfer function of this difference equation is:

$$H(z) = \frac{\beta_0 + \beta_1 z^{-1} + \beta_2 z^{-2}}{1 + \alpha_1 z^{-1} + \alpha_2 z^{-2}}. \tag{4.9}$$

In the following sections expressions for these coefficients are given. These coefficients are for the ramp invariant filter and are tabulated in the ISO standard.

4.2.1 Absolute Acceleration SRS

The absolute acceleration SRS is arguably the most common SRS. The absolute acceleration SRS, as the name indicates, represents the extremal (positive, negative, or maxi-max) absolute acceleration of the single degree-of-freedom oscillator mass. Acceleration spectra are a logical result of measuring shock data with accelerometers—acceleration being the most common type of shock measurements made today. Acceleration also translates directly to force through Newton's law.

The absolute acceleration SRS is almost always plotted in terms of the maxi-max absolute acceleration, usually abbreviated as the MMAA response. An example of the absolute acceleration SRS was shown in Fig. 4.7 for the haversine shock pulse shown in Fig. 4.2.

The denominator coefficients of the transfer function, Eq. 4.9, are based on the equation of motion and they are independent of the output quantity so they are the same in the recursive digital filter equations. For the SDOF oscillator equations of motion, Eq. 4.7 (and Eq. 3.3):

$$\alpha_1 = 2e^{-A}\cos(B),$$
$$\alpha_2 = e^{-2A}. \tag{4.10}$$

The constants A and B along with an additional constant C that will be used later are given by Eq. 3.73:

$$A = \frac{\omega_n T}{2Q},$$
$$B = \omega_n T\sqrt{1 - \frac{1}{4Q^2}}, \tag{4.11}$$
$$C = \frac{\frac{1}{2Q^2} - 1}{\sqrt{1 - \frac{1}{4Q^2}}}.$$

where Q is the quality factor, defined in Eq. 3.8. Rewriting Eqs. 4.11 (Eqs. 3.73) in terms of ζ yields:

$$A = \zeta\omega_n T,$$
$$B = \omega_n T\sqrt{1 - \zeta^2}, \tag{4.12}$$
$$C = \frac{2\zeta^2 - 1}{\sqrt{1 - \zeta^2}}.$$

The ramp invariant digital filter's numerator coefficients for the absolute acceleration SRS are defined by the ISO standard as:

$$\beta_0 = 1 - e^{-A}\left[\frac{\sin(B)}{B}\right]$$
$$\beta_1 = 2e^{-A}\left[\frac{\sin(B)}{B} - \cos(B)\right] \tag{4.13}$$
$$\beta_2 = e^{-2A} - e^{-A}\left[\frac{\sin(B)}{B}\right].$$

4.2.2 Relative Displacement SRS

The relative displacement SRS is not frequently used in traditional shock analysis per se. However, the relative displacement SRS can be used to gain insight into the expected extremal displacements from a shock event. This is important for quantifying shock isolation needs and equipment sway space requirements.

Figure 4.15 shows a plot of the relative displacement SRS for the example haversine shock pulse shown in Fig. 4.2. Normally, the low-frequency portion of a relative displacement SRS would trend to a flat line representing the approximate maximum system displacement from the shock event. However, in the case of the ideal haversine, there is a net velocity change between the initial and final system velocities. This resulting ΔV yields a continuously increasing system displacement and thus there is no low-frequency plateau in the SRS.

Perhaps a more interesting and realistic example is shown in Fig. 4.16. This figure shows the relative displacement SRS calculated from the May 1940 El Centro, California North-South ground motion shown in Fig. 2.12. From this SRS plot it is apparent that the maximum ground motion was in the 20–40 cm range. Quite significant when one considers the common phrase "fixed to ground."

One interesting insight from the SRS presentation method is information on how rapidly system displacements decrease with increasing frequency. Small displacements inherently lead to small strains. As a result, it becomes very difficult to induce damage in ductile structural members at high frequencies simply because

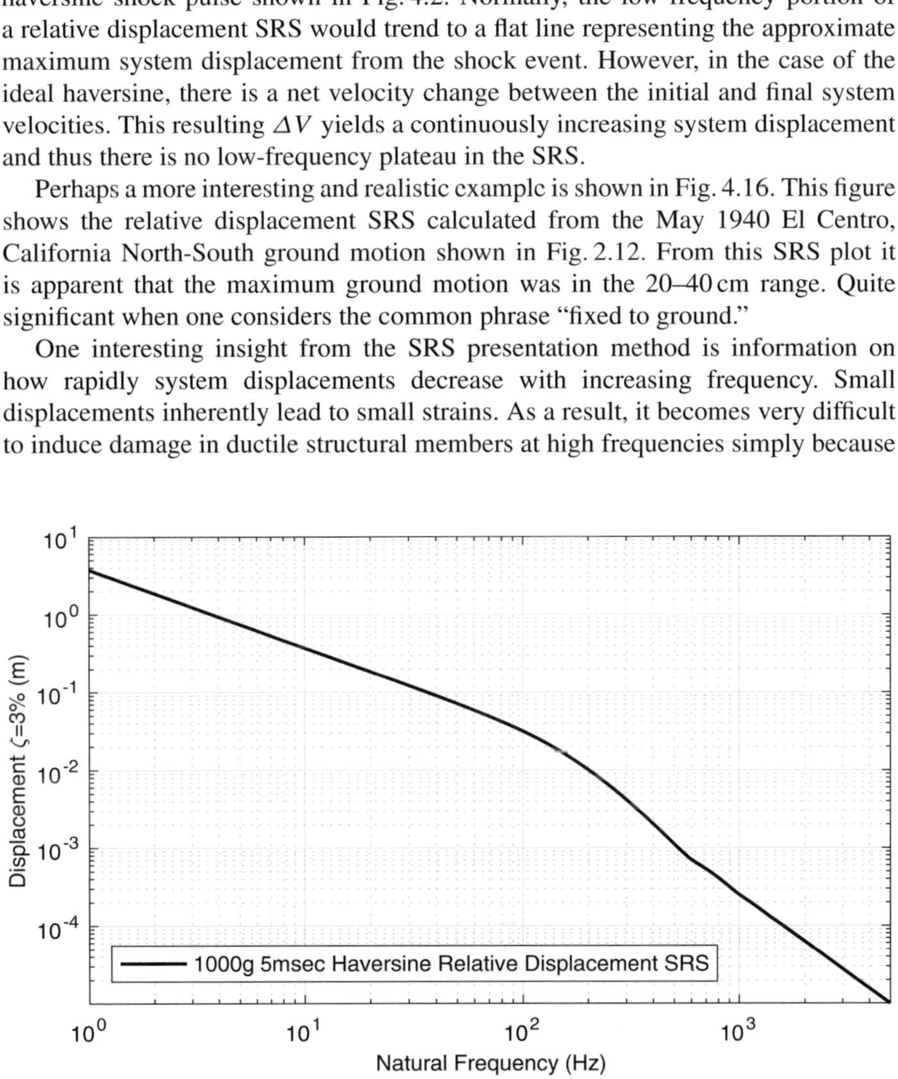

Fig. 4.15 Plot of the relative displacement SRS plots for a 1000 g 5 ms duration haversine shock pulse

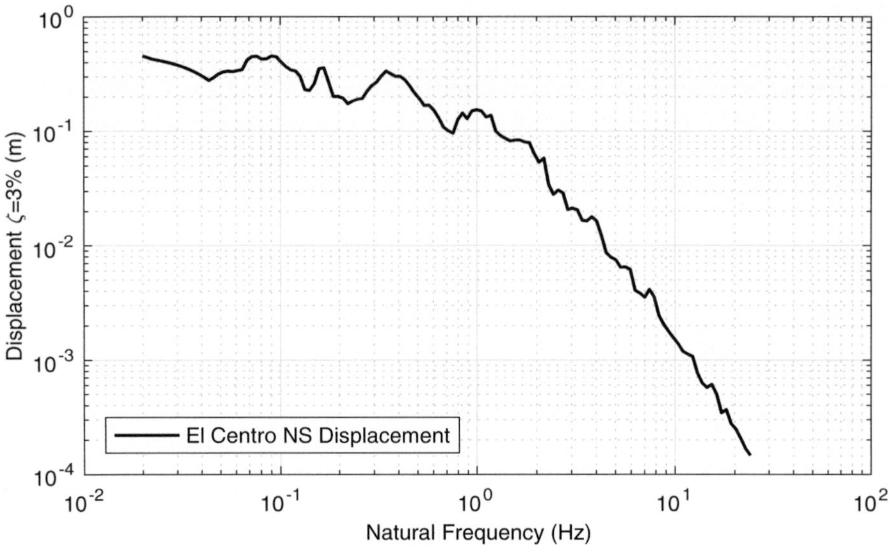

Fig. 4.16 Plot of the relative displacement from the El Centro, California 18 May 1940 North-South earthquake ground response

there is little or no strain developed in the materials. Weld failures, bolt failures, gross yielding, and any other failures requiring significant deformation can be attributed almost exclusively to low-frequency loading.

The ramp invariant filter numerator coefficients for the relative displacement SRS are defined by the ISO standard as:

$$\begin{aligned}\beta_0 &= \frac{1}{\omega_n^3 T}\left[\frac{1-e^{-A}\cos(B)}{Q} - Ce^{-A}\sin(B) - \omega_n T\right]\\ \beta_1 &= \frac{1}{\omega_n^3 T}\left[2e^{-A}\cos(B)\omega_n T - \frac{1-e^{-2A}}{Q} + 2Ce^{-A}\sin(B)\right]\\ \beta_2 &= \frac{1}{\omega_n^3 T}\left[-e^{-2A}\left(\omega_n T + \frac{1}{Q}\right) + \frac{e^{-A}\cos(B)}{Q} - Ce^{-A}\sin(B)\right].\end{aligned} \tag{4.14}$$

The denominator filter coefficients were given previously in Eq. 4.10 and the constants A, B, and C were defined in Eqs. 4.11 and 4.12.

4.2.3 Pseudo-Velocity SRS

The other common SRS calculation is the pseudo-velocity SRS. Almost all early SRS calculations were presented in terms of relative velocity or pseudo-velocity and pseudo-velocity is still the preferred presentation in the U.S. Navy and in seismic communities. Pseudo-velocity will be close but slightly low compared to the relative velocity at low frequencies and it will be higher at high frequencies due to its frequency dependence.

The pseudo-velocity shock response spectrum is derived by taking the relative displacement SRS and multiplying by the SDOF oscillator natural frequency, ω_n. The pseudo-velocity SRS, like the absolute acceleration SRS, is almost always plotted in terms of the maxi-max absolute pseudo-velocity. An example of the pseudo-velocity SRS was shown in Fig. 4.8 for the haversine shock pulse shown in Fig. 4.2. The pseudo-velocity SRS is frequently plotted on special graph paper known as tripartite graph paper or sometimes referred to as four-coordinate paper. This paper contains logarithmic scales for frequency, displacement, velocity, and acceleration such that all three dynamic properties can be read simultaneously as a function of frequency. The details of this presentation method will be presented in Sect. 4.4.

The ramp invariant digital filter's numerator coefficients for the pseudo-velocity SRS are defined by the ISO standard as:

$$\begin{aligned}
\beta_0 &= \frac{1}{\omega_n^2 T}\left[\frac{1-e^{-A}\cos(B)}{Q} - Ce^{-A}\sin(B) - \omega_n T\right] \\
\beta_1 &= \frac{1}{\omega_n^2 T}\left[2e^{-A}\cos(B)\omega_n T - \frac{1-e^{-2A}}{Q} + 2Ce^{-A}\sin(B)\right] \\
\beta_2 &= \frac{1}{\omega_n^2 T}\left[-e^{-2A}\left(\omega_n T + \frac{1}{Q}\right) + \frac{e^{-A}\cos(B)}{Q} - Ce^{-A}\sin(B)\right].
\end{aligned} \tag{4.15}$$

These are just the relative displacement coefficients from Eq. 4.14 multiplied by ω_n.

4.2.4 Relative Velocity SRS

The relative velocity SRS is not frequently used in traditional shock analysis, which is heavily biased toward the absolute acceleration and the pseudo-velocity SRS. However, relative velocity has an elegant relationship with stress which will be discussed further in Sect. 7.3 making it a very important method for presenting shock response data. The relative velocity SRS is also a very efficient intermediate step for calculating energy response spectra. As such, the appropriate filter coefficients are defined here and their incorporation into the energy response spectra calculations will be covered in Chap. 12.

While not in common use outside of U.S. Navy applications, velocity gauges have been used to measure shock input in numerous tests. A velocity SRS is a natural extension of measuring velocity directly. Velocity gauges are still used today in many large-scale U.S. Navy shock tests. However, velocity gages, unlike modern accelerometers, are generally quite large and are thus only adaptable to testing large equipment as opposed to accelerometers, which can be quite small.

While the relative velocity SRS may not be commonly used, it is interesting to see how it compares to the more common pseudo-velocity SRS. Figure 4.17 is a plot comparing the relative velocity SRS with the pseudo-velocity SRS for the example haversine shock pulse shown in Fig. 4.2. The relative velocity SRS is slightly higher than the pseudo-velocity SRS at the very low frequencies and then is only slightly below the pseudo-velocity SRS up until the 200 Hz shock pulse frequency. Above the shock pulse frequency, the relative velocity SRS falls off more rapidly than the pseudo-velocity SRS.

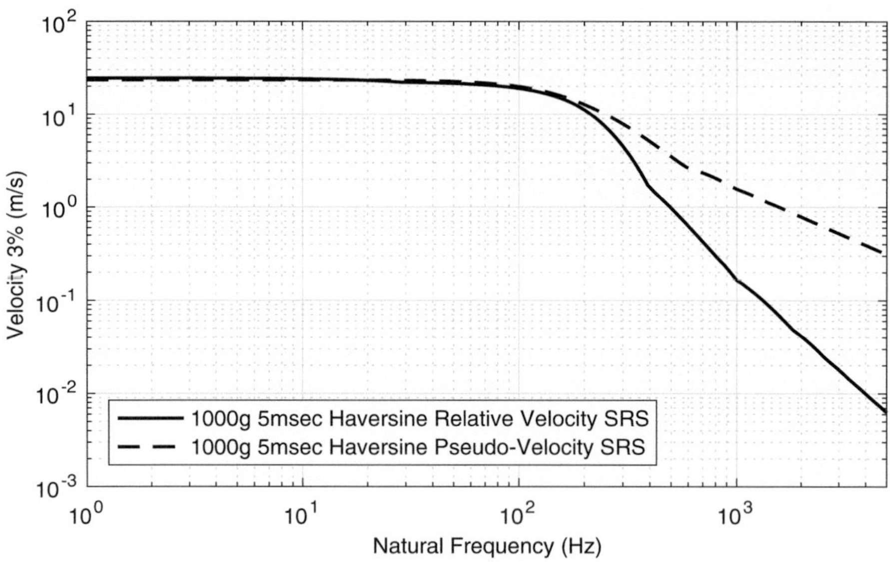

Fig. 4.17 Comparison of the relative velocity and pseudo-velocity SRS plots for a 1000 g 5 ms duration haversine shock pulse

The ramp invariant digital filter numerator coefficients for the relative velocity SRS are defined by the ISO standard as:

$$
\begin{aligned}
\beta_0 &= \frac{1}{\omega_n^2 T}\left[-1 + e^{-A}\cos(B) + \frac{e^{-A}\sin(B)}{\sqrt{4Q^2-1}}\right] \\
\beta_1 &= \frac{1}{\omega_n^2 T}\left[1 - e^{-2A} - \frac{2e^{-A}\sin(B)}{\sqrt{4Q^2-1}}\right] \\
\beta_2 &= \frac{1}{\omega_n^2 T}\left[e^{-2A} - e^{-A}\cos(B) + \frac{e^{-A}\sin(B)}{\sqrt{4Q^2-1}}\right].
\end{aligned}
\tag{4.16}
$$

4.3 Coding an SRS Function

Writing computer code to calculate shock response spectra is not particularly difficult once the methodology is understood. Here again, almost all SRS calculations are performed using the digital filter method with the ISO standard ramp invariant filter coefficients discussed in the previous sections. However, it is also possible to write a code using the direct integration method and Duhamel's integral. The discussion here will only address the ramp invariant filter approach.

Sample pseudo-code is provided in Table 4.3 that outlines the steps to calculate the SRS. The first step after the time history is obtained and converted to the appropriate units is to select the SRS type, natural frequencies, and damping coefficient to be used. The code then loops over all the specified natural frequencies, calculating the filter weights as described in Sects. 4.2.1–4.2.4. The time history must be zero padded to include sufficient time for one full oscillatory cycle—necessary to get the correct residual response. The zero padded signal is then filtered

Table 4.3 SRS pseudo-code

```
Determine the type of SRS desired
Determine the frequencies for the response calculations
Determine the damping coefficient to be used
Loop over all the selected frequencies
    Calculate the constants A, B, and C from Eq. 4.11 or 4.12
    Calculate the denominator filter weights from Eq. 4.10
    Calculate the numerator filter weights β1, β2, and β3
    Pad the time history with zeros to cover the residual response
    Filter the time history
    Search result for primary and residual minimums and maximums
    Combine primary and residual to get the maxi-max spectrum data point
Loop end
Plot results
```

using the calculated filter weights and the digital filter described in Eqs. 4.8 and 4.9. The SDOF response from the filter operation is then searched for the maximum and minimum responses during the primary and residual time windows. These results are then compiled into the SRS data point at that natural frequency and the process is repeated for all selected frequencies.

An alternative to zero padding the time history can be used to calculate the residual response. Given the position, velocity, and acceleration of the last point in the filtered time history, the maximum responses can be predicted because lightly damped sinusoidal motion is well understood. This approach is entirely appropriate when computing resources are more limited since this estimation is computationally faster than padding the time history with sufficient zeros to cover one oscillatory cycle. There are two primary disadvantages to predicting the residual response rather than calculating the residual response. The first is that the prediction must be calculated at both the first quarter peak of the sinusoidal response as well as the three-quarters peak and then logic statements included to ensure the maxima are saved. The second disadvantage is that higher damping values tend to move the predicted response maxima further from the actual response maxima. This is generally not a problem with the low-damping values typically used.

4.4 Pseudo-Velocity SRS on Tripartite Paper

While it is very common to simply plot shock response spectra on traditional log–log paper, there is another type of logarithmic paper that is frequently used for displaying velocity SRS data. Tripartite paper (sometimes referred to as four-coordinate paper) has a traditional log–log arrangement along the abscissa and ordinate for the natural frequency and velocity, respectively. In addition, there are two additional logarithmically scaled sets of gridlines corresponding to displacement and acceleration. Thus, with a single plot, it is possible to read estimates of displacement, velocity, and acceleration simultaneously. This type of plot has always been very popular among U.S. Navy researchers. The biggest drawback to this type of plot is that most analysis packages do not include this plotting capability and many researchers are left with no easy options for displaying their test data in this format.

One of the advantages of the tripartite paper is that since the primary axis is velocity, the relationship between stress and velocity is highlighted as a fundamental property of the SRS. The relationship between stress and velocity is well established in the literature [12–14], and the relationship will be covered in detail in a later chapter. The displacement and acceleration responses are provided but held more in context.

Figure 4.18 shows a plot of the May 1940 El Centro North-South ground motion response pseudo-velocity SRS plotted on tripartite paper. Reading from this plot shows quickly that the maximum ground motion velocity was about 0.9 m/s, the maximum acceleration was about 1 g, and the peak displacement was about 0.4 m.

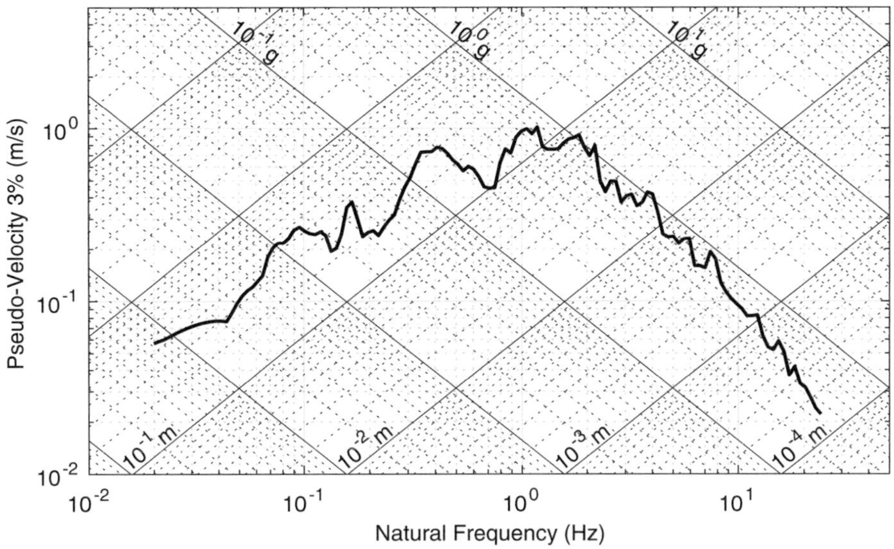

Fig. 4.18 Pseudo-velocity SRS of the El Centro, California 18 May 1940 North-South earthquake ground response plotted on tripartite paper ($\zeta = 3\%$)

It may be remembered that the actual peak acceleration from the time history plot shown in Fig. 2.12 was actually only about 0.25 g. This is a common occurrence with the SRS calculation when many cycles occur in rapid succession. The SDOF system motion calculation can respond sometimes two to four times higher than the signal used to calculate the SRS.

4.5 Non-uniqueness of the SRS

One of the fundamental premises of the SRS is that any two curves with the same SRS will have the same damage potential. This hypothesis has held up to testing for many years and it is generally true that time histories with similar SRS curves produce approximately similar damage statistics. This in turn leads to one of the chief criticisms of the SRS—that it is a non-unique operator. It is theoretically possible to derive an infinite number of time history signals that will yield nominally the same SRS. This non-uniqueness is possible because the SRS is an incomplete transform, containing significantly less data than the original time history. In contrast, the Fourier transform is a complete transform since it contains the same quantity of data as the original signal. Therefore the Fourier transform is reversible while the SRS is not.

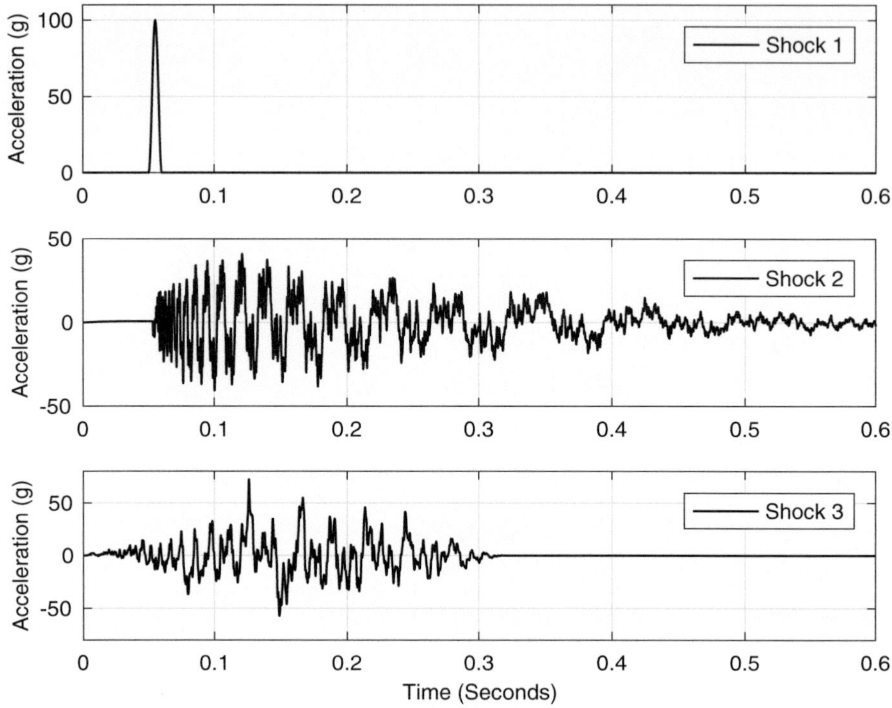

Fig. 4.19 Three sample time histories that yield nominally the same SRS

To illustrate this point, Fig. 4.19 shows three sample time histories with different amplitudes, durations, and waveform characteristics. The first waveform is a classical haversine pulse, the second was created using a summation of exponentially decaying sinusoids, while the third was generated using a summation of half-cycle sinusoids multiplied by a half-sine window. It is obvious from a cursory examination of the three curves that they are quite different from each other. Figure 4.20 shows the MMAA SRS for each of the three shock time histories overlaid for comparison. In this figure, it is immediately obvious that the three curves have essentially the same SRS. Furthermore, one could readily imagine that, with more effort, the minor deviations between the three SRS curves could be reduced. The fact that three time histories of such disparate length and form can yield the same nominal SRS is the essence of the concept of non-uniqueness of the SRS transform.

While the shock time histories and SRS transformations shown in Figs. 4.19 and 4.20 are valid, they are also somewhat contrived. If the SRS calculation is extended beyond the frequency range shown in Fig. 4.20, for which the shock curves were designed, the three SRS curves begin to diverge. Therefore, the non-uniqueness claim is somewhat limited to the defined frequency range. While it is certainly possible to match the SRS over a greater frequency range, the exercise presented here is typically also focused on generating a shock profile that can

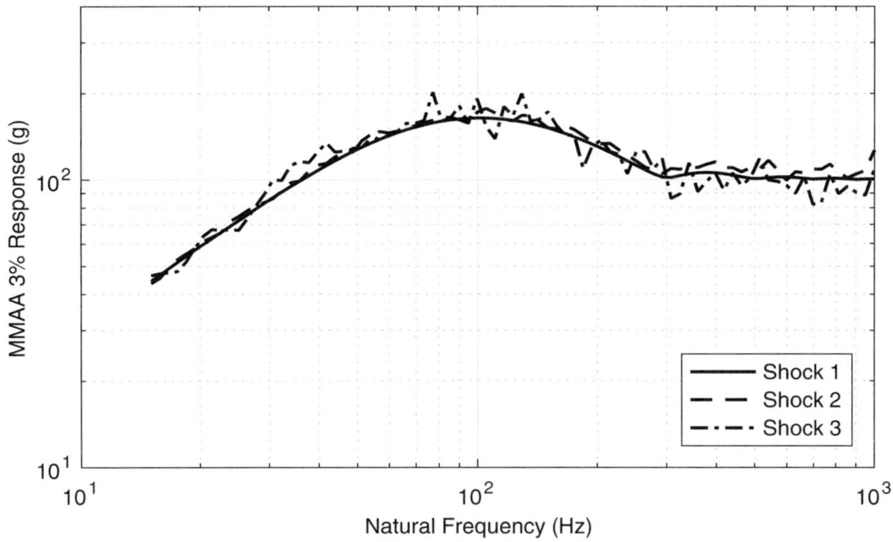

Fig. 4.20 MMAA shock response spectra for the three sample time histories

be tested on an electrodynamic or hydraulic shaker system. These systems have obvious, practical limitations on maximum displacement, velocity, and acceleration. Thus, even though the SRS is nominally the same over the defined frequency range and the shock damage potential should be equivalent, the reality is different. Outside of the defined frequency range, the shocks do differ substantially and if those frequencies are responsible for component damage, then an SRS that is equivalent over a narrower frequency range may not be equally damaging in practice, where all frequency ranges are inherently present.

As a result of the simple example shown in Figs. 4.19 and 4.20, it is often necessary or desirable to specify information about the shock in addition to the SRS. Some measure of the shock pulse duration or form is frequently added to the SRS data. The velocity change is another characteristic that is used to describe the type of shock. This helps ensure that laboratory tests agree with the spirit of the measured field data more precisely than an SRS alone can guarantee.

4.6 Summary

This chapter presented some history of the development of shock response spectra as well as the basics of the how SRS are calculated. The examples should serve to underscore some of the nuances of the SRS and their calculation. The ramp invariant filter coefficients are provided along with sample pseudo-code which can be used by the reader to create their own SRS codes with relative ease.

Problems

4.1 Following the examples provided in this chapter, use a readily available time history and calculate the SDOF response at several frequencies, low, mid-range, and high relative to your shock pulse. Determine where the maximum positive and negative response occurs. Are these maximum in the primary or residual portion of the SDOF response?

4.2 Take one or more readily available time histories and calculate the positive and negative SRS responses. How do the positive and negative responses compare for different types of input excitations? Would there be an advantage to designing with both positive and negative spectra?

4.3 Calculate the pseudo-velocity SRS for several shock time histories from different sources. Compare the width of the velocity SRS plateau from different types of excitations. What can be said about the relationship of the SRS plateau width to the displacement? What about the SDOF acceleration?

4.4 Calculate the FFT and the pseudo-velocity SRS from several shock time histories and compare the results from the two methods. How do they compare? What are the likely advantages and disadvantages of both methods for analysis? For design?

4.5 Much has been said about the non-uniqueness of the SRS transformation. One method for approximating an SRS is to use a summation of decaying sinusoidal curves. This is usually accomplished by assuming a modest number of sinusoidal tones of varying amplitudes and phases for the excitation. A simple optimization routine is then used to tweak the amplitudes and phases such that the resulting SRS matches the desired SRS. Try to develop two different acceleration time histories that provide the same SRS. Do you think they will produce the same damage potential?

4.6 As a follow-on to Problem 4.5. Use a finite element package to model a simple structure such as a cantilever beam with a lumped mass at the free end. Apply the two time histories derived in the previous problem and see if the resulting strain predictions are equivalent.

4.7 Use the pseudo-code provided in this chapter along with the ISO standard filter coefficients to write your own SRS function. Use the examples provided here to compare the output of your function to results presented here.

4.8 Write a similar SRS generating function but replace the ISO standard filter weights with a traditional ordinary differential equation solver to produce the same results. How do the answers compare from these two methods? Are they the same? At what frequencies do they begin to differ and how do they differ?

References

1. McNeil, S. I. (2008). Implementing the fatigue damage spectrum and fatigue damage equivalent vibration testing. In *Proceedings of the 79th Shock and Vibration Symposium*, Orlando, FL.
2. Biot, M. A. (1932). *Transient oscillations in elastic systems*. Ph.D. Thesis, California Institute of Technology.
3. Biot, M. A. (1933). Theory of elastic systems vibrating under transient impulse with an application to earthquake-proof buildings. *Proceedings of the National Academy of Sciences, 19*(2), 262–268.
4. Biot, M. A. (1941). A mechanical analyzer for the prediction of earthquake stresses. *Bulletin of the Seismological Society of America, 31*(2), 151–171.
5. Housner, G. W. (1941). *An investigation of the effects of earthquakes on buildings*. Ph.D. Thesis, California Institute of Technology.
6. DeVries, P. L. (1994). *A first course in computational physics*. New York: Wiley.
7. Fourier, J. (1822). *Théorie Analytique de la Chaleur* (in French). Paris: Firmin Didot Pére et Fils.
8. ISO 18431-4. (2007). *Mechanical vibration and shock — Signal processing — Part 4: Shock-response spectrum analysis*. Geneva: International Standard.
9. Smallwood, D. O. (1985). The shock response spectrum at low frequencies. In *Proceedings of the Shock and Vibration Symposium*, Monterey (pp. 279–288).
10. Harris, C. M., & Crede, C. E. (1961). *Shock and vibration handbook* (Vol. 2, p. 23). New York: McGraw Hill.
11. Gertel, M., & Holland, R. (1967). Data analysis and instrumentation, effect of digitizing detail on shock and Fourier spectrum computation of field data. *Shock and Vibration Bulletin, 36*(Part 6), 1–20.
12. Gaberson, H. A. (2012). The pseudo velocity shock analysis stress velocity foundation. In *Proceedings of the 30th International Modal Analysis Conference*, Jacksonville.
13. Hunt, F. V. (1960). Stress and strain limits on the attainable velocity in mechanical vibration. *Journal of the Acoustical Society of America, 32*(9), 1123–1128.
14. Gaberson, H. A., & Chalmers, R. H. (1969). Modal velocity as a criterion of shock severity. *Shock and Vibration Bulletin, 40*(Part 2), 31–49.

Chapter 5
Classical Shock Theory

Classical shock pulses are typically defined by a single, one-sided shock pulse with a non-zero velocity change. Classical shock pulses are often referred to as simple shocks. There are typically five classical shocks: haversine, half-sine, initial and terminal peak saw-tooth, and trapezoidal [1]. Each of these shocks is easy to describe mathematically and can be well replicated using shock test machines. Classical shocks stand in contrast to the more mathematically complex pyroshock, ballistic shock, earthquake shock, and other waveforms usually described by summations of damped sinusoids or wavelets.

In this chapter, the characteristics of shock response spectra of classical shocks are described. The SRS of the classical shocks are quite similar. Specifically, we explain:

1. the characteristic low-frequency slope of 6 dB/octave in maxi-max absolute acceleration SRS
2. how the SDOF oscillator properties contribute to features in the SRS
3. how to read the velocity change from the SRS
4. how to determine the bandwidth of the shock from the SRS
5. how to interpret the effects of the shock transient from the features of the positive, negative, primary, and residual SRS

5.1 Classical Shocks Have Similar SRS

One interesting feature of the classical shock pulses is that the SRS for these shocks, which are distinctly different in the time domain, are generally quite similar in the spectral domain. To demonstrate this, acceleration time histories and maxi-max acceleration SRS are presented here for the five classical shock profiles. The haversine shock has already been discussed at length and will continue to be

C. Sisemore, V. Babuška, *The Science and Engineering of Mechanical Shock*,
https://doi.org/10.1007/978-3-030-12103-7_5

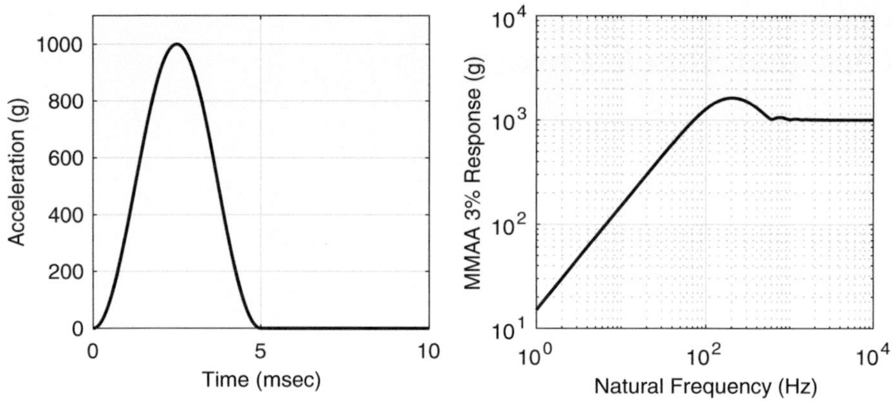

Fig. 5.1 Acceleration time history and SRS for 1000 g 5 ms haversine shock

discussed due to its importance in the mechanical shock field. A haversine shock pulse is an offset cosine waveform described by the equation:

$$\ddot{w}(t) = \frac{A}{2}\left[1 - \cos\left(\frac{2\pi t}{T}\right)\right], \tag{5.1}$$

where A is the haversine amplitude and T is the shock duration. Figure 5.1 shows a time history of a 1000 g haversine shock with a 5 ms duration along with the maxi-max acceleration SRS. Haversine shocks are easily generated on a commercial drop table and are used frequently in test laboratories and test specification development. The pulse shape is also advantageous with transitions to and from the shock pulse being smooth with continuous functions describing the displacement, velocity, and acceleration.

The second classical shock is the half-sine shock. The half-sine shock is exactly as described—the first half of a sinusoidal waveform. The shock pulse is described by the equation:

$$\ddot{w}(t) = A \sin\left(\frac{\pi t}{T}\right). \tag{5.2}$$

Here again, A and T are the shock amplitude and duration, respectively. Figure 5.2 shows a time history of a 1000 g half-sine shock with a 5 ms duration along with the maxi-max acceleration SRS. While the time history plots are somewhat similar between Figs. 5.1 and 5.2, the SRS plots are almost indistinguishable. The SRS for the half-sine is shifted slightly to the left compared to the haversine SRS. The reason for this shift is that a half-sine shock with the same amplitude and duration as a haversine encloses more area on the acceleration versus time plot and thus the resulting velocity change is greater with a comparable amplitude half-sine shock than with a haversine shock.

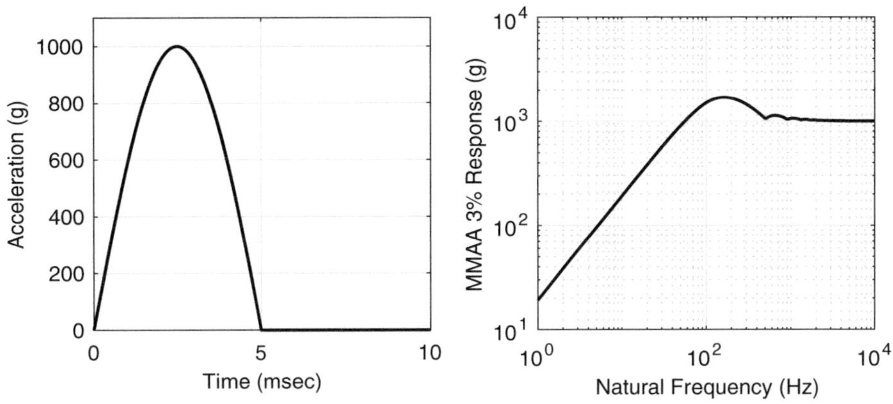

Fig. 5.2 Acceleration time history and SRS for 1000 g 5 ms half-sine shock

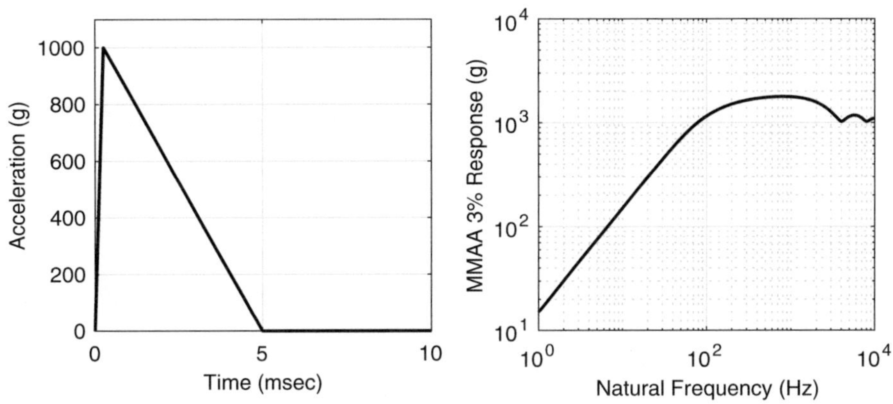

Fig. 5.3 Acceleration time history and SRS for 1000 g 5 ms initial peak saw-tooth shock

There are two classical saw tooth shocks: the initial peak saw-tooth and the terminal peak saw-tooth. Both of these shocks are triangular shocks skewed either to the left or right. The initial peak saw-tooth shock is generally described by the equation:

$$\ddot{w}(t) = A\frac{T - t}{T}, \tag{5.3}$$

where A and T are again the amplitude and pulse duration. The time history and acceleration SRS are shown in Fig. 5.3. In Fig. 5.3, a slight ramp has been incorporated into the rise, consisting of approximately 5% of the total duration. The slight ramp is added because sharp discontinuities do not occur in the laboratory and the digital filter method of SRS calculation does not do well with discontinuities.

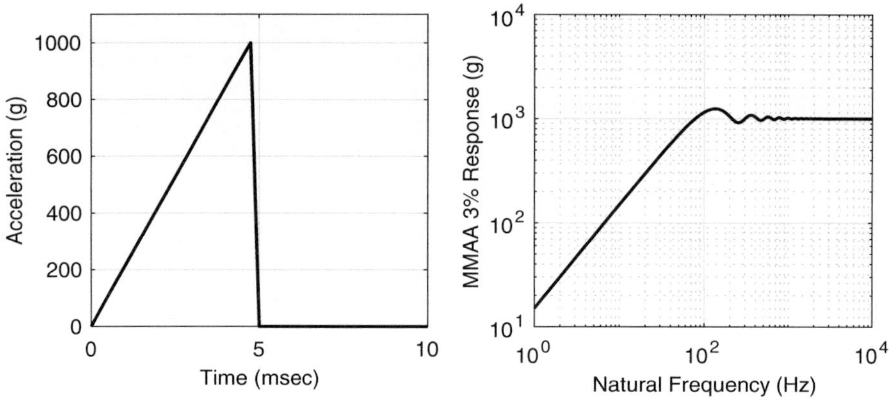

Fig. 5.4 Acceleration time history and SRS for 1000 g 5 ms terminal peak saw-tooth shock

The initial peak saw-tooth shock SRS is similar to the haversine and half-sine SRS at low frequencies but has a longer plateau at the shock pulse frequency.

The terminal peak saw-tooth shock represents the other extreme of the triangular shocks and is described by the similar equation:

$$\ddot{w}(t) = A\frac{t}{T}. \tag{5.4}$$

An example time history and acceleration SRS are shown in Fig. 5.4. Here again, a slight ramp (5% of the total duration) was incorporated to the trailing edge of shock profile to eliminate the discontinuity in the acceleration time history. In contrast to the initial peak saw-tooth, the terminal peak saw-tooth SRS is very similar to both the haversine and half-sine SRS, nearly identical at the low frequencies and a slightly smaller bump at the SRS peak.

The last of the classical shocks described here is the trapezoidal shock. The trapezoidal shock is similar to a square wave but has a sloping rise and fall as shown in Fig. 5.5. The shock shown here has a 10% rise time and 10% fall time with the remaining 80% at the maximum amplitude. The SRS for this shock is also shown in Fig. 5.5. The characteristic bump in the SRS is significantly wider than the bump from the haversine or half-sine shock but the general shape of the SRS is still similar to the other four classical shocks described here.

As a final comparison between the simple, classical shocks, all five maxi-max acceleration SRS curves are plotted together in Fig. 5.6. As can be seen, the five SRS curves have similar shape and magnitude across the frequency spectrum. A closer agreement between the curves could be obtained by adjusting the shock pulse durations and amplitudes between the different waveforms. For this comparison the amplitudes and duration were held constant, resulting in different areas under the acceleration versus time curve. As a result the five shocks impart slightly different velocity changes, ΔV, to the system. The trapezoidal imparts the largest ΔV. This

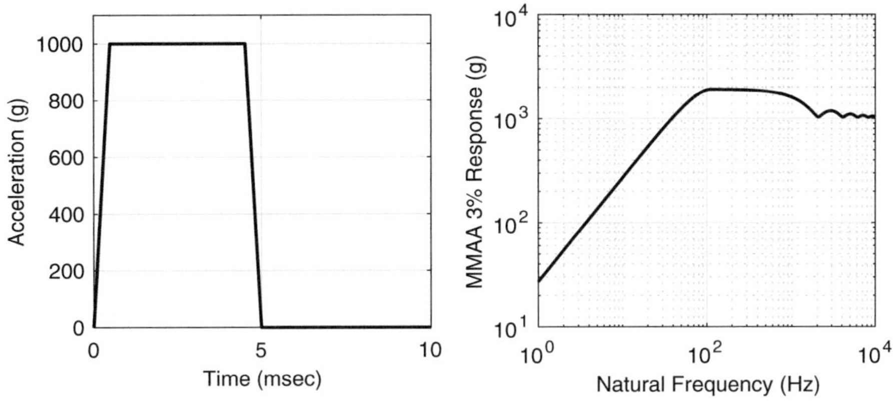

Fig. 5.5 Acceleration time history and SRS for 1000 g 5 ms trapezoidal shock

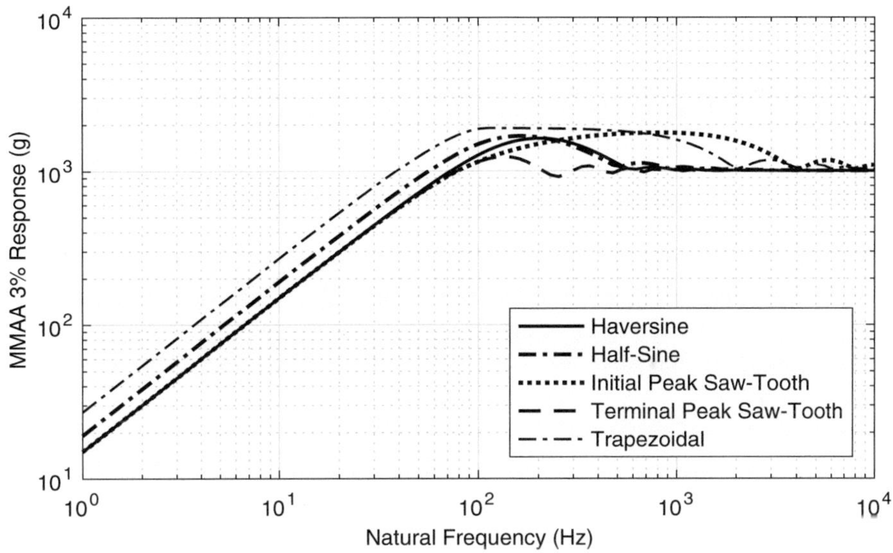

Fig. 5.6 Comparison of maxi-max acceleration SRS for the five 1000 g 5 ms classical shock pulses

coupled with the naturally different shape of the waveforms results in the variations seen. For example, if the duration of the half-sine pulse were reduced from 5 to 4.3 ms, the half-sine and haversine SRS curves would be almost identical. Similar adjustments could be made to all of the shocks but this is left as an exercise for the reader.

Figure 5.7 shows a comparison between the maximum pseudo-velocity SRS for the same five classical shocks on tripartite paper. As was discussed previously, the pseudo-velocity presentation emphasizes the importance of the velocity change in the classical shocks. Since all of the simple shocks are one pulse events representing

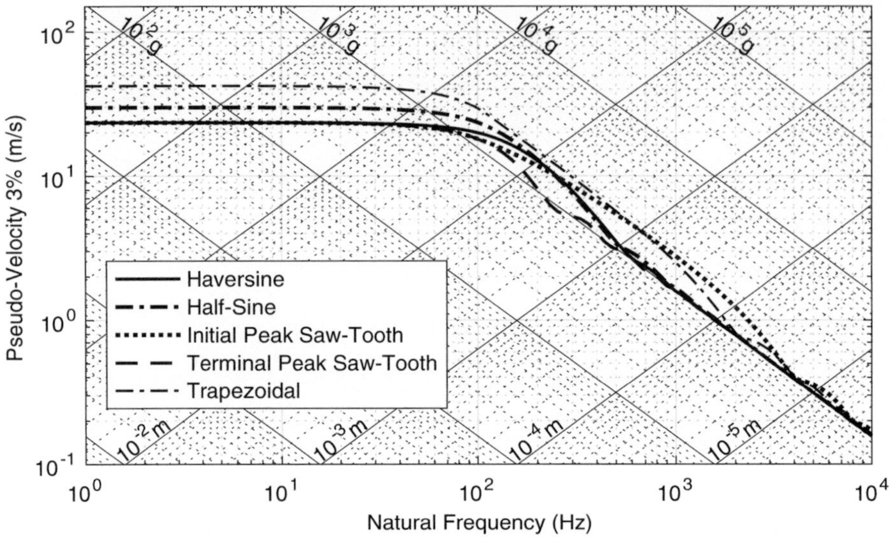

Fig. 5.7 Comparison of maximum pseudo-velocity SRS for the five 1000 g 5 ms classical shock pulses

a velocity change imparted to the system, a velocity presentation may be most appropriate.

The preceding example also shows that the exact form of a classical shock is not necessarily critical for a successful test. Each of the five classical shocks can be tailored to yield essentially the same SRS. This in turn should yield essentially the same damage potential when the shock test is performed on a shock machine.

5.2 Interpretation of Classical Shock SRS

The SRS is typically used to compare the severity of two events or an event with an intended test specification. However, many characteristics of the shock pulse can be read directly from the SRS. The right-hand side of the MMAA SRS tends toward the maximum amplitude of the shock acceleration time history. In other words, as the frequency increases, the SRS amplitude settles at the maximum amplitude in the time history. This was shown in Chap. 4 where the high-frequency SDOF oscillator is stiff and the motion of the SDOF oscillator tracks with the base input motion.

Figure 5.6, showing the MMAA SRS from the five classical shocks, shows all the 1000 g classical shock spectra are at 1000 g by about 10 kHz. This is not equal to the SRS maximum. The peak of a classical shock SRS is between one and two times the peak acceleration amplitude. The exact SRS maximum depends on the shape of the shock pulse and the damping valued used.

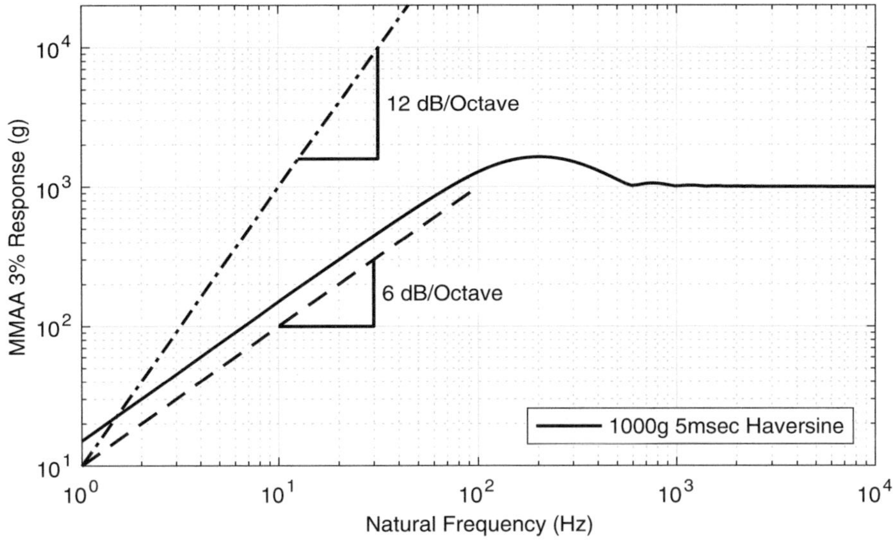

Fig. 5.8 Maxi-max acceleration SRS for a classical shock showing the 6 dB/octave low-frequency slope

The left-hand side of the MMAA SRS defines the type of shock and the velocity change associated with the shock event. The low-frequency slope on a MMAA SRS plot should fall between a low of 6 dB per octave and a high of 18 dB per octave. A 6 dB per octave slope is a slope of one on log–log paper as will be demonstrated in Sect. 5.2.1. Likewise, 12 and 18 dB per octave is a slope of two and three on log–log paper, respectively. Classical shocks all have a 6 dB per octave low-frequency slope indicative of a velocity change associated with the shock event. Figure 5.8 shows a 6 dB per octave line plotted alongside the MMAA SRS for reference. The higher 12 dB per octave slope is indicative of a pyroshock type event characterized by a propagating stress wave without an overall system velocity change.

5.2.1 Slopes in dB per Octave

Slopes on an SRS plot are typically given in units of decibels (dB) per octave. However, when the SRS is plotted on logarithmic paper, the units do provide some simple relationships as will be shown here.

An octave is a frequency interval defined by a doubling of the base frequency regardless of the starting frequency. Thus, 1–2 Hz is an octave just as 100–200 Hz is an octave. The number of octaves between two frequencies is given by a base two logarithm calculation as:

$$N \text{ octaves} = \log_2 \left(\frac{f_2}{f_1} \right). \tag{5.5}$$

Equation 5.5 can also be written in terms of a base ten logarithm as:

$$N \text{ octaves} = \frac{\log_{10}(f_2/f_1)}{\log_{10}(2)}. \tag{5.6}$$

The decibel is a logarithmic unit that defines a ratio between two quantities. Thus, the decibel is usually expressed as a change from some baseline value. For example, +3 dB is a 3 dB increase over a baseline value; −3 dB is a 3 dB reduction from the baseline value; and the term 0 dB is often used to refer to the baseline value. The difference between two SRS amplitudes in terms of dB is obtained by the expression:

$$n \text{ dB} = 20 \log_{10}\left(\frac{A_2}{A_1}\right), \tag{5.7}$$

where A_1 and A_2 are the two amplitudes to be evaluated. Equation 5.7 is applicable to any amplitude related quantity but is not applicable to power or energy quantities. The dB relationship for power quantities replaces the 20 multiplier with 10. This change is due to the quadratic relationship between power and amplitude [2, 3].

As a result of the relationships in Eqs. 5.6 and 5.7, the slope on the plot is given by:

$$\text{Slope} = \frac{n \text{ dB}}{N \text{ octaves}} = \frac{20 \log_{10}(A_2/A_1)}{\log_{10}(f_2/f_1)} \log_{10}(2). \tag{5.8}$$

So, a slope of one on log–log paper corresponds to $A_2/A_1 = 10$ while $f_2/f_1 = 10$ which is equal to

$$\text{Slope} = \frac{20 \log_{10}(10)}{\log_{10}(10)} \log_{10}(2) = 20 \log_{10}(2) = 6.02 \text{ dB/octave}. \tag{5.9}$$

This is commonly rounded down and stated that a slope of one on log–log paper is equal to 6 db/octave. Likewise, a slope of two on log–log paper corresponds to an amplitude ratio of $A_2/A_1 = 100$, which substituted into Eq. 5.8 yields approximately 12 dB/octave.

5.2.2 *Why the Low-Frequency Slope is 6 dB/Octave*

It was stated earlier that the MMAA SRS low-frequency slope from a classical shock should tend toward a constant value of 6 dB/octave. Figure 5.8 shows this quite clearly for the classical haversine shock. But why does this happen?

For an undamped system, pseudo-velocity and relative velocity are the same in the residual vibration era. Since the velocity change imparted by a classical shock in this era is independent of the SDOF natural frequency, the velocity SRS is a flat line equal to the velocity change from the base input shock pulse for the residual portions of the relative velocity SRS. This is true regardless of the type of classical shock, where the velocity change monotonically increases, and is illustrated in Fig. 5.7.

The maximum relative velocity in the free vibration era is the velocity change imparted by the shock:

$$\Delta V = \int_0^{T_d} \ddot{w}(t)dt. \tag{5.10}$$

For example, the velocity change of a half-sine pulse with amplitude A and duration T is:

$$\Delta V = \int_0^{T} A \sin\left(\frac{\pi \tau}{T}\right) d\tau \tag{5.11}$$

$$= -\frac{AT}{\pi} \cos\left(\frac{\pi \tau}{T}\right)\Big|_0^T \tag{5.12}$$

$$= -\frac{2AT}{\pi}. \tag{5.13}$$

Solving this equation using the parameters from Fig. 5.2, $A = 1000\,\mathrm{g}$ and $T = 0.005\,\mathrm{s}$, gives a maximum velocity of 31.2 m/s, nearly the same as shown in Fig. 5.7. The slight difference in velocity between the value calculated here and the value read from the plot is that the derivation here assumed an undamped system where the SRS calculations presented previously assumed a small damping in the SDOF oscillator.

In Chap. 4, we showed that the low-frequency portion of the MMAA SRS of a classical shock is defined by the residual spectra. In Chap. 3 we showed that the response to an impulse is equivalent to the free vibration response for an initial velocity. This means we can determine the slope of the MMAA SRS by looking at the MMAA SRS of an impulse. Using the expressions derived in Chap. 3, the relative displacement impulse response is

$$z(t) = \frac{\Delta V}{\omega_n} \sin(\omega_n t), \tag{5.14}$$

so the maximum absolute value is just

$$Z(\omega_n) = \frac{\Delta V}{\omega_n}. \tag{5.15}$$

In Chap. 3 we also showed that the absolute acceleration is related to relative displacement for an undamped SDOF oscillator by

$$\ddot{x}(t) = -\omega_n^2 z(t), \tag{5.16}$$

so the MMAA SRS is

$$\text{SRS}_{\text{MMAA}} = Z(\omega_n) = \omega_n \Delta V. \tag{5.17}$$

Since the net velocity change is constant in a classical shock, the MMAA SRS is linear in ω_n which gives the 6 db/octave slope.

Another way to understand the low-frequency characteristics of the MMAA SRS of classical shocks is to use Duhamel's integral. Referring back to the SDOF oscillator described in Chap. 3, and shown in Fig. 3.3, the response of the SDOF mass, $x(t)$, to an arbitrary base excitation force $m\ddot{w}(t)$ can be obtained using Duhamel's integral as:

$$z(t) = \int_0^t m\ddot{w}(\tau)h(t-\tau)d\tau. \tag{5.18}$$

If we ignore the damper in Fig. 3.3 and assume an undamped system, then the impulse response function is given by:

$$h(t-\tau) = \frac{1}{m\Omega}\sin[\Omega(t-\tau)]. \tag{5.19}$$

Thus, the response of the SDOF oscillator mass after the shock pulse has passed, assuming the system is initially at rest with $z(0) = 0$ and $\dot{z}(0) = 0$, is given by

$$z(t) = \frac{1}{\Omega}\int_0^t \ddot{w}(\tau)\sin[\Omega(t-\tau)]d\tau. \tag{5.20}$$

Taking the derivative of Eq. 5.20 with respect to t (not τ) to get velocity yields:

$$\dot{z}(t) = -\int_0^t \ddot{w}(\tau)\cos[\Omega(t-\tau)]d\tau. \tag{5.21}$$

We first expand the angle subtraction terms, $\sin[\Omega(t-\tau)]$ and $\cos[\Omega(t-\tau)]$ in the displacement and velocity relationships in Eqs. 5.20 and 5.21 using the angle difference identities.

$$\sin[\Omega(t-\tau)] = \sin[\Omega t]\cos[\Omega\tau] - \cos[\Omega t]\sin[\Omega\tau], \tag{5.22}$$

$$\cos[\Omega(t-\tau)] = \cos[\Omega t]\cos[\Omega\tau] + \sin[\Omega t]\sin[\Omega\tau]. \tag{5.23}$$

Next, we can express $\ddot{w}(t)$ in terms of its Fourier spectrum at the system natural frequency. The Fourier transform is traditionally given as:

$$X(\Omega) = \int x(t)e^{-i\Omega t}dt. \tag{5.24}$$

However, the exponential term can be replaced by sines and cosines using Euler's formula as:

$$e^{-i\Omega t} = \cos(\Omega t) - i\,\sin(\Omega t). \tag{5.25}$$

Substituting Eq. 5.25 and the trigonometric angle expansions into Eqs. 5.20 and 5.21 yields:

$$\Omega z(t) = [\mathrm{Re}\; X(\Omega)]\sin(\Omega t) + [\mathrm{Im}\; X(\Omega)]\cos(\Omega t) \tag{5.26}$$

and

$$\dot{z}(t) = [\mathrm{Re}\; X(\Omega)]\cos(\Omega t) - [\mathrm{Im}\; X(\Omega)]\sin(\Omega t). \tag{5.27}$$

For small values of Ω, the maximum response occurs after the shock excitation has ended, in the residual vibration era. For an undamped system, the residual response is a sine wave defined by the initial conditions. If the shock duration is T, then the initial conditions for the free response in the residual time frame are $z(T)$ and $\dot{z}(T)$ and the relative displacement is given by Eqs. 3.25 and 3.26:

$$z(t) = z(T)\cos(\Omega t) + \frac{\dot{z}(T)}{\Omega}\sin(\Omega t),\;\; t > T_d \tag{5.28}$$

and the relative velocity is

$$\dot{z(t)} = -\Omega z(T)\sin(\Omega t) + \dot{z}(T)\cos(\Omega t),\;\; t > T_d. \tag{5.29}$$

The maximum value of the residual response is just the amplitude of the sine wave, which is given by:

$$Z = \sqrt{z^2(T) + \frac{\dot{z}^2(T)}{\Omega^2}}. \tag{5.30}$$

Substituting the relationships from Eqs. 5.26 and 5.27 into Eq. 5.30 and simplifying yields:

$$\Omega Z = \sqrt{[\mathrm{Re}\; X(\Omega)]^2 + [\mathrm{Im}\; X(\Omega)]^2}. \tag{5.31}$$

The left-hand side of Eq. 5.31 is just the pseudo-velocity SRS at the frequency Ω and the right-hand side is the magnitude of the Fourier spectrum at the same frequency. Thus, it follows that:

$$\dot{Z}_{max} = |X(\Omega)|\,. \tag{5.32}$$

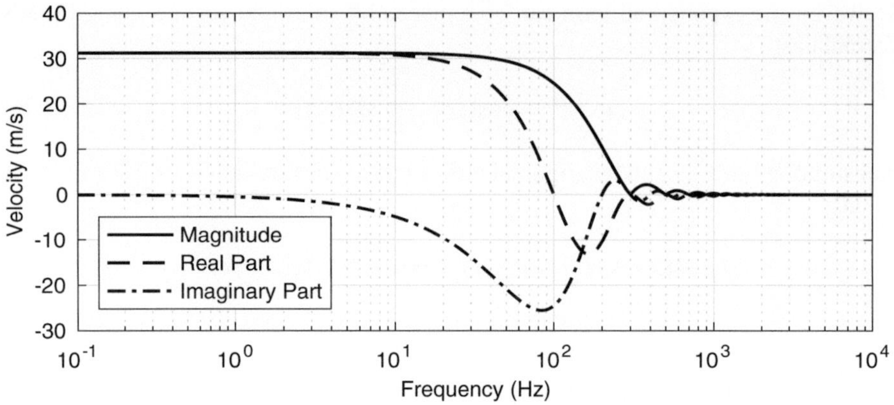

Fig. 5.9 Fourier transform real, imaginary, and magnitude plot for a 1000 g 5 ms half-sine shock pulse

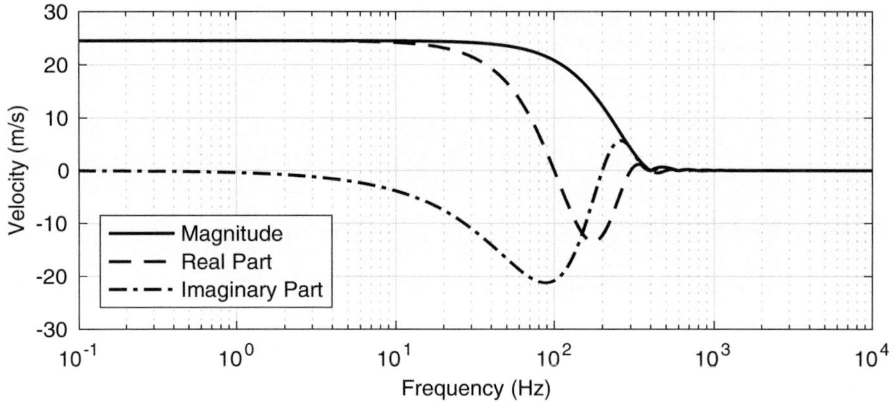

Fig. 5.10 Fourier transform real, imaginary, and magnitude plot for a 1000 g 5 ms haversine shock pulse

This theory is demonstrated with three examples from the classical shock pulses. Figure 5.9 shows the Fourier transform results from the 1000 g 5 ms half-sine shock pulse shown in Fig. 5.2. In this figure, the real and imaginary parts of the Fourier spectra are plotted along with the magnitude. The low-frequency portion is relatively flat and the magnitude tends to the real-valued portion of the Fourier spectrum as the imaginary part tends to zero at low frequencies. The magnitude is not a true zero slope line but is smooth and asymptotically flat as the frequency goes to zero. Likewise, a similar result is shown in Fig. 5.10 for the haversine shock pulse from Fig. 5.1. Finally, Fig. 5.11 shows the same low-frequency, flat Fourier spectrum trend for the initial peak saw-tooth shock pulse of Fig. 5.3. This relationship to the Fourier spectrum defines why the classical shock pulses tend to a flat line in the velocity SRS and a 6 dB/octave slope in the MMAA SRS. This also shows why

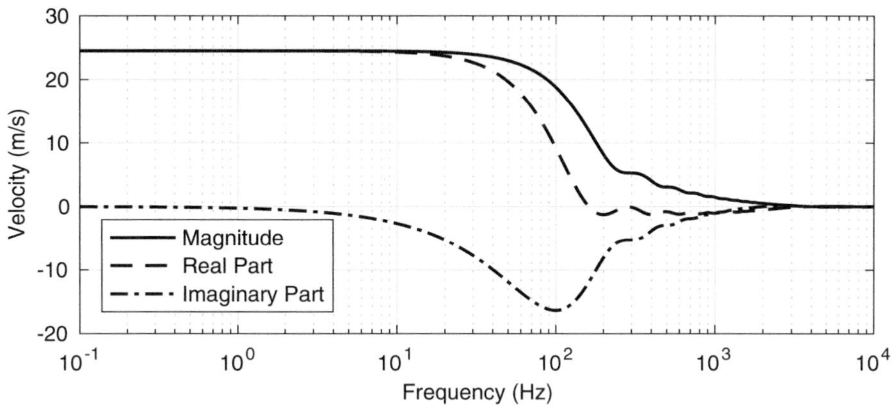

Fig. 5.11 Fourier transform real, imaginary, and magnitude plot for a 1000 g 5 ms initial peak saw-tooth shock pulse

shock pulses that do not have a net velocity change, such as a pyroshock event, have steeper slopes—their Fourier transforms should start with a $\Delta V = 0$ instead of the impact velocity.

The derivation given here shows why the low-frequency portion of the velocity SRS is flat for a classical shock pulse. This derivation applies for any shock excitation. Once the SRS is defined by the residual response, the slope of the MMAA spectra will always tend toward 6 dB/octave.

5.2.3 *Estimating Velocity from the SRS*

When a velocity SRS is used instead of the acceleration SRS, the low-frequency line tends to a flat line equal to the system velocity change associated with the shock event when the SDOF oscillator damping is zero. For SRS calculations with non-zero damping, the velocity will trend to a value slightly less than the actual velocity change. This can be seen in Fig. 5.7. As was stated earlier, the five classical shocks were all plotted with the same amplitude and duration; however, the different shapes enclose differing areas on the acceleration versus time plot leading to different velocity changes from the shock pulses.

It is also possible to obtain the velocity change from the acceleration SRS plot with a simple calculation. The conversion from acceleration to velocity in the frequency domain is simply a division by the circular frequency and a unit conversion from g to velocity units. The equation is given by

$$\Delta V = \frac{Ag}{2\pi f}, \tag{5.33}$$

where A is the acceleration read from the MMAA SRS plot, g is the conversion factor to engineering units (1 g = 9.81 m/s in SI units), and f is the frequency where the acceleration was read. The true system velocity change is technically defined by the acceleration at the very low frequencies which should approach a 6 dB per octave straight line as shown in Fig. 5.8. Since the slope should be constant at low frequencies, it is only necessary to select a representative point at a very low frequency to extract the velocity. For example, in Fig. 5.8, the acceleration at 1 Hz is 15.025 g; inserting this point into Eq. 5.33 yields a velocity change of 23.46 m/s. This is the same velocity read directly from Fig. 5.7. Note that Eq. 5.33 is exactly the expression for pseudo-velocity of an undamped SDOF system derived in Chap. 3, Eq. 3.21.

This relationship intrinsically means that as the low-frequency line on the MMAA SRS moves left and up, the velocity change increases. Likewise, as the low-frequency line moves right and down, the velocity change decreases. In most cases, the velocity change of a classical shock is more closely correlated with damage than the maximum acceleration, which is discussed in Chap. 7. As a result, a higher velocity shock is considered to be more severe than a lower velocity shock, regardless of the peak acceleration in the MMAA SRS plot. The exception is if one shock excited a fundamental resonant mode where another shock did not.

Figure 5.12 shows a plot of a MMAA SRS from an actual drop test. Since this is real test data, the resulting SRS curves are not always clean and well-defined due to non-linear effects and interactions. The drop test performed here was actually a destructive test with multiple internal component collisions coupled with material

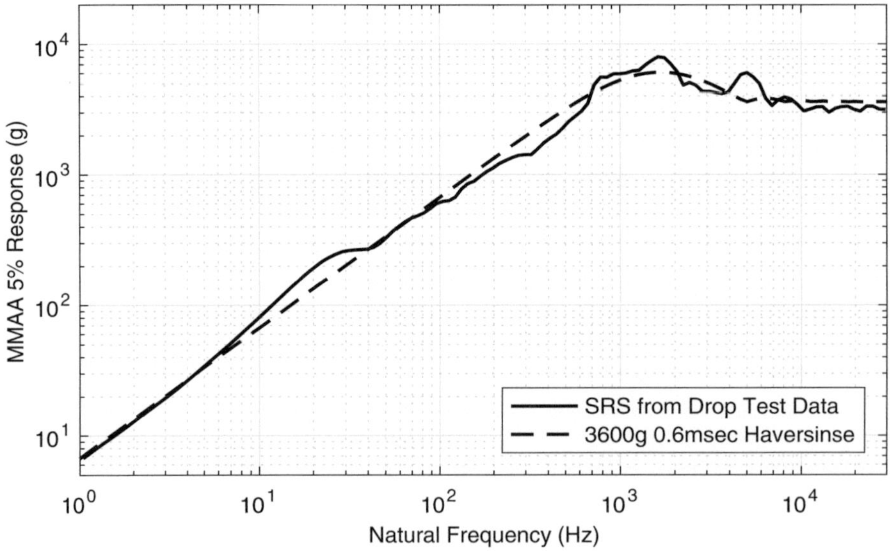

Fig. 5.12 Maxi-max acceleration SRS data from an actual drop test overlaid with a best-fit haversine SRS

yield and fracture. Figure 5.12 also shows an overlay of a best-fit classical haversine. The low-frequency portion of the original SRS overlays well with the classical fit by design. Extracting the velocity change from these data shows that the data correspond to an approximate 10.5 m/s velocity change. Physically, this corresponds to about a 5.6 m drop height. The drop height was calculated using the standard conservation of mechanical energy relations.

5.2.4 *Effect of Damping on the SRS*

While a damping estimate is not as easily extracted from the plot as the peak acceleration, an understanding of the damping can be gleaned. Low system damping will result in a higher, sharper peak in the SRS. High system damping will yield an SRS with a lower, more flattened SRS peak.

Figure 5.13 shows test data from a single system level drop test with multiple internal components. One of the components has relatively little internal damping as shown by the higher peak in the SRS. In contrast, the SRS from the second component shown in Fig. 5.13 indicates high internal damping. This is apparent from the low, flattened SRS peak. Since both component responses were recorded from a single test, the low-frequency slope line converges to show a common impact velocity as should be expected.

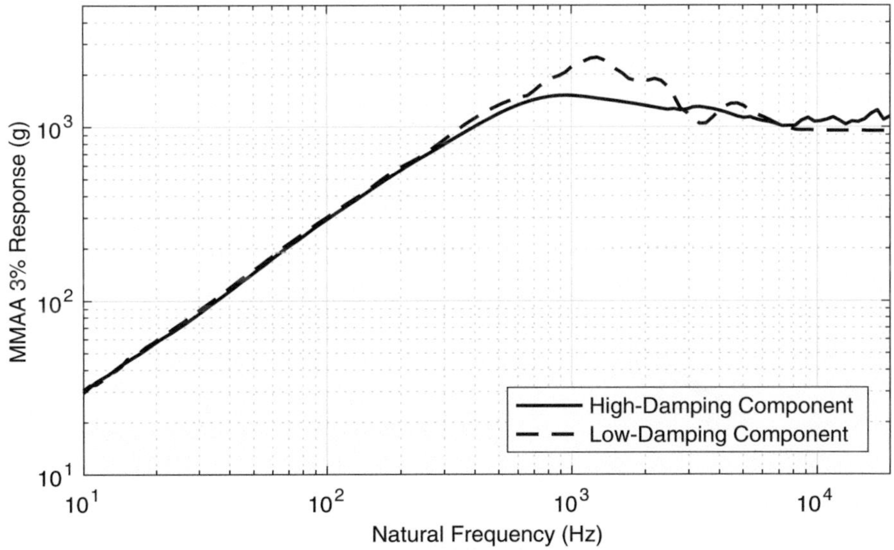

Fig. 5.13 Maxi-max acceleration SRS from a test of a low-damping component and a high-damping component

One of the more problematic aspects of analyzing real test data can also be seen in Fig. 5.13. Real test data often include multiple resonances and often includes the effects of interaction between multiple components in a system. This frequently appears in the high-frequency portion of the SRS as seen here. This is in contrast to the idealized shape of the mathematically true haversine shock. A good approach for analyzing real test data are to overlay the data with an ideal haversine as a way of seeing through the noise of the actual system and understanding the actual physics.

In addition to the changes in the SRS shape with the intrinsic system damping, the damping selected while calculating the SRS will also have an effect on the resulting SRS. Figure 5.14 shows the MMAA SRS calculated using the same 1000 g 5 ms haversine shock pulse with no damping ($\zeta = 0$) and with 10% damping ($\zeta = 0.1$). Most SRS calculations are performed with 5% or less damping. SRS curves calculated using intermediate damping values will fall proportionately between the zero and 10% curves shown here. The first change between the two curves is the height of the SRS peak. The SRS maximum for the undamped case is 1.7 times greater than the peak acceleration in the time history. The SRS calculated with 10% damping is only 1.5 times greater than the acceleration time history. For comparison, an SRS calculated with 3% damping will have a peak at 1.64 times the maximum acceleration, while an SRS with 5% damping will have a peak at 1.6 times the maximum acceleration.

More interesting perhaps is the obvious change in the velocity shown in Fig. 5.14. The two plots show that the acceleration is the same since both SRS curves end at 1000 g at high frequencies. However, the velocity change is significantly different.

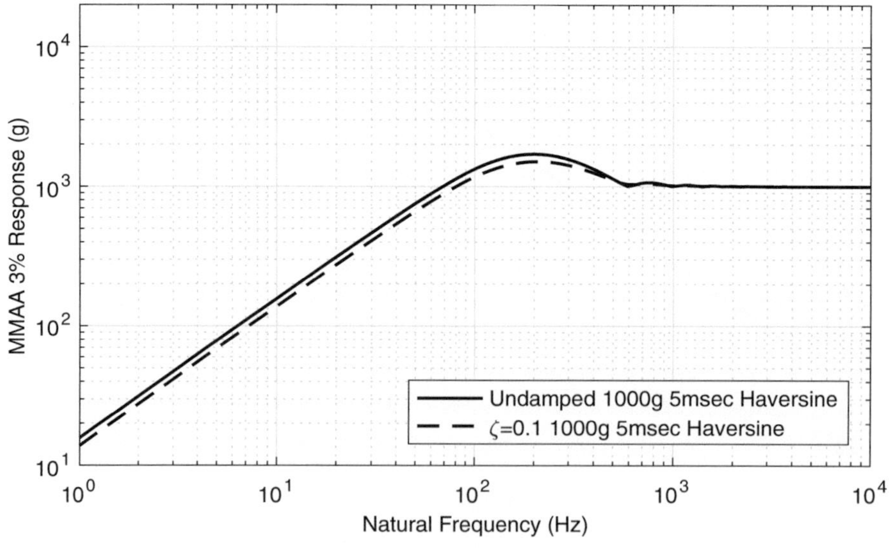

Fig. 5.14 Maxi-max acceleration SRS for a haversine calculated with no damping and 10% critical damping

Using Eq. 5.33, the predicted velocity change using the undamped model is 24.5 m/s while the velocity change from the 10% damped mode is 21.6 m/s. This is somewhat problematic since a change in the SDOF oscillator model effectively changes the apparent rigid body motion of the impacting system. As a result, care should be taken using larger damping values to ensure that the resulting SRS curves are appropriate to the problem at hand.

5.3 Shock Bandwidth

One of the potential problems with using acceleration SRS is the common misconception that shock energy is constant at high frequencies. This erroneous interpretation arises since the acceleration SRS tends to a flat line at the shock acceleration amplitude and remains there with increasing frequency. Figure 5.6 shows this clearly with all of the shock amplitudes trending toward 1000 g at high frequencies. If the plot had been continued past 10 kHz, the MMAA SRS would remain flat. This is not the case with the PVSRS plot shown in Fig. 5.7. The pseudo-velocity SRS falls off rapidly above about 200 Hz. While there is an obvious difference between the two presentation methods, the shock pulse is the same. Intuition says that the shock energy is not infinite and must taper off at some point.

The Fourier transform magnitude helps us understand the energy contained in the shock pulse. Figure 5.15 shows the Fourier transform magnitude of the 1000 g

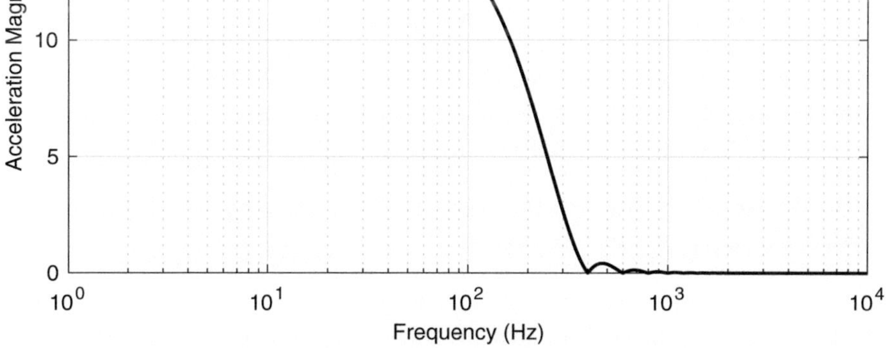

Fig. 5.15 Fourier transform magnitude for 1000 g 5 ms haversine shock

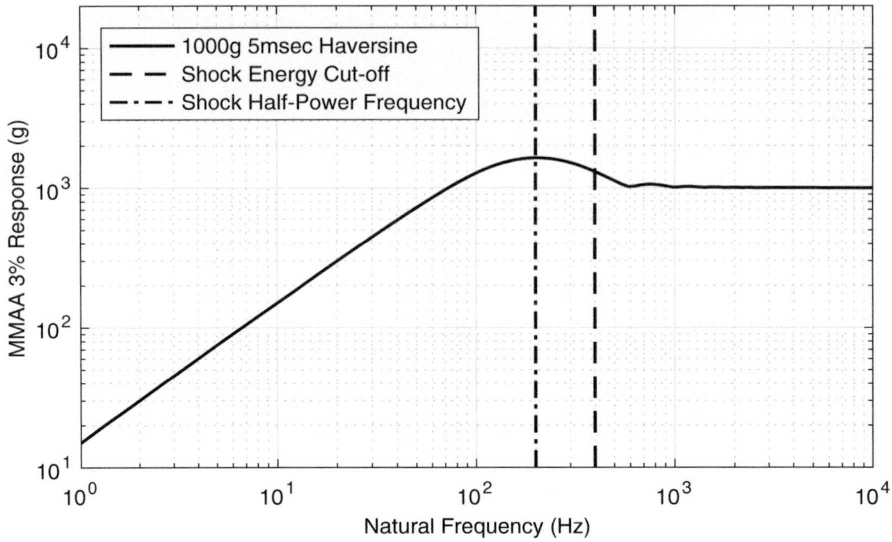

Fig. 5.16 MMAA SRS plot for 1000 g 5 ms haversine shock showing energy cut-off frequencies

5 ms haversine pulse shown in Fig. 5.1. Parseval's identity states that the energy contained in the square of the time history $[x(t)]^2$ is equal to the square of the energy in the frequency response $[X(\omega)]^2$. Figure 5.15 shows that the frequency response effectively drops to a negligible value at 400 Hz in this example or twice the frequency of the 5 ms (200 Hz) shock pulse. The energy is reduced by half at the frequency of the shock pulse, 200 Hz in this example. The locations of these frequencies with respect to the MMAA SRS are shown in Fig. 5.16.

The consequence of this energy cut-off is that the classical shock pulses will not significantly excite system resonances above the shock pulse frequency. Further, the pulse may not sufficiently excite system frequencies approaching the shock pulse frequency since the energy is already beginning to diminish rapidly. This is much more apparent with the pseudo-velocity SRS presentation shown in Fig. 5.17. As a result, care should be taken when interpreting the energy input to a system based solely on the MMAA SRS as shown here. The peak in the MMAA haversine SRS occurs at the half-power frequency. Beyond the MMAA SRS peak there is essentially no energy being imparted to the system. As a result, the system is simply tracking the input motion as opposed to being dynamically excited by the input motion.

The results shown here for the haversine are equally applicable to the other classical shock pulses since the classical shock pulse response spectra are similar as demonstrated previously. The classical shocks are all one-sided shocks in the time domain and exhibit similar high-frequency energy roll-off above the excitation frequency. For this reason, care should be taken when designing a test to ensure that the system resonant frequencies of interest are fully excited when using classical shocks.

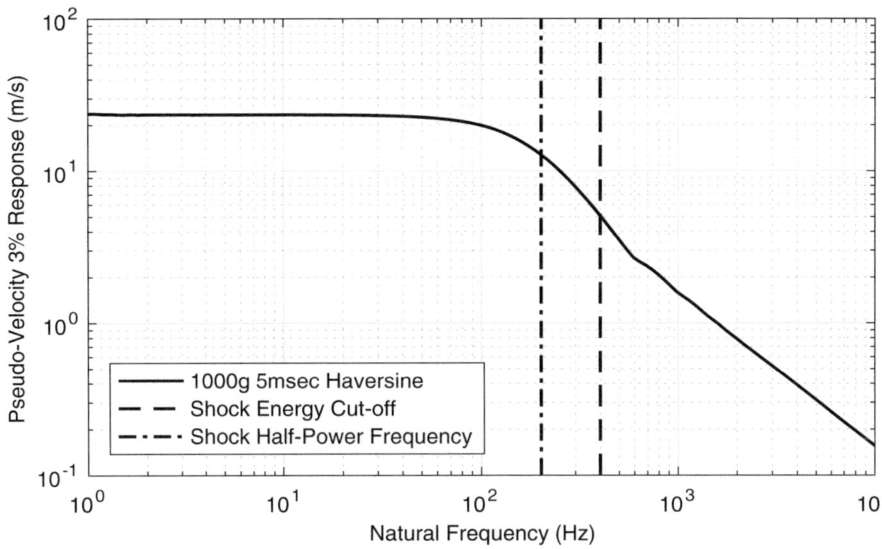

Fig. 5.17 Pseudo-velocity SRS plot for 1000 g 5 ms haversine shock showing energy cut-off frequencies

One shock that is unique in regard to its bandwidth is the ideal impulse. An impulse is a Dirac delta function at $t = 0$. This means that its bandwidth is infinite (i.e., the Fourier magnitude is constant to $f = \infty$). Consequently, the SRS is defined entirely by the residual response, so the MMAA SRS is a line with slope 6 dB/octave and the SRS should not have a high-frequency plateau. While the Dirac delta function may have infinite energy in theory, it cannot happen in practice. In real-world excitations, the shock pulse will always have a finite amplitude and a measurable duration, unlike the Dirac delta function. However, a Fourier analysis of the Dirac delta function does reveal an interesting limitation of the ramp invariant method for calculating SRS.

When the ramp invariant digital filter is used to compute the SRS, the Dirac delta impulse is eventually transformed into a triangle pulse with duration equal to twice the sample rate, $2\Delta t$. This is a result of having a finite sampling rate. The resulting SRS is illustrated in Fig. 5.18. The sample rate for the digital filter was selected as $f_s = 2$ kHz and the SRS is calculated up to 1 kHz. The figure shows that the discretization affects the SRS above 200 Hz. This illustrates the rule of thumb mentioned in Chap. 3, Sect. 3.4.2.4 that the sampling rate should be at least ten times the highest SDOF natural frequency if the peak dynamic response must be estimated accurately. Conversely, one should be careful when computing the SRS at resonant frequencies greater than about 10% of the sample frequency.

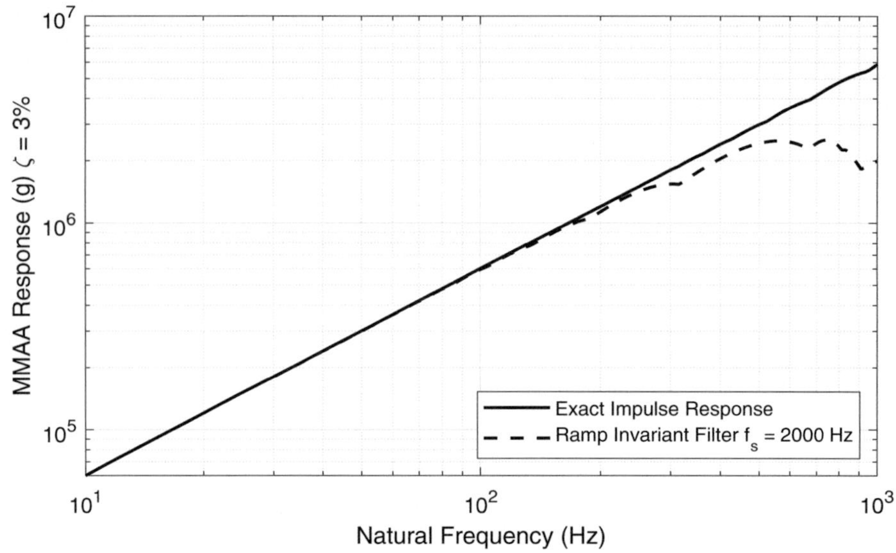

Fig. 5.18 Maxi-max absolute acceleration SRS of a 1000 g impulse shock

5.4 Positive and Negative, Primary and Residual

While most SRS plots will show the maxi-max absolute acceleration or pseudo-velocity, the positive, negative, primary, and residual SRS plots can be helpful in understanding the shock transient. Figure 5.19 shows a maximum positive and negative SRS plot of the same 1000 g 5 ms haversine pulse overlaid with the shock energy cut-off lines from the Fourier spectrum data. The maximum negative SRS shows a very distinct tapering off of the shock response at the same frequencies as the first minimum in the Fourier spectrum. The reason for the disparity between the positive and negative responses at high frequency is that the positive SRS is following the positive haversine shock pulse whereas the negative response is not being excited because there is negligible relative motion between the base and SDOF oscillator. Hence there is a negligible oscillation occurring on the return portion of the shock. This result is identical to the Fourier spectrum results shown in Fig. 5.15.

Figure 5.19 also shows that the negative SRS is slightly less than the positive SRS below the half-power frequency. This difference is a result of the SRS being calculated with a small but non-zero damping coefficient. Thus, the damping in the SDOF oscillator requires that the second half-cycle have a slightly smaller amplitude than the first half-cycle. If the positive and negative SRS are calculated using an undamped SDOF oscillator, the two SRS will be identical until just before the half-power frequency.

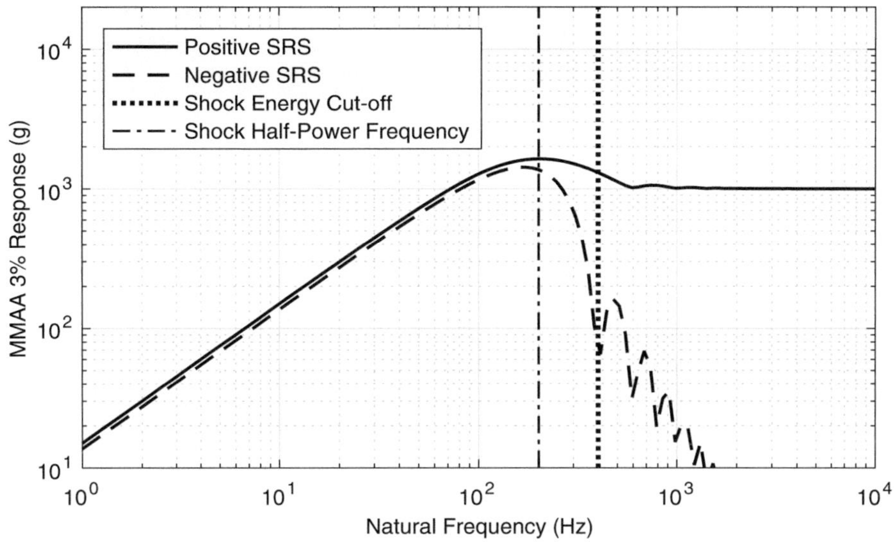

Fig. 5.19 Maximum positive and negative acceleration SRS plot for 1000 g 5 ms haversine shock showing energy cut-off frequencies

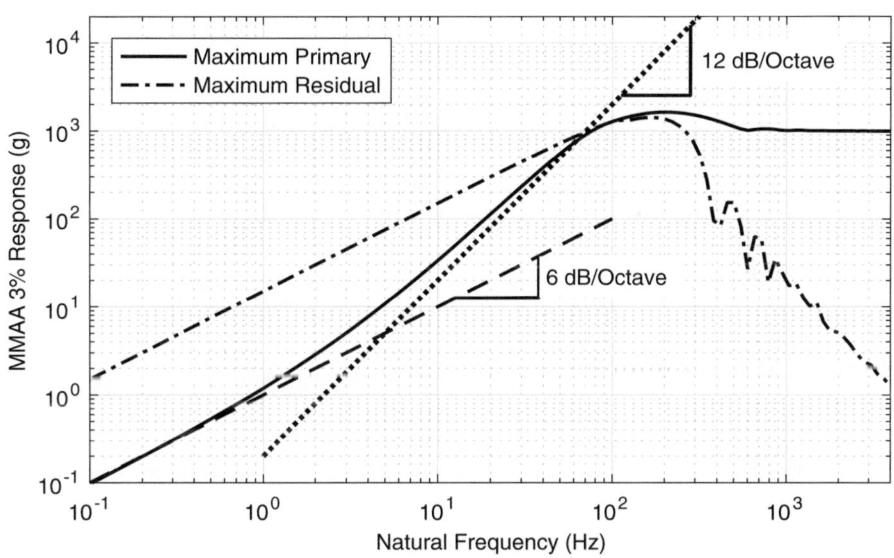

Fig. 5.20 Primary and residual acceleration SRS plot for 1000 g 5 ms haversine shock showing low-frequency slopes

Figure 5.20 shows the absolute maximum primary and residual SRS from the same 1000 g 5 ms haversine pulse. As was discussed previously in Chap. 4, the residual response dominates the low-frequency portion of the MMAA SRS while the primary response defines the high-frequency portion of the SRS. The residual response is derived from the motion after the shock transient has past, while the primary response is derived from the SDOF oscillator response during the time of the shock pulse. Figure 5.8 shows that the low-frequency portion of the MMAA SRS has a 6 dB per octave slope and Fig. 5.20 shows that this corresponds to the maximum residual response.

Figure 5.20 shows that the primary response falls off more rapidly as the frequency decreases from the shock pulse frequency. The slope is almost double the 6 dB/octave slope initially with an eventual transition to the expected 6 dB/octave slope at very low frequencies. The plot shows that the residual is not only the driving portion of the SRS but is substantially more driving than the primary response. As the frequency decreases, the primary portion of the response essentially becomes negligible in the overall shock response.

Figure 5.21 shows a −12 dB/octave slope plotted alongside the high-frequency portion of the maximum residual SRS. As was stated previously, the residual response falls away similar to the Fourier magnitude spectrum. The −12 dB/octave slope is an indication of how rapidly the energy diminishes past the shock pulse frequency. Therefore, classical one-sided shock pulses should not be relied upon to excite system resonances above the excitation bandwidth. The half-power bandwidth is likely a more appropriate limit for realistic resonant frequency excitation.

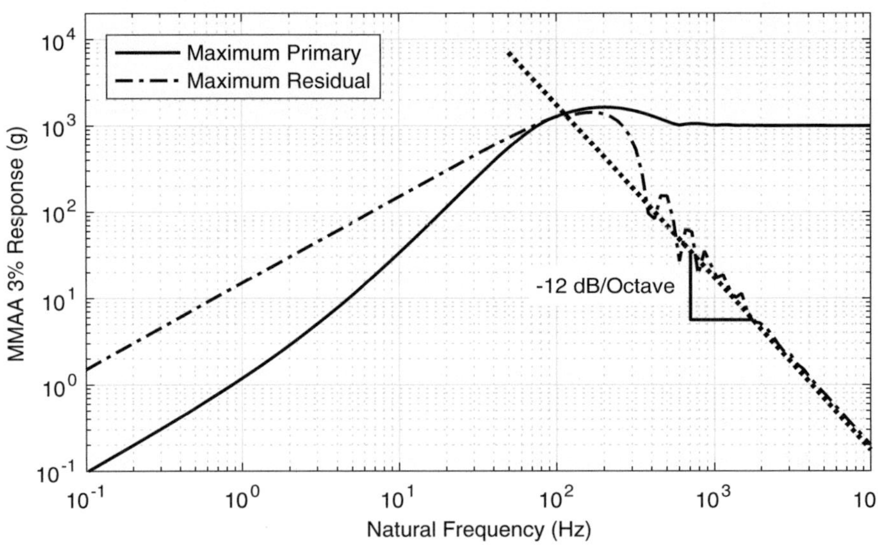

Fig. 5.21 Primary and residual acceleration SRS plot for 1000 g 5 ms haversine shock showing high-frequency slope

5.5 Summary

This chapter has discussed many of the characteristics of classical shock pulses in relation to their energy content and SRS representation. The five classical shock pulses have similar SRS with some mild variations largely resulting from the difference in the area enclosed on the acceleration versus time plot. The data presented here show that the classical shock pulses exhibit a substantial velocity change as a result of the shock event. This velocity change has a corresponding momentum change that is capable of significant damage to systems and components. However, it was also shown that the energy contained in the shock diminishes rapidly past a frequency equal to about one-half the shock pulse frequency. As a result, system resonances above about half the shock excitation frequency will not be well excited by a classical shock pulse.

Problems

5.1 This chapter described and compared the five classical shock types. However, any largely one-sided shock pulse with a finite velocity change will result in similar SRS properties. Generate several one-sided shock pulses and calculate their SRS. Compare your results to the classical results presented here. How are they similar? How are they different?

5.2 The step function is not considered a classical shock. What properties of the step function make it a poor representation of a shock event?

5.3 The initial peak and terminal peak saw-tooth shocks in Sect. 5.1 were plotted with a slight slope to the leading and trailing edges, respectively. How does changing that slope alter the resulting SRS? What do the changes to the SRS mean physically?

5.4 The trapezoidal shock in Sect. 5.1 is defined with a 10% rise and fall time. How does changing the rise and fall time alter the resulting SRS? What happens if the rise time and fall time are not equal?

5.5 Use the closed-form equations in Chap. 3 to compute the MMAA SRS of a 1000 g 5 ms step function. How does the SRS compare to the SRS of the trapezoidal shock?

5.6 Use the closed-form equations in Chap. 3 to compute the MMAA SRS of a 1000 g 5 ms descending ramp function. The descending ramp function starts with a 1000 g impulse. How does the SRS compare to the SRS of the initial peak saw-tooth shock?

5.7 Use the closed-form equations in Chap. 3 to compute the MMAA SRS of a 1000 g 5 ms ascending ramp function. The trailing edge slope of the ascending ramp is infinite. How does the SRS compare to the SRS of the terminal peak saw-tooth shock?

5.8 Section 5.2 shows that the velocity change from the five classical shocks differs considerably. All of the classical shocks in this section had the same peak acceleration and pulse duration. What causes the difference in the velocity changes? Derive acceleration amplitudes and pulse durations for four of the five classical shocks such that all five pulse styles have the same velocity change.

5.9 Section 5.2.2 demonstrates how the SRS low-frequency slope relates to the Fourier transform of the shock pulse. Calculate the Fourier transform for several sets of measured shock events to see how the low-frequency slope compares. Shocks with a high velocity change will have a long, flat low-frequency portion whereas shocks with a low velocity change should trend down and level out closer to zero. Note that it may be necessary to process a lot of data after the shock in order to obtain the low-frequency resolution in the Fourier transform to see these trends.

5.10 Section 5.2.3 describes the process for estimating the impact velocity from test data. Use the techniques from this section to estimate impact velocities for several sets of experimental data from tests you have collected. How do the estimates from the SRS compare to the known impact velocities?

5.11 Figure 5.15 shows the Fourier transform magnitude for a classical haversine shock pulse. The purpose of this plot is to identify the half-power frequency of the shock. Create similar plots for the other four classical shock pulses to see if or how the half-power frequency changes with different shock pulses.

5.12 It was shown that the low-frequency SRS slope tends to approximately 6 dB/octave. At what relative location to the shock pulse frequency does the SRS trend to a relatively constant slope? How is this related to the bandwidth of the shock pulse?

5.13 Compare the low-frequency SRS slopes of the ideal classical shock pulses with the low-frequency slopes of some readily available drop shock test data. How do the slopes compare? How does the high-frequency portion of the SRS compare?

5.14 If the velocity change of the shock can be estimated from the low-frequency portion of the SRS, where will the SRS tend to if the defined shock has no net velocity change? Demonstrate this with an example.

References

1. Gaberson, H. A. (2011). Simple shocks have similar shock spectra when plotted as PVSS on 4CP. In *Proceedings of the 82nd Shock and Vibration Symposium*, Baltimore.
2. ISO 80000-3. (2006). *International Standard, Quantities and units—Part 3: Space and time.*
3. ISO 80000-8. (2007). *International Standard, Quantities and units—Part 8: Acoustics.*

Chapter 6
Oscillatory and Complex Shock Theory

Oscillatory shock pulses are defined as transient waveforms with positive and negative amplitudes. They are more complex waveforms, often built up of multiple periodic, decaying harmonic functions, but they can also be made up of a series of classical shocks. Oscillatory shocks have a sharp distinction from classical shocks in that there is typically no net velocity change associated with the shock event. While these shocks have no net velocity change, they may or may not have a displacement change. For example, an earthquake shock will have no velocity change; however, it is possible for the ground to shift resulting in a net displacement. Pyroshock, ballistic shocks, and earthquake excitation are examples of oscillatory shocks. An example time history of an oscillatory shock is shown in Fig. 6.1.

One of the fundamental problems with oscillatory shocks is that they are difficult to reproduce in the laboratory. Oscillatory shocks are typically produced by some sort of explosive device or an earthquake, so they are essentially decaying random excitations. While explosive testing can be done, it is expensive and fraught with safety concerns. Reproducing earthquake motion is a more tractable problem but requires specialized equipment to generate the large displacements typically associated with an earthquake event.

Complex shocks are a combination of classical shocks, perhaps more than one in series, with an oscillatory shock superimposed. This is typical in structures where the energy from an impact or pyrotechnic event excites the vibration modes of the structure. Figure 6.2 shows an example of how this may occur. A structure is subjected to a classical shock at its base. The shock-sensitive component is located within the structure away from the base. In this example, the shock-sensitive component is on the third level. The shock environment that the component experiences is the response of the system to the base excitation measured at the base of the component. In this case, we have made the implicit assumption that the component is relatively small compared to the overall system. The structure between the base, where the classical shock is applied, and the shock-sensitive component acts as a complex mechanical filter. The component shock excitation will therefore

C. Sisemore, V. Babuška, *The Science and Engineering of Mechanical Shock*,
https://doi.org/10.1007/978-3-030-12103-7_6

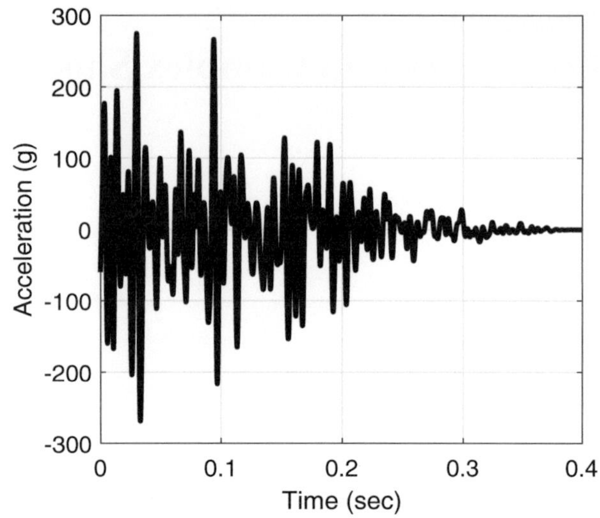

Fig. 6.1 Example of an oscillatory shock

be a combination of the classical shock and the dynamic response of the intervening structure—it will be a complex shock.

The theory and discussion presented in this chapter will focus on the general characteristics of oscillatory and complex shocks. Oscillatory shocks are typically high acceleration, high-frequency events with negligible velocity, or displacement changes. The information is equally applicable to earthquake shocks and pyroshocks. Specific characteristics of these shocks are highlighted and discussed. In this chapter we explain:

1. the SRS characteristics associated with oscillatory shocks including effects of bandwidth and duration;
2. interpretation of oscillatory shock SRS inflection points;
3. the characteristic low-frequency slope of 12 dB/octave for shocks with no velocity change;
4. how to interpret the SRS for two-sided pulses and complex shocks.

6.1 Oscillatory Shock Waveform

Oscillatory shocks can take numerous forms, ranging from relatively simple to extremely complex. The most basic oscillatory shocks are essentially decaying random excitations. The spectral content of a decaying random signal is difficult to specify, so in general, an oscillatory waveform is represented as a summation of damped harmonics. The simplest waveform can be expressed as a single decaying sine tone as shown in Fig. 6.3. This figure shows a 2 kHz decaying sinusoid with an initial amplitude of approximately 3000 g and a duration of about 8 ms. The

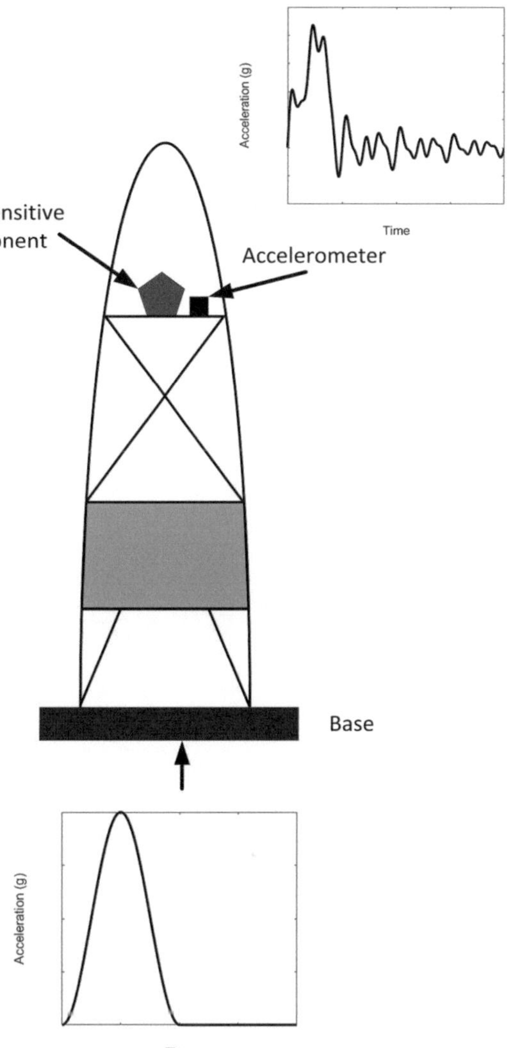

Fig. 6.2 Classical shock at the base of a structure becomes a complex shock at a shock-sensitive component

decaying sinusoid is not a true oscillatory shock in the sense that there is a small but finite velocity change associated with this signal due to the fact that the positive peaks are all slightly higher than their corresponding negative peaks. Nevertheless, the decaying sinusoid is a reasonable approximation of an oscillatory shock.

Figure 6.4 shows the calculated MMAA SRS from the oscillatory shock waveform given in Fig. 6.3. This SRS highlights many of the distinctive differences between the oscillatory shock SRS and the classical shock SRS plots presented in Chap. 5. The first significant difference is the height of the maximum SRS peak. The SRS peak is approximately 15,000 g from a 3000 g shock pulse, a five times amplification. This is in sharp contrast to the 1.64 times amplification seen with

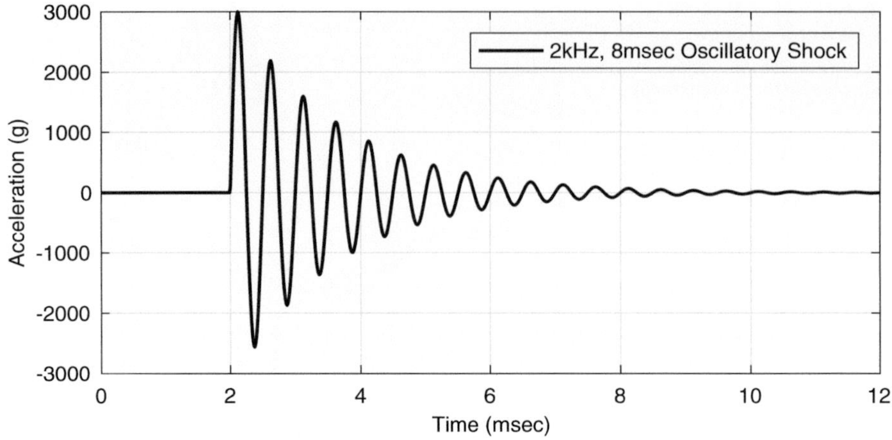

Fig. 6.3 Acceleration time history of a 3000 g 2 kHz oscillatory shock

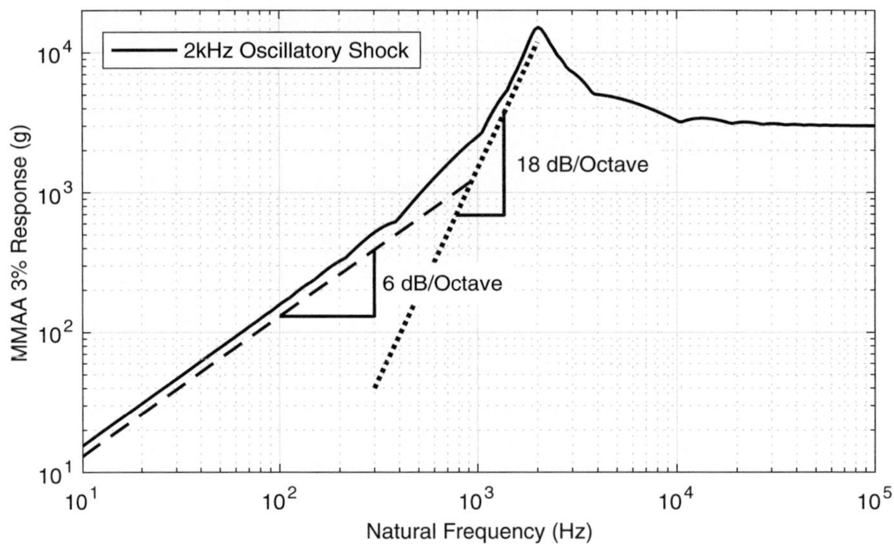

Fig. 6.4 MMAA SRS of a 3000 g 2 kHz oscillatory shock showing slope lines

the classical haversine shock. The high-frequency portion of the spectrum still trends to the maximum acceleration, 3000 g in this example, the same as with the classical shocks. Below the shock frequency, 2 kHz in this example, the initial slope is approximately 18 dB/octave. The low-frequency slope becomes more shallow at about half the shock frequency and tends to the same 6 dB/octave slope of the classical shocks at the lowest frequencies. The system velocity change can be calculated from the 6 dB/octave portion of the SRS using Eq. 5.33 which yields a velocity change of about 2.4 m/s in this example. Integrating the acceleration time history gives a velocity change of 2.52 m/s which is slightly larger than the velocity change calculated from the SRS. The difference is due to the inclusion of 3% damping in the SRS calculation. If the SRS calculation is repeated with zero damping, Eq. 5.33 yields a velocity change prediction of 2.52 m/s as it should to match the numerical integration. The velocity change from the oscillatory shock is a result of the non-symmetric nature of the example acceleration waveform.

Figure 6.5 shows a plot of the maximum primary and residual SRS from the example waveform. This plot is fundamentally different from the classical shocks. The low-frequency portion of the classical shock SRS was driven by the residual response where the high-frequency portion was dictated by the primary SRS. In contrast, Fig. 6.5 shows that for an oscillatory shock, the primary response is equal to the MMAA response everywhere of interest. The residual portion of the SRS does become greater than the primary at about 20 Hz in this example, a point in the SRS of little interest when the peak occurs at 2 kHz. The residual portion of the spectrum is quite interesting in the region from about 400 Hz up to nearly 2 kHz. The apparent flat spectrum is actually a result of the length of the oscillatory signal. Given that the waveform is known and decaying, if the waveform is truncated early it will not have fully decayed to zero. Likewise, a long window will essentially have zero padding at the end. Figure 6.6 shows the same primary and residual SRS calculated with various lengths of the waveform shown in Fig. 6.3. This plot shows that the shorter waveforms, those truncated before the wave completely decays, have a higher residual response across the entire frequency spectrum. Likewise, the spectra calculated with longer time windows have lower residual responses. This is a result of the excitation being allowed to decay to a lower value before

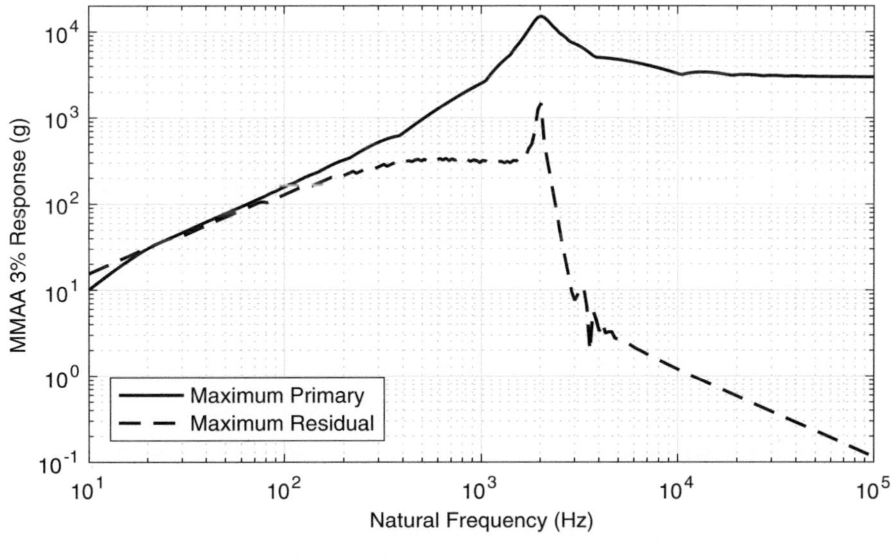

Fig. 6.5 Maximum primary and residual SRS of the 3000 g 2 kHz oscillatory shock

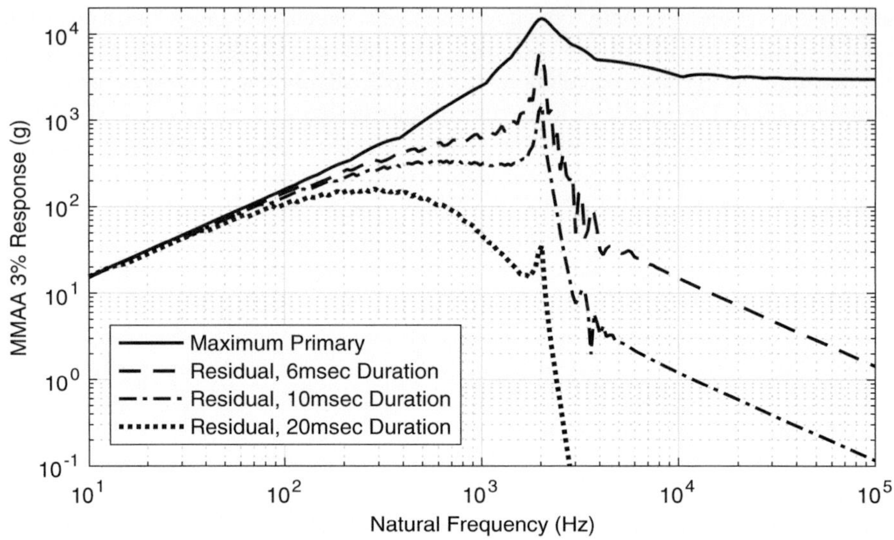

Fig. 6.6 Maximum primary and residual SRS of the 3000 g 2 kHz oscillatory shock calculated with different waveform time windows

the free vibration portion of the calculation begins. As a result, care should be taken when attempting to interpret the residual portion of the SRS for a oscillatory shock event. It is important to note the extremely rapid decay of the residual SRS above the primary shock frequency but the exact amplitude of the response may not necessarily be correct.

In contrast to the primary and residual plots, the maximum positive and negative SRS are very similar. This is also a fundamental difference between oscillatory shocks and the classical shocks. Figure 6.7 shows plots of the maximum positive and negative SRS from the example waveform shown in Fig. 6.3. As can be seen here, the two SRS are nearly identical with the positive SRS being slightly higher than the negative SRS almost universally. The similarity is a result of the two-sided nature of the oscillatory shock event. The positive and negative oscillations are similar in magnitude and duration. The fact that the negative SRS is almost universally lower than the positive SRS is a result of the damped sinusoid calculation where the negative pulse amplitude is always a slightly lower absolute magnitude than the corresponding positive pulse.

6.2 Shock Bandwidth

The Fourier transform magnitude also can be used to understand the energy contained in the shock pulse according to Parseval's identity. Figure 6.8 shows a plot of the Fourier magnitude spectrum for the example shock pulse shown in Fig. 6.3. The Fourier spectrum for the oscillating shock differs from the classical

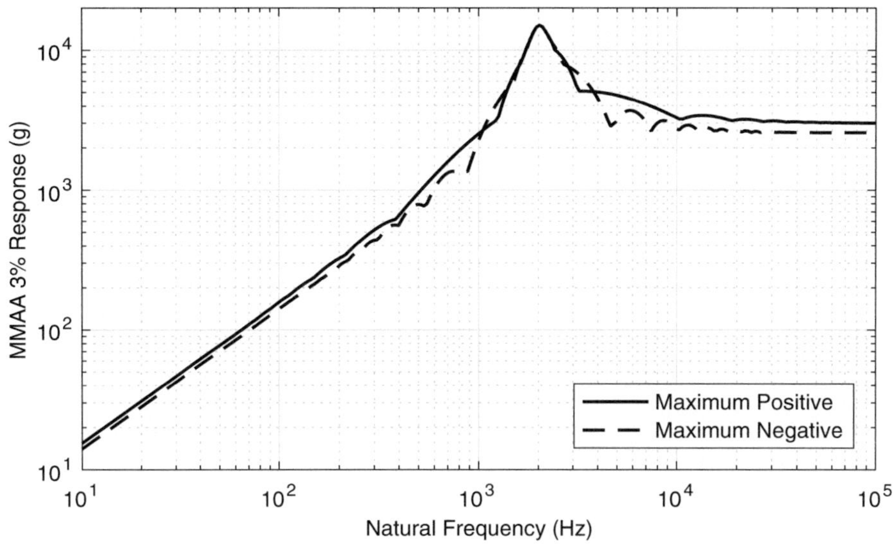

Fig. 6.7 Maximum positive and negative SRS of the 3000 g 2 kHz oscillatory shock

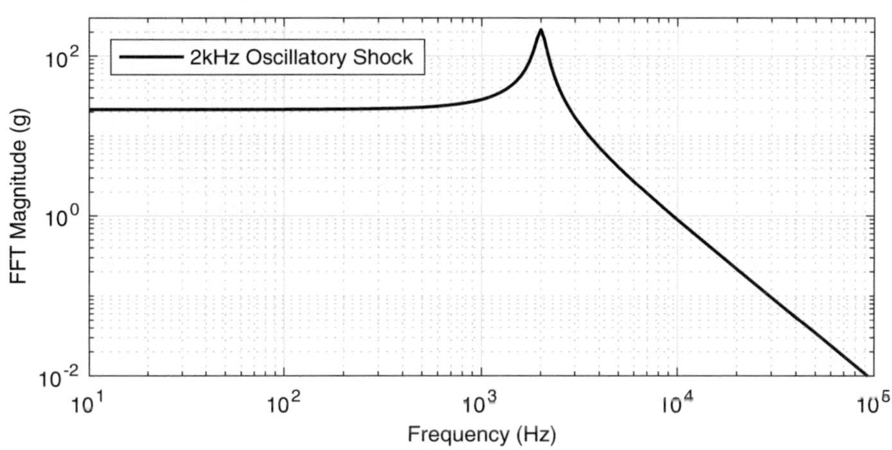

Fig. 6.8 Fourier transform magnitude for the 3000 g 2 kHz oscillatory shock

shocks in that the resonance is quite pronounced in the Fourier spectrum. However, there is significant similarity in that the energy falls of rapidly beyond the shock pulse frequency. This is more easily seen when the SRS is plotted in terms of pseudo-velocity as shown in Fig. 6.9. In the pseudo-velocity plot, it is inherently more obvious where the energy begins to diminish in the shock signal. It should not be expected that resonances significantly above the shock pulse frequency will be excited by the shock event. Here again, it should be intuitive that the shock energy input to the system is finite and diminishes at high frequency.

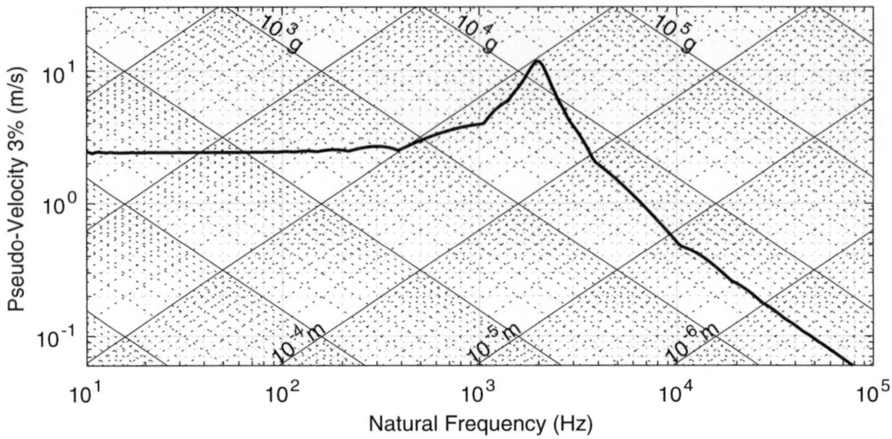

Fig. 6.9 Pseudo-velocity SRS for the 3000 g 2 kHz oscillatory shock

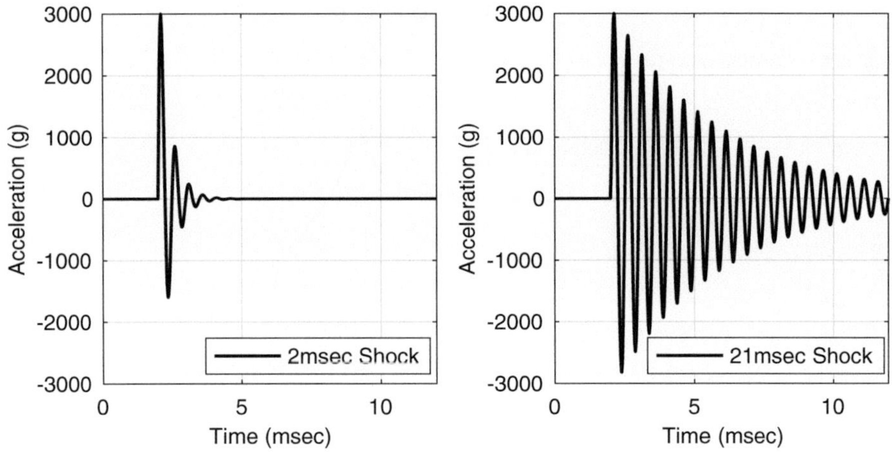

Fig. 6.10 Acceleration time histories of a 3000 g 2 kHz oscillatory shocks with short and long oscillatory times

6.3 Effects of Shock Length

The length of the shock excitation also plays a significant role in the energy transferred to the system being excited. The energy contained in a shock pulse is proportional to the area enclosed by the acceleration versus time curve. As such, a longer shock pulse imparts more energy to the system than a shorter duration pulse with the same amplitude. Figure 6.10 shows two oscillatory shock time histories both similar to the one shown previously in Fig. 6.3. The shock on the left side of Fig. 6.10 depicts a very short duration, rapidly decaying shock pulse where the pulse on the right-hand side is the same frequency but with a much longer ring-

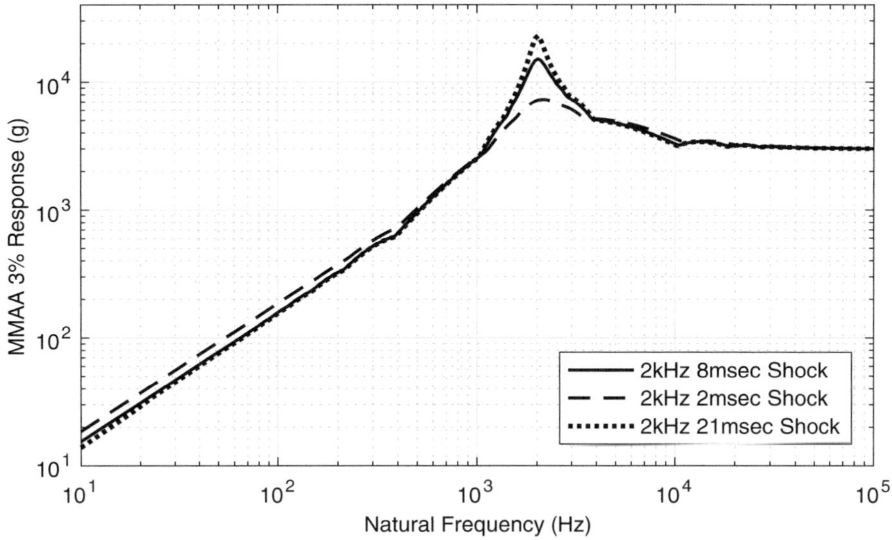

Fig. 6.11 MMAA SRS for three 3000 g 2 kHz oscillatory shocks with different oscillatory times

down time. The original oscillatory shock shown in Fig. 6.3, at the beginning of this chapter, has a duration between these two shocks. Figure 6.11 shows the MMAA SRS for all three of these oscillatory shocks. The SRS from all three shocks has the same high-frequency line at 3000 g, representing the maximum amplitude of the shocks. The low-frequency line is almost identical for the medium and long duration oscillatory shock and slightly higher for the short duration shock. Since the short duration shock decays very rapidly, the area under the positive side of the time history curve is sufficiently greater than the area under the negative side of the curve that the velocity change from this shock is slightly higher than the other two curves. Velocity changes calculated from the MMAA SRS curves are 3.02, 2.41, and 2.22 m/s for the 2, 8, and 21 ms shock pulses, respectively.

Figure 6.11 also shows an interesting effect on the SRS slope near the primary shock frequency. In contrast to the classical shocks where the low-frequency slope was 6 dB/octave and begin to become more shallow as it rolled over near the pulse frequency, the oscillatory shock slopes become significantly steeper as the SRS approaches the shock frequency. The slope steepness increases with the excitation length—equivalent to the number of oscillatory cycles. While it is often stated that the low-frequency SRS slope of an oscillatory shock is 12 dB/octave, this is only true for shocks with a modest number of oscillatory cycles. The slope of the short duration shock in Fig. 6.11 is about 9 dB/octave near the peak, while the 8 ms shock has a slope previously shown to be approximately 18 dB/octave. The long duration shock SRS slope is almost 24 dB/octave near the peak. With each of these slope changes comes a corresponding increase in the peak SRS amplitude. The peak amplitudes are approximately 7300, 15,000, and 23,000 g for the 2, 8, and 21 ms shock pulses, respectively. The peak acceleration for all three shock pulses was actually 3000 g as indicated by the time histories.

Figure 6.11 shows the significance of the number of oscillations in a shock pulse. The number of oscillations is used in much the same way that vibrating a system at its resonant frequency significantly amplifies its response. The SRS is calculated using an SDOF oscillator response and increasing the number of shock cycles near the SDOF oscillator resonance will naturally build up the response to a greater level than just one or two cycles. The real concern here is that the SDOF oscillator may not be a particularly good representation of the physical system. Physical systems have a tendency to display significant non-linearities in their response as the excitation levels increase. Thus, a high-peaked SRS generated by a long duration oscillatory shock input in the laboratory may not actually be as severe as anticipated because of non-linearities in the responding components.

6.4 Understanding the SRS Inflection Points

A careful examination of the SRS plots shown in Figs. 6.4, 6.9, and 6.11 indicates a series of inflection points in the SRS both below and above the primary shock frequency. On the low-frequency side, there appear to be inflection points at approximately one-fifth and one-half of the shock pulse frequency. Symmetric inflection points are seen on the high-frequency side at approximately twice and five-times the shock frequency—the inverse of the low-frequency points. Here again, these characteristics are not present in the classical shock pulses.

To understand the source of these inflection points, it is useful to plot the response of the SDOF oscillator to the shock input in question at a few distinct frequencies. One of the advantages of the SRS is the ability to distill a large quantity of complex data, the acceleration time history, into a simpler, more readable presentation. However, sometimes it is necessary to look again at the underlying data and the intermediate calculation results to gain more understanding. Figure 6.12 shows the calculated SDOF acceleration response to the 8 ms 2 kHz oscillatory shock in the upper plot and the Fourier magnitude spectrum of the SDOF response in the lower plot. The dashed line shows the original shock pulse and the solid line shows the response. At about 2.8 and 3.6 ms the SDOF response clearly shows the effects of multiple harmonic oscillations added together. The Fourier spectrum also clearly shows this with a dominant peak at 800 Hz, the frequency of the SDOF oscillator, and a secondary peak at 2 kHz, the frequency of the excitation. The classical shocks do not show this characteristic because they do not drive the SDOF oscillator at the oscillating shock frequency. The continuous driving of the SDOF oscillator at the shock frequency means that the shock energy is distributed between two natural frequencies. Since the SRS is calculated by looking for the maximum responses, and the waveform is not a single frequency waveform, the resulting SRS reflects this composite response.

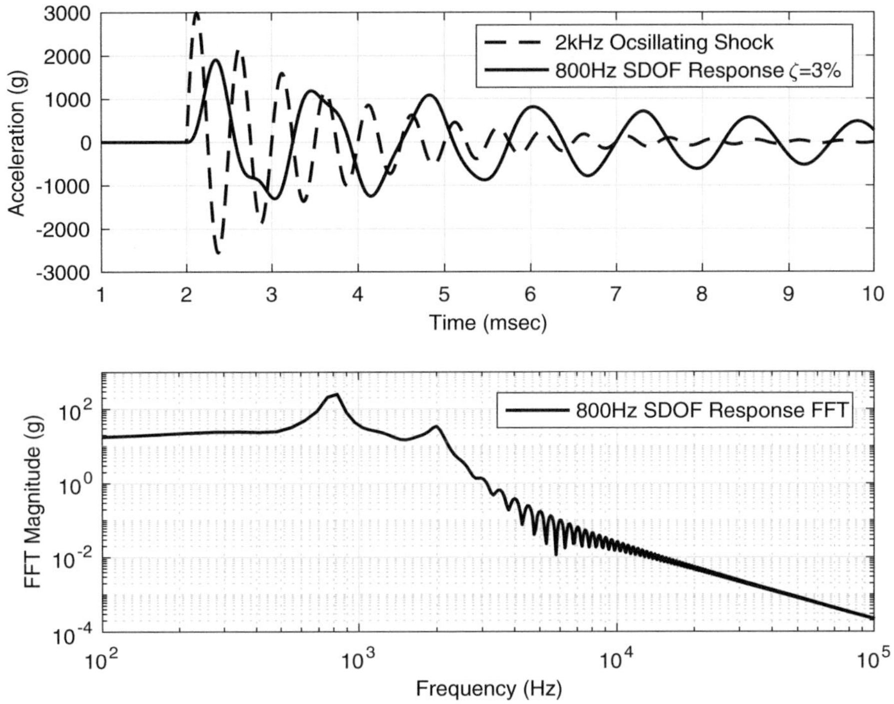

Fig. 6.12 Acceleration time history and Fourier magnitude spectrum of a 800 Hz SDOF system to the 3000 g 2 kHz oscillatory shock

Figure 6.13 shows the oscillatory shock temporal response and Fourier magnitude spectrum with a 2 kHz SDOF oscillator. In this example, the shock frequency and SDOF oscillator frequency are the same so there is no interference between the input and response, allowing the response oscillations to build up over several cycles. The 2 kHz SDOF oscillator actually reaches its maximum response on the fifth cycle. Here again, this is a model form assumption that may or may not be reflective of a physical system. Inherently relying on five oscillations to reach an assumed level of damage potential may be unrealistic given the potential nonlinearities in a physical system under high strain excitation.

Figure 6.13 only shows one dominant peak in the Fourier spectrum corresponding to the SDOF oscillator and the shock excitation frequency. Since the energy is concentrated at one frequency and not split between two frequencies, the FFT magnitude is much higher. This behavior yields much higher SRS peaks from an oscillatory shock as compared to a classical shock.

An interesting comparison can also be made from the responses of the three different duration shock pulses shown previously in Fig. 6.11. The peak amplitudes of 7300, 15,000, and 23,000 g for the three shocks correspond to multipliers of 2.4, 5.0, and 7.7 of the peak temporal acceleration. Looking at Fig. 6.13 it was noted

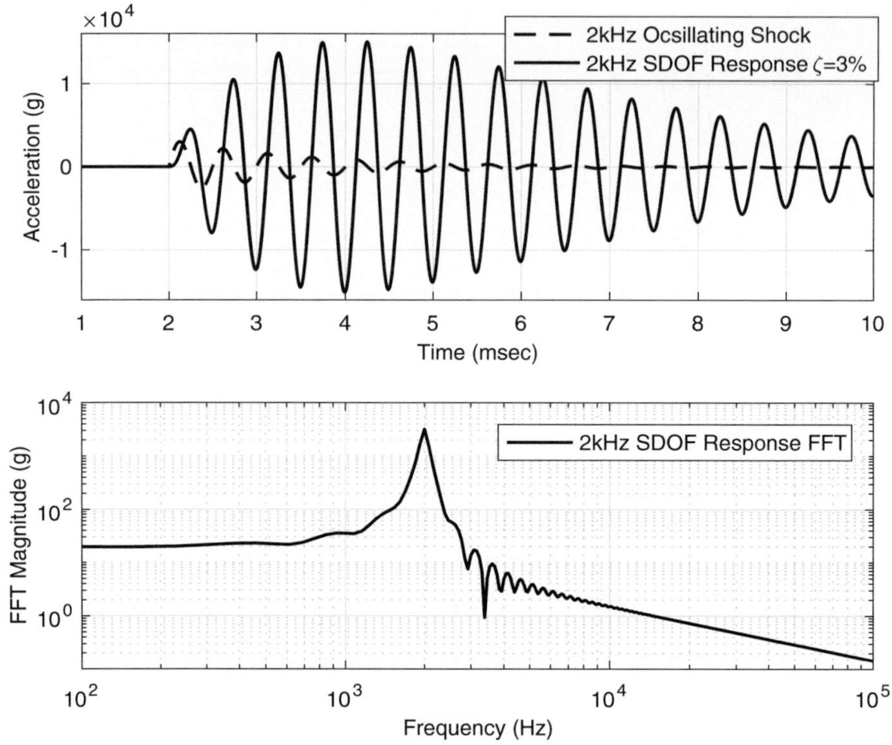

Fig. 6.13 Acceleration time history and Fourier magnitude spectrum of a 2 kHz SDOF system to the 3000 g 2 kHz oscillatory shock

that it took five oscillations to reach the peak amplitude. Likewise, a similar plot could be generated for the short and long duration shock and the number of cycles to reach the maximum absolute amplitude is 2.5 cycles for the short duration shock and seven cycles for the long duration shock. Thus, the number of cycles required to reach the SRS peak is approximately equal to the amplification factor between the maximum temporal acceleration and the SRS peak.

Finally, Fig. 6.14 shows the oscillatory shock temporal response and Fourier magnitude spectrum with a 5 kHz SDOF oscillator. In this example, the SDOF oscillator is effectively stiff compared to the frequency of the input shock and there is little relative motion between the SDOF oscillator mass and ground. This is evident in the Fourier spectrum where the peak response occurs at the shock pulse frequency, 2 kHz, and the response at the 5 kHz SDOF oscillator frequency is about one-fourth of the shock frequency response.

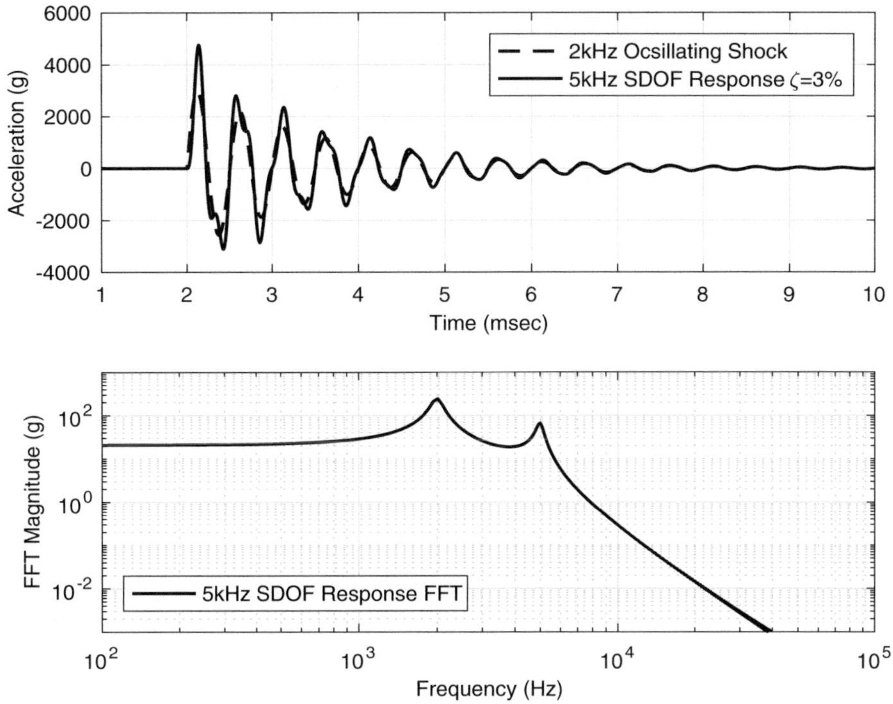

Fig. 6.14 Acceleration time history and Fourier magnitude spectrum of a 5 kHz SDOF system to the 3000 g 2 kHz oscillatory shock

6.5 Two-Sided Shock Pulse

The preceding discussion has focused on an idealized oscillating shock defined by a decaying sinusoidal excitation. There is another shock form that is frequently and somewhat erroneously considered an oscillating shock. The two-sided shock pulse shown in Fig. 6.15 is a single sine cycle containing both a positive and negative shock pulse of equal amplitude and duration. This is technically not an oscillating shock because there is only a single cycle, but it is also not considered a classical shock.

The pulse shown in Fig. 6.15 is a 2 kHz sine pulse with a 3000 g amplitude. Unlike the oscillating shocks discussed previously, this shock pulse integrates to a zero velocity change because the positive and negative shock pulses are equal and opposite. The right-hand plot in Fig. 6.15 shows the MMAA SRS of the two-sided pulse. Here, as in the oscillating shock SRS plots, the high-frequency line levels off at the maximum acceleration amplitude of 3000 g. The peak is not as sharp as seen with the true oscillating shocks, and the low-frequency slope is approximately 12 dB/octave over a large portion of the low-frequency spectrum. The slope does start to become slightly shallower at the lowest frequencies, tending

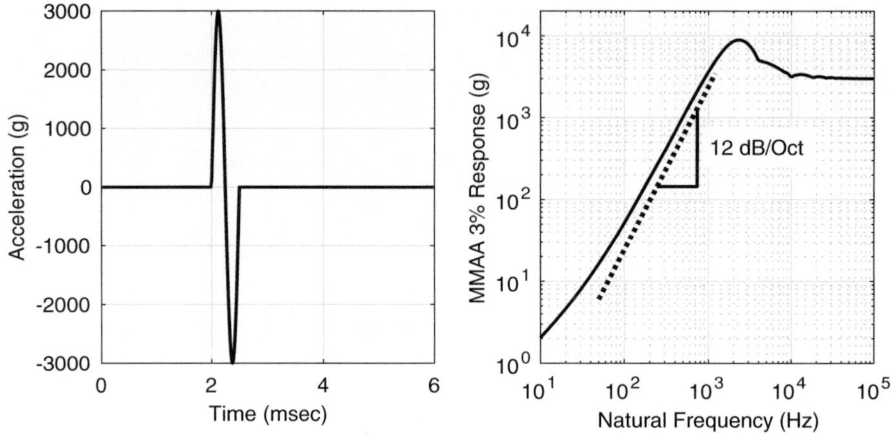

Fig. 6.15 Acceleration time history of a 3000 g 2 kHz two-sided shock pulse with the corresponding MMAA SRS

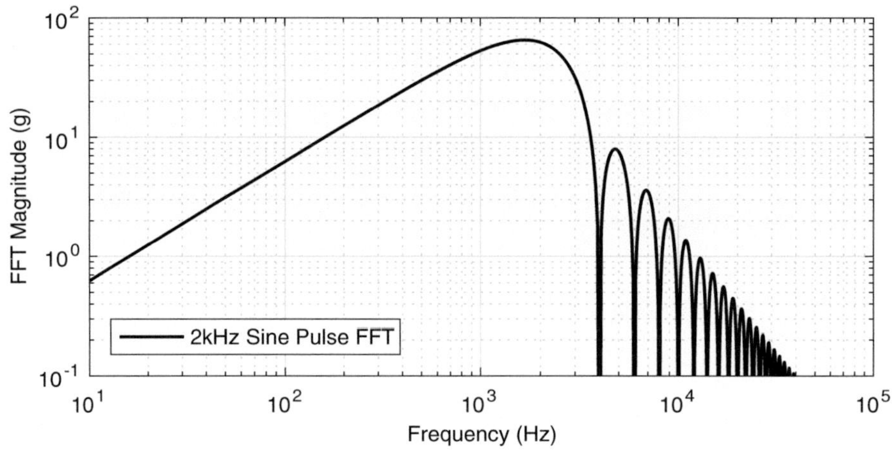

Fig. 6.16 Fourier magnitude spectrum of the 3000 g 2 kHz two-sided shock pulse

toward approximately 9 dB/octave but not yet reaching the 6 dB/octave seen in the oscillating shocks or classical shocks.

Figure 6.16 shows the Fourier magnitude spectrum of the two-sided shock pulse shown in Fig. 6.15. The Fourier spectrum shows that the energy contained in the shock falls off rapidly above the shock excitation frequency, 2 kHz in this example. Just as with the classical shock pulses, the shock energy is negligible at twice the shock frequency as shown in Fig. 6.16. Several smaller bands of energy occur in integer increments of the shock excitation frequency, although the magnitudes are diminishing rapidly and these should not be counted on to sufficiently excite higher frequency resonances in the component under test.

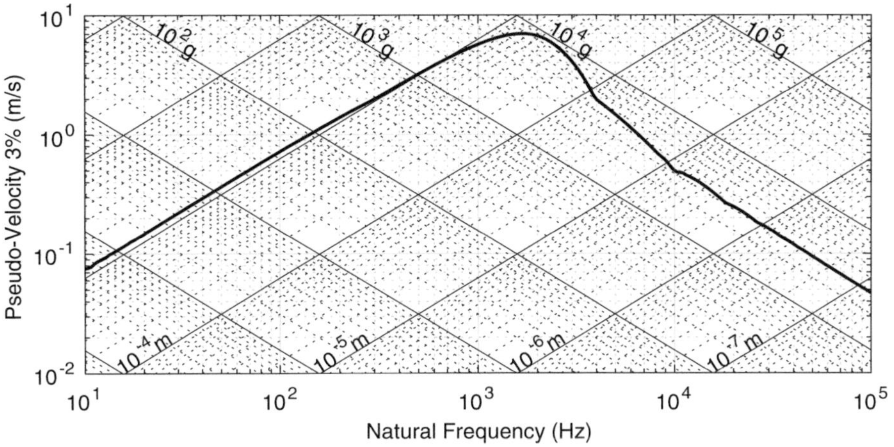

Fig. 6.17 Pseudo-velocity SRS of the 3000 g 2 kHz two-sided shock pulse

In contrast to the classical shocks, there is also very little energy contained in the shock pulse at low frequencies. The Fourier spectrum is steadily decreasing with decreasing frequency down to zero at 0 Hz. The half-power bandwidth for this shock runs from 500 Hz up to about 3 kHz. Frequencies outside of this range will be poorly excited and frequencies within the range but away from the peak may be inadequately excited. The limited energy at low frequencies is a direct result of the zero velocity change. This is shown in the pseudo-velocity SRS given in Fig. 6.17. The pseudo-velocity SRS also shows a steady downward trend at low frequencies in sharp contrast to the flat low-frequency profile of the oscillatory shock shown in Fig. 6.9 or the classical shock pseudo-velocity spectra shown in Chap. 5.

An additional difference between the two-sided pulse and the true oscillating shock is apparent in the MMAA SRS. Comparing the MMAA SRS in Fig. 6.15 right-hand plot with the typical oscillatory shock MMAA SRS given in Fig. 6.4 shows that there are no discrete inflection points apparent in the low-frequency slope portion of the two-sided shock pulse MMAA SRS. To see the difference, we will again plot the response of the SDOF oscillator to the shock input. Figure 6.18 shows the calculated SDOF acceleration response to the two-sided shock pulse in the upper plot and the Fourier magnitude spectrum of the SDOF response in the lower plot. The dashed line shows the original shock pulse and the solid line shows the response of an 800 Hz SDOF oscillator. In contrast to the time history responses seen with the oscillating shock in Fig. 6.12, the SDOF response here is a single frequency response. The Fourier spectrum also shows that this is a single frequency response at the SDOF oscillator frequency. The Fourier magnitude spectrum of the oscillatory shock indicated two response frequencies—the SDOF oscillator response and the driving frequency of the shock.

Figure 6.19 shows the calculated SDOF acceleration response to the two-sided shock pulse and the Fourier magnitude spectrum of a 5 kHz SDOF oscillator response. In this example, the SDOF oscillator response is amplified over the input

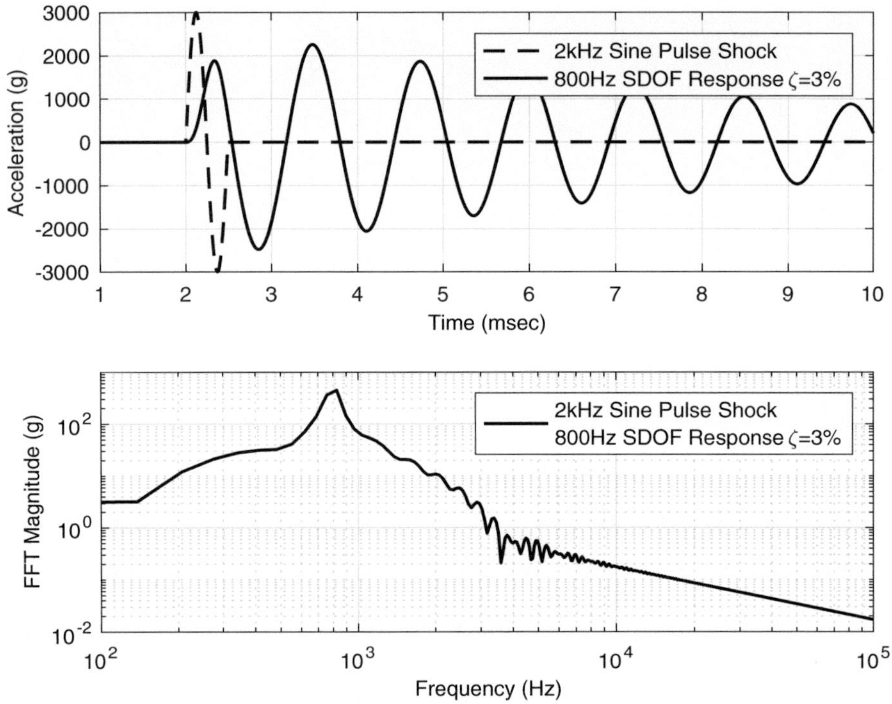

Fig. 6.18 800 Hz SDOF oscillator response to the 3000 g 2 kHz two-sided shock pulse and the Fourier magnitude spectrum of the SDOF response

shock acceleration. In addition, the Fourier spectrum shows both the shallow peak at the 2 kHz shock frequency and a sharper, taller peak at the 5 kHz frequency of the SDOF oscillator. The inflection points seen on the high-frequency side of Fig. 6.15 are a result of the transition between the maximum acceleration occurring at the driving frequency or the SDOF oscillator response frequency and the natural zeros in the Fourier amplitude spectrum that occur at integer intervals of the shock frequency. The two primary high-frequency inflection points in Fig. 6.15 occur at 4 kHz and 10 kHz. Both of these frequencies correspond to zero energy frequencies in Fig. 6.16.

6.6 Complex Shocks

The classical shock discussion from Chap. 5 and the preceding development of oscillatory shock theory has always assumed that the shocks occur independently. However, this is often not the case in real-world systems. Many times the measured shock event and the resulting SRS is a compilation of a complex series of events often comprised of more than one distinct shock event. Figure 6.20 shows a

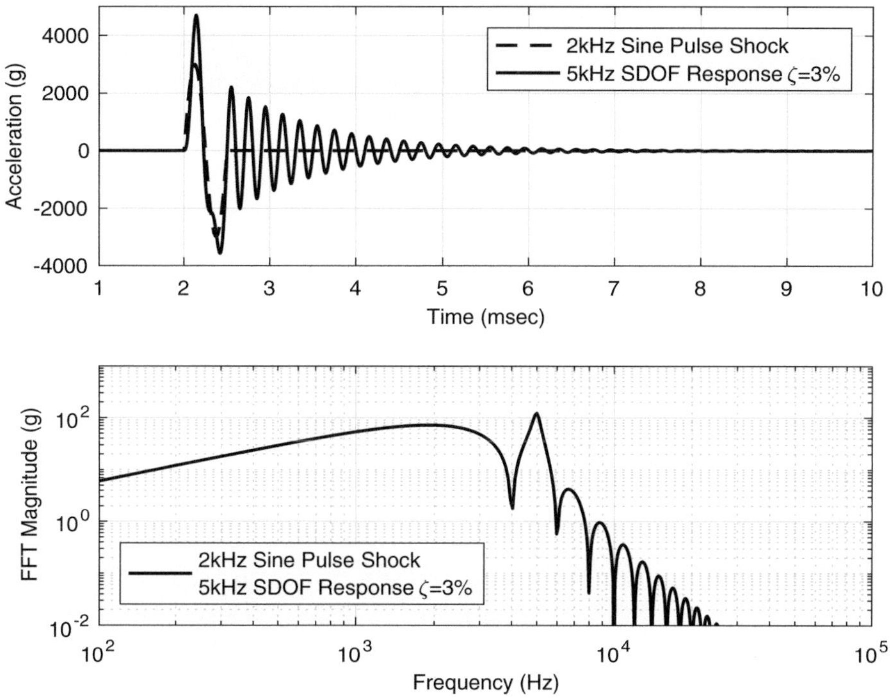

Fig. 6.19 5 kHz SDOF oscillator response to the 3000 g 2 kHz two-sided shock pulse and the Fourier magnitude spectrum of the SDOF response

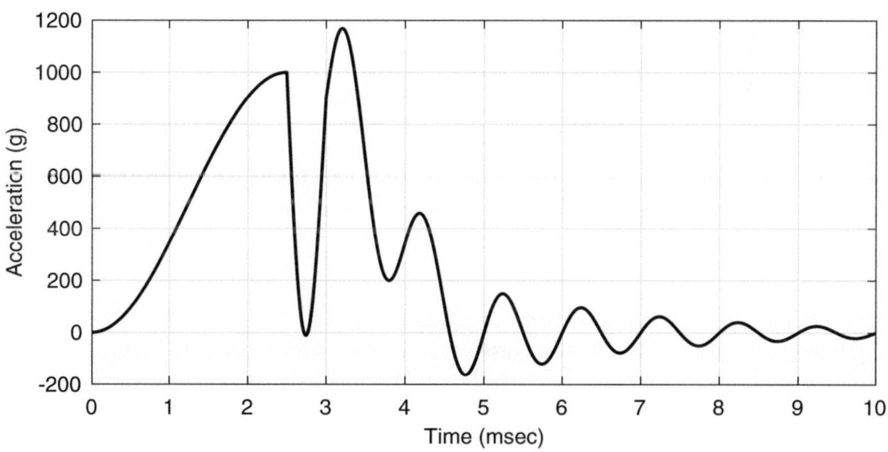

Fig. 6.20 Time history response of a hypothetical complex shock waveform

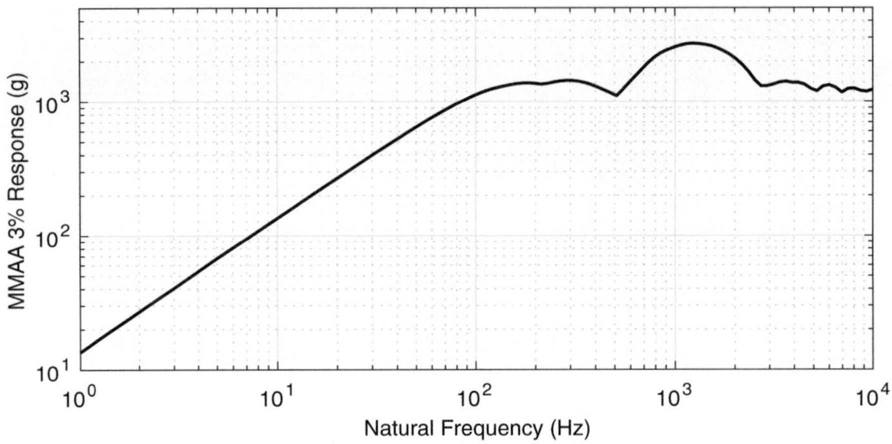

Fig. 6.21 Absolute acceleration SRS of the hypothetical complex shock waveform

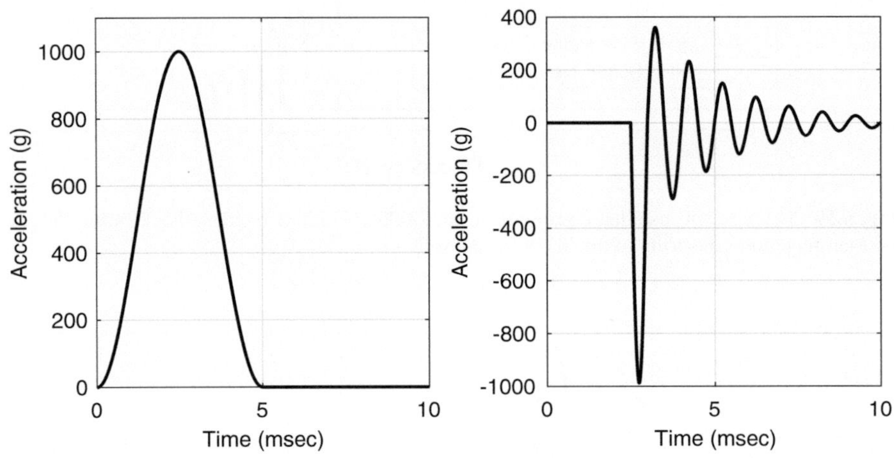

Fig. 6.22 Acceleration time history plots of the two underlying shocks used to derive the complex shock waveform: a classical 1000 g 5 ms haversine on the left and a 1 kHz damped sinusoid on the right

hypothetical time history response of a complex shock waveform. The time history plot appears to show the rise of a classical haversine shock prior to the introduction of a more complex waveform at about 2.5 ms. Figure 6.21 shows the resulting maximax absolute acceleration SRS from the complex waveform of Fig. 6.20.

The SRS in Fig. 6.21 shows the expected 6 dB/octave slope of the classical haversine as well as a portion of the haversine-like hump at about 200 Hz. In contrast to the single event shocks, there is a second, higher hump that peaks at about 1.2 kHz as well as some fluctuations in the SRS above 3 kHz. The time history plot shown in Fig. 6.20 was created as the sum of a classical haversine and a damped sinusoidal response. The two individual responses are shown in Fig. 6.22. The haversine shock, shown on the left-hand side, is simply a classical haversine with a 1000 g amplitude and a 5 ms duration. The damped sinusoidal response, shown on the right-hand side, is offset and inverted such that it begins at the haversine peak and a non-linear damping is used such that the first pulse is significantly higher than the remaining oscillations.

While the shock time history shown here was created from a summation of analytical waveforms, it is representative of certain real-world phenomena. For example, if a welded steel component is tested on a drop table at a relatively high acceleration and a weld were to fail, the accelerometer would record the drop shock as well as the shock created by the weld failure. Furthermore, the weld failure is most likely to occur when the component reaches its maximum deflection—occurring at approximately the same time as the maximum acceleration. In addition, a weld failure releases a stress wave propagating through the material but with little momentum change, similar to a pyrotechnic shock event. As was stated earlier, pyrotechnic shock events are characterized by a propagating stress wave but with relatively little momentum change, approximated well by a damped sinusoidal waveform.

Figure 6.23 shows the same SRS of the complex waveform shown in Fig. 6.21 along with the SRS from the two underlying waveforms from Fig. 6.22. Here the SRS of the classical haversine is clearly seen slightly above the haversine shape of the complex shock. The actual shock's SRS is slightly lower than the classical haversine because the oscillatory shock component subtracts area from under the haversine curve, resulting in less net velocity change than the pure haversine

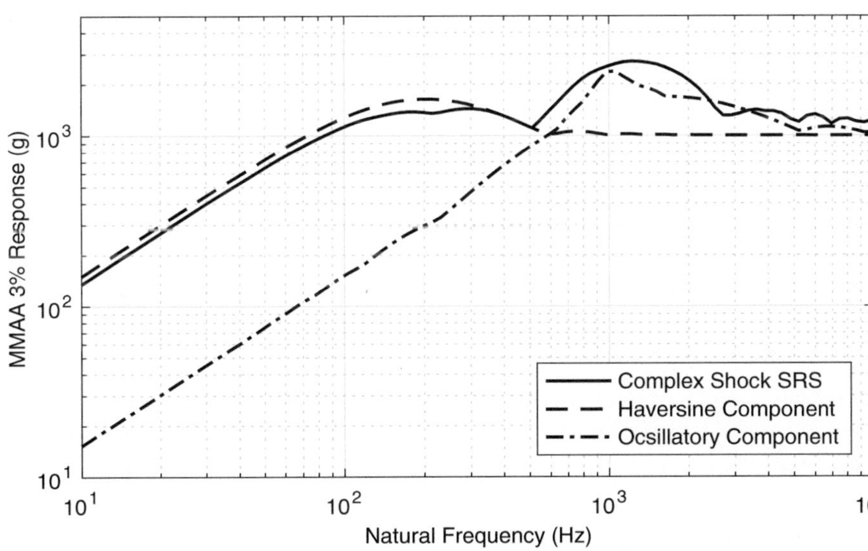

Fig. 6.23 Absolute acceleration SRS of the hypothetical complex shock waveform overlaid with the SRS from the two underlying shocks

alone. The SRS hump corresponding to the oscillatory component is not nearly as recognizable as the haversine portion. This difference is due to the fact that the oscillations are occurring about the falling portion of the haversine response and not about zero as in the theoretical cases.

6.7 Summary

This chapter has presented the primary characteristics of oscillatory shock pulses as well as aspects of a classical two-sided shock pulse. Oscillatory shocks are similar in some respects to classical shocks and distinctly different in other areas. The small velocity change associated with the oscillatory shock is perhaps the most significant difference between these and the classical shocks. Small velocity changes also yield small momentum changes.

Oscillatory shocks typically have a much higher peak in the MMAA SRS as compared to the classical shocks. The reason for this is that the classical shocks are a single pulse whereas the oscillatory shocks excite the system for several oscillations allowing the shock to build up a resonance response in the system, thereby increasing its amplitude.

It was also shown that the energy contained in the shock diminishes rapidly past about one and a half times the primary shock frequency, becoming negligible at double the shock pulse frequency. As a result, system resonances above about 1.5 times the shock excitation frequency will not be well excited by the oscillatory shock pulse.

Problems

6.1 Oscillatory shocks are often associated with pyroshock events although they can occur from other sources as well. Pyroshocks are distinctive for having a zero velocity change where a normal oscillatory shock usually has some small velocity change associated with it. How can you tell whether or not the oscillatory shock has a velocity change associated with it from looking at the MMAA SRS? It is obvious from the time history as well?

6.2 Why is the SRS from an oscillatory shock essentially unaffected by the residual spectrum? What does it mean if the residual spectrum does have a substantial effect on the SRS?

6.3 What are some of the differences apparent in the SRS between an oscillatory shock with a small number of oscillations and one with a large number of oscillations?

6.4 Calculate the SRS from some of your own experimental oscillatory shock data and look for the inflection points to the left and right of the SRS peak. Are the inflection points obvious? Are they in a similar location to the ones shown in Sect. 6.4?

6.5 The two-sided shock pulse shown in Sect. 6.5 is a special case of the oscillatory shock because it only has a single forward and reverse pulse. Does this shock have a velocity change associated with it? Can you tell this from the SRS plot? Can you tell from the Fourier transform magnitude?

6.6 The complex shock waveform shown in Sect. 6.6 is a linear combination of a classical shock pulse and an oscillatory shock pulse. What do each part of the complex shock represent physically? What does the combination physically represent? Try to find a set of measured test data with two distinct pulses in the SRS and separate the events into the two underlying events.

6.7 Many of the example oscillatory shocks in this chapter used a single decaying sinusoid tone. Calculate the SRS for some other example oscillatory shocks, and compare the results. For example, a summation of decayed sinusoidal tones, a summation of two tones with differing frequencies, or a more complex waveform. Are the SRS characteristics similar?

6.8 In Sect. 6.6 a complex shock was created by the superposition of a damped sinusoid onto a haversine. The first peak of the damped sinusoid is negative as shown in Fig. 6.22. Compute the MMAA SRS for this complex shock but with the first peak of the damped sinusoid positive. Compare your result to Fig. 6.23.

6.9 Consider the complex shock comprised of two haversines separated in time as shown in Fig. 6.24. The first haversine is a 1000 g 1 ms haversine. The second, which starts 4 ms later, is a −500 g 2 ms haversine.

1. What is the velocity change associated with the complex shock?
2. Plot the MMAA SRS of each haversine and of the complex shock. How does the trailing haversine affect the SRS? Can the velocity change be discerned from the SRS of the complex shock?
3. Vary the time between the pulses and compute the SRS. How does the time gap between the pulses affect the SRS?

6.10 A complex shock consists of a haversine with two small amplitude, high-frequency sinusoids as shown in Fig. 6.25. The haversine is a 1000 g 5 ms pulse. The parameters of the two sinusoids are:

- $A_1 = 200\,\mathrm{g}$
- $f_1 = 1000\,\mathrm{Hz}$
- $\zeta_1 = 0.02$
- $A_1 = 100\,\mathrm{g}$
- $f_1 = 2000\,\mathrm{Hz}$
- $\zeta_1 = 0.01$

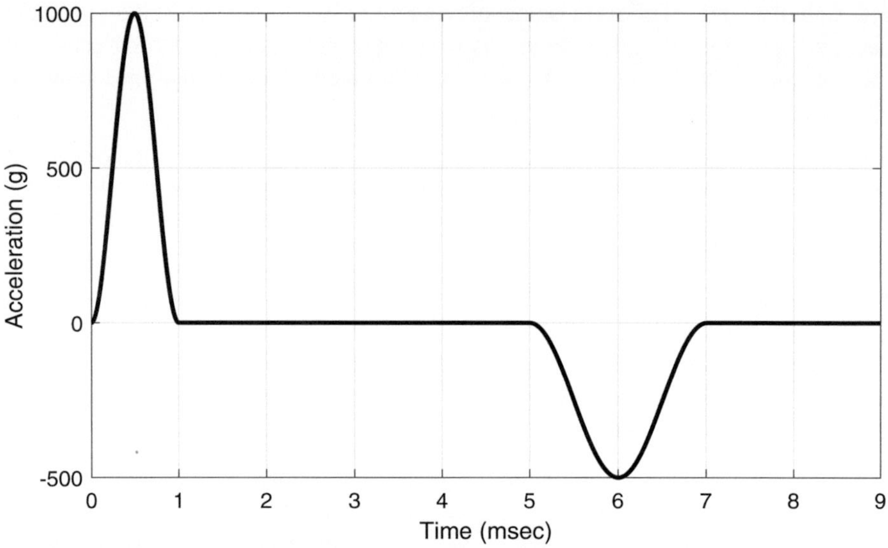

Fig. 6.24 Figure problem 9—complex shock waveform comprised of two haversines separated in time

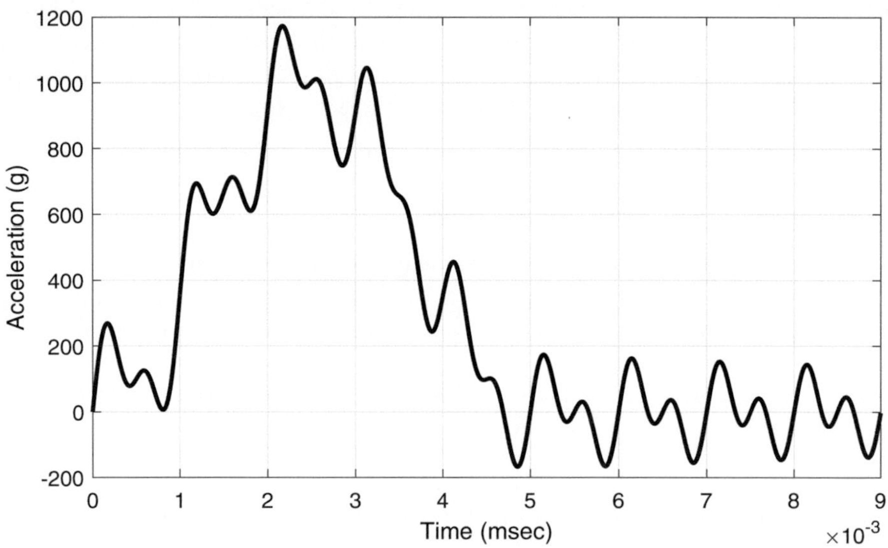

Fig. 6.25 Figure problem 10—complex shock waveform comprised of a haversine and two high-frequency damped sinusiods

Plot the SRS of the haversine, the two sinusoids, and the complex shock. How do the sinusoids affect the SRS from the haversine?

Chapter 7
Design for Shock with SDOF Spectra

When designing a structure, the designer must understand the environment in which the structure will be used. Much has been written about the properties, utility, and limitations of shock spectra; but, how are shock spectra used in design? Shock spectra are often used to compare the severity of two shock events. It is very common to see the shock spectra obtained from a test overlaid with shock spectra from a requirement or shock spectra from two tests overlaid to see how the two tests compare. Making an overlay plot is a relatively straightforward task; however, interpreting that plot can often involve considerable engineering judgment and differences of professional opinion. Likewise, if a designer is given an SRS, how can that be used for design or in a finite element simulation to evaluate a design?

The designer must also understand the functional requirements for the system in the shock environment. Does the system have to function during the shock event or only after the shock event? Electronic components on aerospace systems are frequently required to operate through a shock event. In other fields, it may be acceptable if the system returns to operation within a defined time window after the shock. Is it acceptable for the system to suffer some amount of damage? In naval applications, plastic deformation or buckling of struts supporting equipment may be acceptable in severe shock events. Likewise, structural damage may be permitted in buildings during a severe earthquake as long as the structure does not completely collapse. Shock spectra are useful tools for a designer when the failure mode of interest is over-stress. They are not very useful for designing structures for fatigue.

This chapter presents some basic information for designing systems and components to survive shock testing and shock exposure. The intention here is to provide insight to design so that engineers will be able to design their systems to pass shock. In this chapter, we explain:

1. how to tell if two SRS curves are nominally equivalent (how close is close enough when comparing SRS data);
2. tips for translating SRS data into values for finite element analysis;

C. Sisemore, V. Babuška, *The Science and Engineering of Mechanical Shock*,
https://doi.org/10.1007/978-3-030-12103-7_7

3. a derivation of the stress–velocity relationship;
4. sizing structural members to absorb strain energy;
5. general design guidance for shock.

7.1 Use of the SRS for Design

The SRS provides information about the spectral content of the transient excitation. Other information, such as velocity change, can also be obtained from an SRS. If we are analyzing an existing system, the system parameters—mass, stiffness, and damping—are fixed. On the other hand, when designing a system, we have the ability to select or tailor the system parameters so that the system response to a particular shock falls within an acceptable range. If a structure can be adequately modeled as an SDOF oscillator (not an unreasonable assumption during a project's conceptual or preliminary design phases), one would naturally tailor the design such that the natural frequency is in the region where the SRS is low, and avoid frequency ranges where the shock excitation has the greatest damage potential—frequencies where the SRS is high.

The base excitation example shown in Fig. 4.1 has the largest damage potential between 200 and 600 Hz. Certain low frequencies are also potentially damaging, specifically around 30 Hz. As such, we would tailor the system design with resonances outside of these high SRS regions. How is this accomplished? The natural frequency of the SDOF oscillator is given by $\omega = \sqrt{k/m}$. Thus, if the design needs to have a higher natural frequency, the stiffness, k, can be increased, the mass, m, decreased, or a combination of both can be pursued. Likewise, if the design needs to be moved to a lower natural frequency, the mass can be increased, the stiffness decreased, or both. While changes in the mass are effective, they are not always practical. Increasing the mass of a structure may be an issue for aerospace systems but not for civil structures.

Stiffness can be increased with the use of gussets, stiffeners, braces, or changing cross-sections. While adding structural stiffeners increases the mass, the penalty is usually small compared to the effect on stiffness. Here again, this is an inherent flexibility available to designers that is not available when simply analyzing an existing system. Basic handbooks for calculating the stiffness and natural frequency for generic simple shapes can prove extremely valuable at this stage in the design. An excellent reference for calculating frequencies for common structural shapes was written by Blevins [1].

Sometimes, it may not be possible to tailor the structure's properties so that its response to the shock is relatively benign. In these situations, the designer has two options:

1. increase the damping in the structure so that the shock is attenuated before it reaches shock-sensitive components;
2. isolate the structure or the shock-sensitive parts in it.

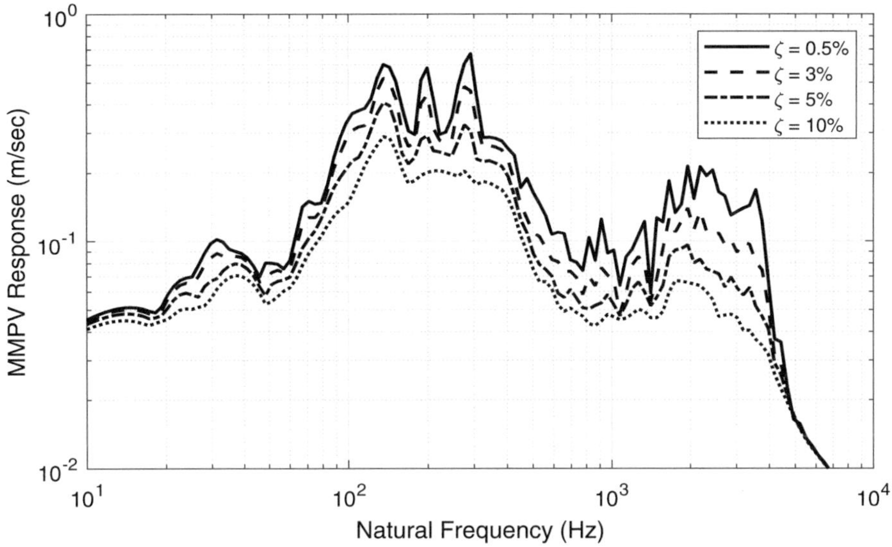

Fig. 7.1 Pseudo-velocity shock response sensitivity to damping ratio, ζ

Damping does not change the overall shape of the SRS; however, it smooths the features and generally lowers the overall level. That can be enough to enable the system to operate successfully in the shock environment if the system damping can be increased. The effect of increasing the damping in a structure is shown in Fig. 7.1, assuming again that the structure can be adequately modeled by the SDOF system. The amplitude of the response spectrum decreases as the damping increases, and the spectrum is smoothed. In order to appreciably reduce the peak response, the damping must be increased substantially. In Fig. 7.1, the peak response of a system with 10% damping is about 30% of the peak response of a system with 3% damping. However, it should be cautioned that very high damping values can increase the response since they allow the damper to become an alternate load path for the shock.

The damping in a structure can be increased with the use of damping treatments, but the effects may be too small. Mechanical isolators, which are usually placed at the interface of the structure to the mounting surface, can be a more effective solution. They protect the structure from the shock by absorbing and dissipating its energy before it is transmitted into the structure. The shock absorber on an automobile is a classic example of a shock isolator. Isolators are mechanical low pass filters. They pass the low-frequency portion of the shock, with amplification around the isolator's resonant frequency, and they attenuate the shock at high frequencies. Isolators may increase the compliance of the structure, which could lead to larger relative displacements, so the designer must ensure that there is adequate sway space and strain relief in cables that cross the isolated interface. An example of a passive shock isolator is the wire rope isolator. Wire rope isolators tend to be much stiffer in tension than compression and have a modest stroke so they are best suited for use with single-sided, high-frequency shock environments.

7.2 Equivalent SRS

When using an SRS for design, it is important to understand what parts of the SRS are of interest and when two SRS represent essentially the same environment. Figure 7.2 shows two MMAA SRS curves from the same drop shock test overlaid for comparison. Are these two SRS equivalent? Do the environments represented by the two SRS have the same damage potential? In the MMAA SRS presentation, the eye is immediately drawn to the differences in the peaks at about 750 Hz. One might be tempted to say that Location 2 saw a significantly more intense shock than Location 1. After all, at 750 Hz, the Location 1 peak is about 3600 g, while the Location 2 peak is at 6400 g. However, the data were collected from two points on the same part that were only a few centimeters apart. Is this reasonable or is it a property of the SRS? Looking at Fig. 7.2, it is obvious that the velocity change from the impact is the same because the velocity change is read from the low-frequency slope. It also appears that the acceleration time history peaks were probably similar because the high-frequency plateau is trending to approximately the same acceleration magnitude. Then, what causes the difference in the SRS peaks?

Figure 7.3 shows two plots with the same peak acceleration. The left-hand plot is a classical 1000 g half-sine shock pulse. The right-hand plot is the same classical

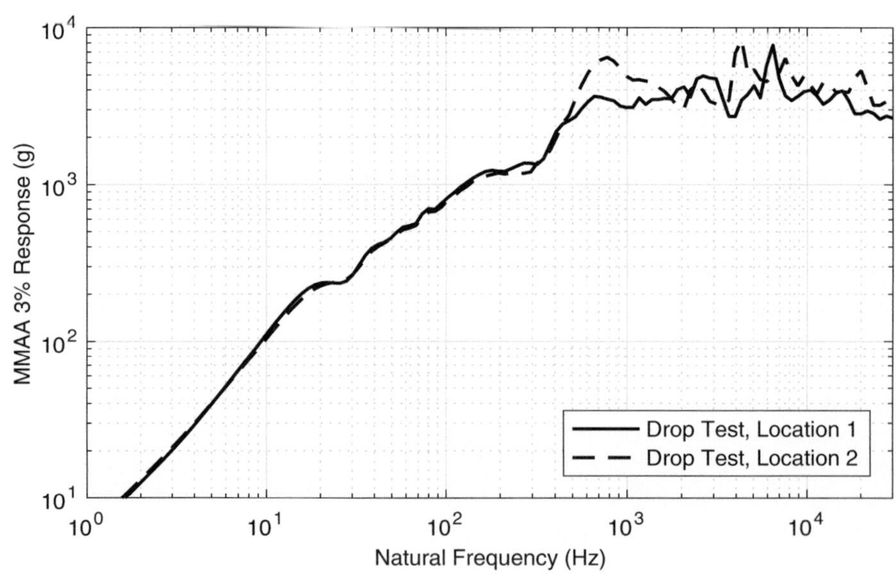

Fig. 7.2 Comparison of MMAA SRS from two nearby locations in the same drop test

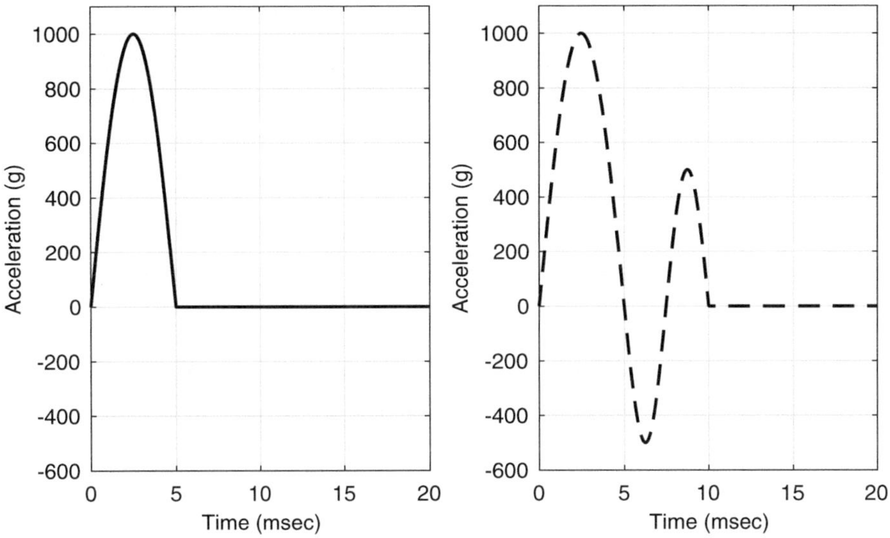

Fig. 7.3 Acceleration time history of 1000 g 5 ms half-sine shock pulse shown on left; same pulse with an additional 500 g sinusoidal oscillation on the right

half-sine shock pulse with an additional sinusoidal cycle at half the amplitude and double the frequency of the first shock pulse. Figure 7.4 shows the resulting MMAA SRS of the two shock pulses overlaid for comparison. Both SRS plots have the same velocity change, as shown by the low-frequency slope, and the same peak acceleration. However, the shock with the additional oscillation has a higher SRS peak than the classical half-sine. In this example, the classical half-sine has a peak SRS amplitude of 1700 g where the more complex waveform has a peak SRS amplitude of 2485 g, a considerable increase.

Returning to the original question: how do we know if two SRS are nominally equivalent? Looking at Fig. 7.4, both shocks have the same velocity change and the same peak acceleration. By these measures, one would say that the two SRS are equivalent even though one has a significantly higher peak than the other. This is similar to the number of cycles in the oscillatory shocks. As was shown in Sect. 6.3, the more cycles in an oscillatory shock, the higher the SRS peak. This is a result of the amplification seen in the response of the SDOF oscillator, not necessarily a result of a physical phenomenon. Returning to the time history plot in Fig. 7.3, if the component survived the strain induced by the 1000 g single shock pulse shown on the left-hand side, it seems unlikely that adding an additional 500 g cycle will generate higher strains in the part than the first pulse. If the strain is less on the second oscillation, then it seems unlikely that the second oscillation could induce a failure if the first pulse did not. The only exception would be if a fundamental resonance of the part occurs at that frequency where the second oscillation may be amplified over the first oscillation. This assumption is the reason the SRS is higher, because the input is driving the SDOF oscillator at its natural frequency.

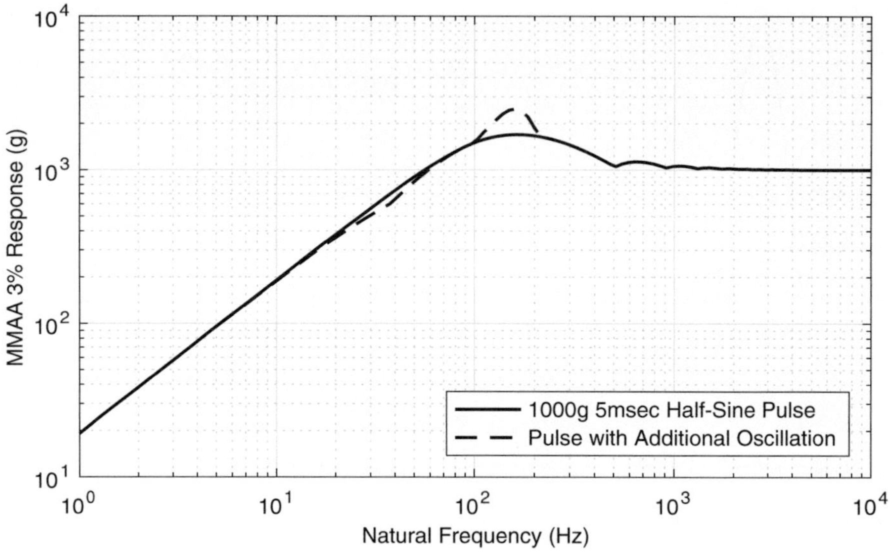

Fig. 7.4 Comparison of MMAA SRS from a half-sine shock pulse and a half-sine shock with an additional oscillation

In general, to determine if two SRS are equivalent, the focus should not be on the small differences but rather on the gross motions that define the SRS. The most significant is to determine which SRS has the highest velocity change. It can be shown that the velocity change is proportional to stress. As a result, if one SRS shows a substantially higher velocity change than another SRS, the shock that creates the higher velocity change is more severe. If both SRS show the same velocity change, then there is a reasonably good chance that the two SRS will produce the same damage potential. The exception to this rule is if one SRS represents a shock with greater bandwidth, that shock may excite a fundamental resonance that the shock with smaller bandwidth cannot.

As will be shown in the next section, peak acceleration by itself is generally a poor indicator of relative severity. If two shocks have the same velocity change but different peak accelerations, then the one with the higher acceleration will have a larger shock bandwidth. This has the potential to make that shock more severe if additional resonances are encompassed but it is not guaranteed to be more severe. Peak acceleration should never be used as the sole indicator of shock severity.

7.3 Relationship of Pseudo-Velocity to Stress

There has been considerable debate within the shock community about the best format for displaying SRS data. Many people prefer the acceleration SRS and many others prefer the pseudo-velocity SRS. Both response spectra contain nominally the same information but the presentations tend to emphasize different aspects of the same event. The actual method of presentation will most likely be determined by the common practice of the local engineers, the company, or the desire of the test laboratory.

It has been the authors' experience that absolute acceleration SRS plots tend to focus attention on the maximum acceleration in the shock. Many people who are minimally skilled in the shock field will naturally focus on the acceleration level obtained and ignore or downplay the frequency content of the pulse. Several examples have been presented here utilizing 1000 g shock pulses. While that sounds severe, it may not be a concern depending on the nature of the event. If someone drops their cell phone on a concrete floor, they can easily generate a 1000 g shock loading for the phone and likely a significant repair or complete replacement. On the other hand, a mechanic might drop a wrench on the same concrete floor multiple times, generating 1000 g shock pulses without any damage at all. The reason is that the impact frequencies may significantly differ and the resulting stress developed in the brittle phone parts may be damaging whereas the stress developed in a ductile steel wrench is insignificant.

The other SRS in common use is the pseudo-velocity SRS. Pseudo-velocity has the units of velocity but is not the actual relative velocity as was demonstrated previously. While pseudo-velocity itself seems a bit more esoteric than maximum acceleration, it has some advantages with evaluating a shock environment for structural design. When plotted on tripartite paper, all of the important effects of the environment on a system—relative displacement, velocity, and acceleration as functions of natural frequency—are accessible to the designer in one place. For these reasons, many people find this to be the preferred SRS for comparison of shock severity.

Another important property of pseudo-velocity is its relationship to the stress developed in a part as a result of a shock event. This is an often underappreciated aspect of the pseudo-velocity SRS. Almost all shock theory has been described using the ideal spring-mass-damper oscillator; however, the relationship between pseudo-velocity and stress is subtle. The relationship derives from the wave equation and modal analysis, which is discussed in Chap. 8. The relationship between pseudo-velocity and stress was initially described by Hunt [2] and elaborated on by Kinsler and Frey [3], and Gaberson [4]. The underlying relationship between pseudo-velocity and stress derives from the fact that the mode shapes are functions of the natural frequencies. This characteristic is illustrated on an axially loaded rod and on a beam in transverse bending in the next sections. However, it applies to any structure because all structures' responses can be expressed in terms of periodic mode shapes.

7.3.1 *Stress–Pseudo-Velocity Derivation for an Axially Loaded Rod*

To illustrate the relationship between pseudo-velocity and stress, we first examine an axially loaded rod. This one-dimensional assumption greatly simplifies the mathematical development, but is readily extendable to two and three dimensions. First, we assume a long, slender, prismatic rod with constant cross-section A and length L oriented such that the x-axis is along the rod's length as shown in Fig. 7.5. Next, consider a short length of the rod, dx, at some distance x from one end as shown in Fig. 7.5. We start by assuming a small axial displacement, u, resulting from the traveling stress wave as shown in the rod segment sketch in Fig. 7.6. The resulting unit elongation at any cross-section is given by:

$$\varepsilon = \frac{\left(u + \frac{\partial u}{\partial x}dx\right) - u}{dx} = \frac{\partial u}{\partial x}, \tag{7.1}$$

which defines the strain in a rod segment. The free body diagram showing the forces acting on the rod segment is shown in Fig. 7.7. In this figure, the rod cross-sectional area is given by A and the axial force is given by $F = \sigma A$. The mass of the rod segment is calculated with the density and volume as $\rho A dx$. Applying Newton's law and using the free body diagram give

$$-\sigma A + \left(\sigma + \frac{\partial \sigma}{\partial x}dx\right)A = \rho A dx \frac{\partial^2 u}{\partial t^2}. \tag{7.2}$$

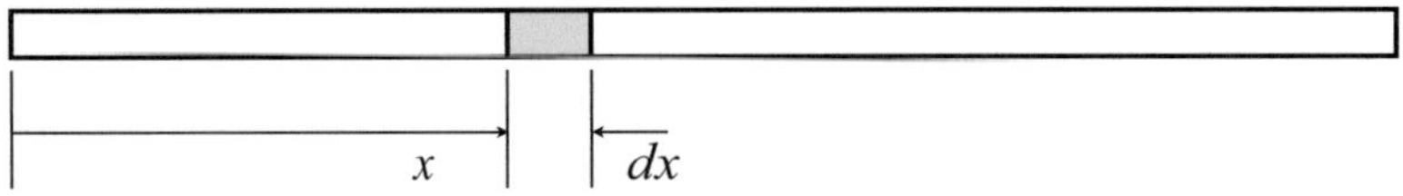

Fig. 7.5 Sketch of a long one-dimensional prismatic rod subject to vibratory motion

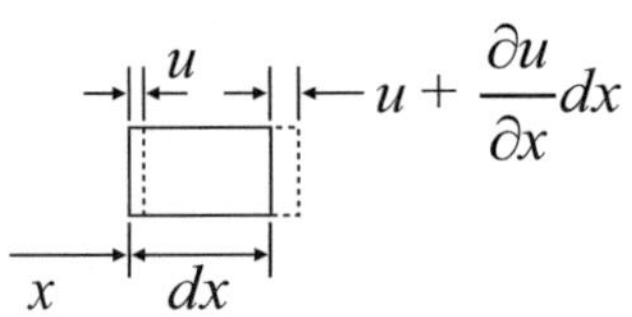

Fig. 7.6 Sketch of unstrained and strained rod segment

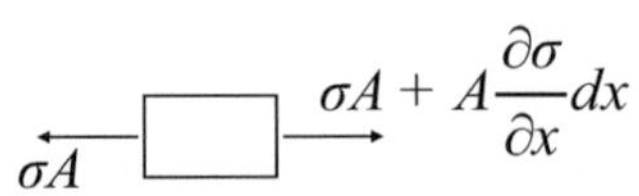

Fig. 7.7 Free body diagram of rod segment

Dividing through by the cross-sectional area and simplifying give

$$\frac{\partial \sigma}{\partial x} = \rho \frac{\partial^2 u}{\partial t^2}. \tag{7.3}$$

Since stress and strain are related through $\sigma = E\varepsilon$ and the strain is given in Eq. 7.1, substituting for σ in the left-hand side of Eq. 7.3 gives

$$\frac{\partial \sigma}{\partial x} = \frac{\partial}{\partial x}\left(E\frac{\partial u}{\partial x}\right) = E\frac{\partial^2 u}{\partial x^2}. \tag{7.4}$$

Rearranging Eqs. 7.3 and 7.4 gives

$$\frac{\partial^2 u}{\partial t^2} = \frac{E}{\rho}\frac{\partial^2 u}{\partial x^2} \tag{7.5}$$

which is the one-dimensional wave equation, usually written as:

$$\frac{\partial^2 u}{\partial t^2} = c^2\frac{\partial^2 u}{\partial x^2} \tag{7.6}$$

where c is the speed of sound in the material and is given by:

$$c = \sqrt{\frac{E}{\rho}}. \tag{7.7}$$

The solution to the partial differential equation of motion in Eq. 7.6 has a simple physical meaning. The solution is an acoustic wave propagation problem represented by a plane wave traveling along the rod's length. The wave travels along the length and is reflected back from the end. This motion continues indefinitely in the undamped case described here. The wave velocity depends only on the material density and modulus of elasticity. This acoustic wave represents a disturbance in the beam geometry resulting in local strain within the bar. This simple relationship defines how quickly stress propagates through a system. Shock information contained in the stress wave is transmitted through a mechanical system at the material's speed of sound. As a consequence of this phenomenon, material changes and interfaces representing impedance mismatches within the system can alter the stress wave as it travels, sometimes significantly.

The general solution to the wave equation derived in Eq. 7.6 is of the form:

$$u(t, x) = f_1(ct - x) + f_2(ct + x), \tag{7.8}$$

which represents a traveling wave. The displacement, u, is a function of two independent variables: the distance along the beam, x, and the time t. For the solution of this equation, we will assume a complex harmonic solution of the form:

$$u(t,x) = Ae^{i(\Omega t - \lambda x)} + Be^{i(\Omega t + \lambda x)}, \tag{7.9}$$

where A and B are complex amplitude constants and λ is the wavelength constant defined by $\lambda = \Omega/c$. This form of the solution is more common in the field of acoustics but it leads to a more insightful conclusion. For boundary conditions, we will assume that the long rod is free at both ends, $x = 0$ and $x = L$. The free end boundary condition requires that there is no internal force at the ends, or $\partial u/\partial x = 0$.

Substituting $\partial u/\partial x = 0$ at $x = 0$ into Eq. 7.9 gives $0 = -A + B$ so that we know $B = A$ and Eq. 7.9 can be simplified as:

$$u(t,x) = Ae^{i\Omega t}\left(e^{-i\lambda x} + e^{i\lambda x}\right). \tag{7.10}$$

This can be rewritten in terms of the cosine function as:

$$u(t,x) = 2Ae^{i\Omega t}\cos(\lambda x). \tag{7.11}$$

The free end boundary condition at $x = L$ requires a solution for the partial derivative of Eq. 7.11 such that

$$\sin(\lambda L) = 0. \tag{7.12}$$

Valid solutions to this equation occur when

$$\lambda_n = \frac{n\pi}{L} \qquad n = 1, 2, 3, \ldots \tag{7.13}$$

and therefore the rod's natural frequencies are given by:

$$\Omega_n = \frac{n\pi c}{L} \qquad n = 1, 2, 3, \ldots \tag{7.14}$$

It can be shown that these frequencies are also identical to those for a rod that is fixed at both ends.

Physical vibrations are defined by the real part of Eq. 7.11, given by:

$$u(t,x) = [C_n\cos(\Omega_n t) + D_n\sin(\Omega_n t)]\cos(\lambda_n x). \tag{7.15}$$

To solve this equation, we assume initial conditions at $t = 0$ such that the displacement, $u = 0$ at one end, with an initial velocity, $\partial u/\partial t = V_0$ applied at the same end. This initial velocity represents a shock pulse applied to one end of the rod. Using the initial displacement condition, it can be readily seen that at $t = 0$ and $x = 0$, C_n must be zero to ensure $u(t,x) = 0$. Thus, the solution is reduced to:

$$u(t,x) = D_n\sin(\Omega_n t)\cos(\lambda_n x). \tag{7.16}$$

Taking the partial derivative with respect to time and solving for the initial velocity condition gives

$$\frac{\partial u(0,0)}{\partial t} = D_n \Omega_n \cos(\Omega_n t)\cos(\lambda_n x) = V_0. \tag{7.17}$$

Substituting $x = 0$ and $t = 0$ and solving for D_n gives

$$D_n = \frac{V_0}{\Omega_n}. \tag{7.18}$$

Thus, the solution to the one-dimensional wave equation for the free-free long rod subjected to an initial velocity at one end is

$$u(t,x) = \frac{V_0}{\Omega_n}\sin(\Omega_n t)\cos(\lambda_n x). \tag{7.19}$$

Substituting the wavelength constant gives

$$u(t,x) = \frac{V_0}{\Omega_n}\sin(\Omega_n t)\cos\left(\frac{\Omega_n}{c}x\right). \tag{7.20}$$

The purpose of this derivation was to show the relationship between stress and pseudo-velocity. Strain is related to stress through Hooke's Law as $\sigma = E\varepsilon$ where E is the modulus of elasticity. The strain is given by Eq. 7.1 as $\varepsilon = \partial u/\partial x$. Thus, the modal strain at any point is given by:

$$\frac{\partial u(t,x)}{\partial x} = -\frac{V_0}{\Omega_n}\frac{\Omega_n}{c}\sin(\Omega_n t)\sin\left(\frac{\Omega_n}{c}x\right). \tag{7.21}$$

However, we are generally interested in the maximum modal strain in the rod which occurs when the trigonometric functions equal unity. Thus,

$$\varepsilon_{max} = \frac{V_0}{c}. \tag{7.22}$$

Likewise, the maximum stress in the rod is given by:

$$\sigma_{max} = \frac{V_0}{c}E. \tag{7.23}$$

Substituting for the speed of sound as defined in Eq. 7.7 gives

$$\sigma_{max} = V_0\sqrt{E\rho}. \tag{7.24}$$

The conclusion from this derivation is that the maximum stress is directly related to the modal velocity. Since we have assumed the motion is sinusoidal or a summation of sinusoids, the velocity here is the modal velocity. The conclusion, however, is very significant in that stress is proportional to the modal velocity alone. Similar stress relationships can be derived for displacement, $x = V/\Omega$, and acceleration, $a = V\Omega$; however, the results are more complicated. In terms of the displacement, the relationship is

$$\sigma_{max} = x_{max}\Omega\sqrt{E\rho}. \tag{7.25}$$

For a specific natural frequency, $\Omega = \Omega_n$, $x_{max}\Omega$ is the maximum pseudo-velocity. In terms of acceleration, the stress relationship is

$$\sigma_{max} = \frac{a_{max}}{\Omega}\sqrt{E\rho}. \tag{7.26}$$

The complication is that in the displacement and acceleration relationships the frequency is required to determine the effect on stress. In other words, it is not adequate to provide a peak displacement or peak acceleration without further context. This is not true for velocity. Is a peak displacement of 0.010 cm acceptable in a vibrating rod? It depends on the frequency. Is a peak acceleration of 1000 g acceptable? It again depends on the frequency. However, a maximum velocity of 7 m/s is going to be either good or bad across all frequency ranges.

The wave equation solution is typically not carried all the way to the stress–pseudo-velocity relationship derived here; however, the insights gained with this derivation are important. A similar derivation can be performed for transverse bending with the same results. It is for this reason that many people find pseudo-velocity to be the preferred method for quantifying shock severity. The derivation presented here is based on a similar derivation by Gaberson [4] as well as work by Hunt [2], Kinsler and Frey [3].

7.3.2 *Stress–Pseudo-Velocity Derivation for a Beam in Bending*

The previous derivation demonstrated the stress–pseudo-velocity relationship for an axially loaded rod. A similar derivation can be performed to demonstrate that the same stress–pseudo-velocity relationship holds for a beam with a transverse loading (Fig. 7.8). To simplify the derivation, we start with the assumption that the wavelength is long with respect to the beam thickness. This assumption allows us to neglect rotary inertia and shear effects. Following the basics of long beam theory, the bending stress is proportional to the bending moment and the beam curvature.

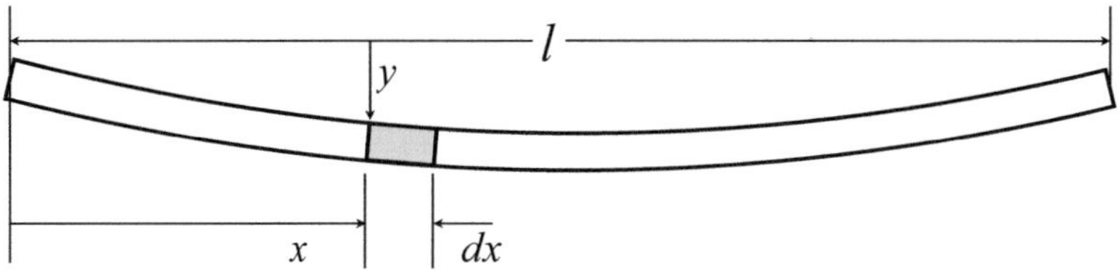

Fig. 7.8 Sketch of a long prismatic beam subject to transverse vibratory motion

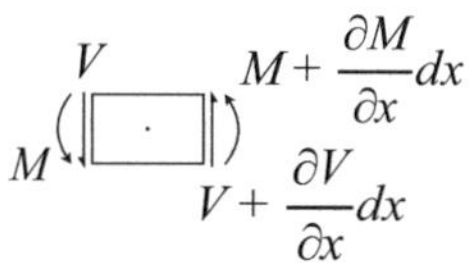

Fig. 7.9 Sketch of loaded beam segment

The bending moment is given by:

$$M = -EI\frac{\partial^2 y}{\partial x^2}, \tag{7.27}$$

where E is the beam's modulus of elasticity, I is the cross-sectional moment of inertia, and y is the transverse displacement. The shear force in the beam is given by the first derivative of the bending moment, $V = \partial M/\partial x$.

To derive the equation of transverse motion, the sum of the vertical forces acting on the beam segment as shown in Fig. 7.9 yields

$$-V + V + \frac{\partial V}{\partial x}dx = \rho A\frac{\partial^2 y}{\partial t^2}dx. \tag{7.28}$$

Simplifying and dividing through by dx give

$$\frac{\partial V}{\partial x} = \rho A\frac{\partial^2 y}{\partial t^2}. \tag{7.29}$$

Substituting for V gives

$$\frac{\partial V}{\partial x} = \frac{\partial}{\partial x}\left(\frac{\partial M}{\partial x}\right) = \frac{\partial^2 M}{\partial x^2} = \rho A\frac{\partial^2 y}{\partial t^2}. \tag{7.30}$$

Substituting Eq. 7.27 into Eq. 7.30 yields

$$-EI\frac{\partial^4 y}{\partial x^4} = \rho A\frac{\partial^2 y}{\partial t^2}. \tag{7.31}$$

Rearranging this equation gives the Euler–Bernoulli beam equation:

$$\frac{\partial^2 y}{\partial t^2} + \frac{EI}{\rho A}\frac{\partial^4 y}{\partial x^4} = 0. \tag{7.32}$$

We will solve Eq. 7.32 using the separation of variables technique as we did for the long rod example in Sect. 7.3.1. Thus, we assume that

$$y(x, t) = Y_x(x)Y_t(t). \tag{7.33}$$

Substituting this equation into Eq. 7.32 and defining $a^2 = EI/\rho A$ yield

$$Y_x\frac{d^2 Y_t}{dt^2} + a^2 Y_t\frac{d^4 Y_x}{dx^4} = 0. \tag{7.34}$$

Separating the variables gives

$$\frac{1}{Y_t}\frac{d^2 Y_t}{dt^2} + \frac{a^2}{Y_x}\frac{d^4 Y_x}{dx^4} = 0. \tag{7.35}$$

Since both halves of Eq. 7.35 cannot be zero, otherwise the solution is trivial and meaningless, both halves must equal a constant which we define in terms of Ω as:

$$\frac{1}{Y_t}\frac{d^2 Y_t}{dt^2} = -\frac{a^2}{Y_x}\frac{d^4 Y_x}{dx^4} = -\Omega^2. \tag{7.36}$$

Thus, the partial differential equation has been reduced to two ordinary differential equations, one in time and a second in space. The temporal equation is given by

$$\frac{d^2 Y_t}{dt^2} + \Omega^2 Y_t = 0, \tag{7.37}$$

and with one notation simplification defined by, $\alpha^4 = \Omega^2/a^2$, the spacial equation is given by:

$$\frac{d^4 Y_x}{dx^4} - \alpha^4 Y_x = 0. \tag{7.38}$$

The solution to the spacial component given in Eq. 7.38 was published by Timoshenko [5], as well as numerous other authors, and is given by:

$$Y_x = C_1 \sin(\alpha x) + C_2 \cos(\alpha x) + C_3 \sinh(\alpha x) + C_4 \cosh(\alpha x), \tag{7.39}$$

where C_1 through C_4 are constants to be determined.

To solve for the constants, we need to define appropriate boundary conditions. For this derivation, we will assume simply supported or pinned-pinned boundary conditions for the transverse beam. Fixed or free boundary conditions could be used with similar conclusions although the mathematics is less accommodating. At the pinned end, the deflection and bending moment must be zero. Therefore, the boundary conditions are

$$Y_x = \frac{d^2 Y_x}{dx^2} = 0 \tag{7.40}$$

at $x = 0$ and $x = L$. Differentiating and solving for the bending moment at $x = 0$ first give

$$\begin{aligned}
\frac{d^2 Y_x}{dx^2} &= 0 = \alpha^2 \left[-C_1 \sin(\alpha x) - C_2 \cos(\alpha x) + C_3 \sinh(\alpha x) + C_4 \cosh(\alpha x)\right] \\
0 &= -C_1 \sin(0) - C_2 \cos(0) + C_3 \sinh(0) + C_4 \cosh(0) \\
0 &= -C_2 + C_4 \\
C_2 &= C_4.
\end{aligned} \tag{7.41}$$

Likewise, solving for the enforced displacement at $x = 0$ gives

$$\begin{aligned}
Y_x = 0 &= C_1 \sin(\alpha x) + C_2 \cos(\alpha x) + C_3 \sinh(\alpha x) + C_4 \cosh(\alpha x) \\
0 &= C_1 \sin(0) + C_2 \cos(0) + C_3 \sinh(0) + C_4 \cosh(0) \\
0 &= C_2 + C_4 \\
C_2 &= -C_4.
\end{aligned} \tag{7.42}$$

The only way that Eqs. 7.41 and 7.42 can both be true is for C_2 and C_4 to both equal zero. Using the pinned boundary condition at $x = L$ and the solutions for C_2 and C_4 just developed, we can solve for C_1 and C_3 by:

$$\begin{aligned}
\frac{d^2 Y_x}{dx^2} &= 0 = \alpha^2 \left[-C_1 \sin(\alpha L) + C_3 \sinh(\alpha L)\right] \\
0 &= -C_1 \sin(\alpha L) + C_3 \sinh(\alpha L).
\end{aligned} \tag{7.43}$$

Since the hyperbolic sine function can only be zero if $L = 0$, which is not a very interesting beam, we find that C_3 must be zero. And finally, we have that

$$Y_x = 0 = C_1 \sin(\alpha L). \tag{7.44}$$

Since the sine term can be zero, we will define the constant C_1 to be the maximum transverse displacement, y_{max}. Equation 7.44 is equal to zero when αL is a multiple of π. Thus, α is restricted to values of

$$\alpha = \frac{n\pi}{L} \qquad n = 1, 2, 3, \ldots \tag{7.45}$$

Since α was previously defined by $\alpha^2 = \Omega/a$, we would like to rewrite Eq. 7.45 in terms of Ω as:

$$\Omega = \alpha^2 a = a\left(\frac{n\pi}{L}\right)^2 = \left(\frac{n\pi}{L}\right)^2 \sqrt{\frac{EI}{\rho A}} \qquad n = 1, 2, 3, \ldots \tag{7.46}$$

Returning to the temporal solution in Eq. 7.37, the solution is known to be of the form:

$$Y_t(t) = C_5 \sin(\Omega t) + C_6 \cos(\Omega t), \tag{7.47}$$

where C_5 and C_6 are again constants to be determined. If we assume an initial condition that the displacement, $y = 0$ at time, $t = 0$, then $C_6 = 0$ by inspection. The remaining constant is rolled into the constant defined by the spacial solution. Thus, the total solution then is given by the product of the temporal solution and the spacial solution as defined previously. Thus,

$$y(x, t) = y_{max} \sin(\alpha x) \cos(\Omega t). \tag{7.48}$$

To complete the stress proof, we return to the equation for the bending moment given in Eq. 7.27 and note that the bending stress is given as a function of the bending moment as:

$$\sigma = \frac{Mh}{I} = -Eh\frac{\partial^2 y}{\partial x^2}, \tag{7.49}$$

where h is the distance from the neutral axis to the beam's outer fiber. Substituting the total solution into the stress solution and differentiating give

$$\sigma = y_{max} Eh\alpha^2 \sin(\alpha x) \cos(\Omega t). \tag{7.50}$$

The maximum stress occurs when the sine and cosine term are maximum, thus:

$$\sigma_{max} = y_{max} Eh\alpha^2. \tag{7.51}$$

Substituting for α yields

$$\sigma_{max} = y_{max} Eh\Omega\sqrt{\frac{A\rho}{EI}} \tag{7.52}$$

Rearranging and substituting for the material's sound speed c from Eq. 7.7 yield

$$\sigma_{max} = y_{max} \Omega\rho c h\sqrt{A/I}. \tag{7.53}$$

The term $h\sqrt{A/I}$ is a beam shape factor since it contains only dimensional parameters. The shape factor will be defined as K_b and the maximum stress is then given by:

$$\sigma_{max} = \rho c K_b \Omega y_{max}. \tag{7.54}$$

Since the vibration is sinusoidal, $\Omega y_{max} = v_{max}$. Thus, the maximum stress from the transverse vibration in a beam is proportional to the maximum modal velocity. The final relationship is

$$\sigma_{max} = \rho c K_b v_{max}. \tag{7.55}$$

As with the axially loaded long rod, there is no frequency dependence in the maximum stress–pseudo-velocity relationship. The equation could be manipulated and defined in terms of maximum deflection or maximum acceleration but the modal frequency, Ω, would be naturally pulled into the expression. Thus, the engineer would be forced to define the peak stress in terms of a peak displacement and a frequency or a peak acceleration and a frequency. However, where pseudo-velocity is defined, the maximum stress is defined independent of frequency. This is the stress–pseudo-velocity proof. The preceding derivation is based on the work of Gaberson and Chalmers [4, 6] as well as work by Hunt [2].

7.3.3 Examples

To illustrate the pseudo-velocity relationship to stress for design, two examples are provided. The first example is a for an axially loaded rod fixed at one end and the second example is for a cantilever beam.

7.3.3.1 Axially Loaded Rod Example

Consider the undamped, slender rod shown in Fig. 7.10, fixed at one end and free at the other. For these boundary conditions, the characteristic equation for free vibration of the bar is

$$\cos\left(\frac{\Omega L}{c}\right) = 0. \tag{7.56}$$

The roots of this equation are

$$\frac{\Omega L}{c} = \frac{\pi}{2}, \frac{3\pi}{2}, \ldots, \frac{(2r-1)\pi}{2}, \tag{7.57}$$

and the natural frequencies are

$$\Omega_r = \frac{(2r-1)c\pi}{2L}. \tag{7.58}$$

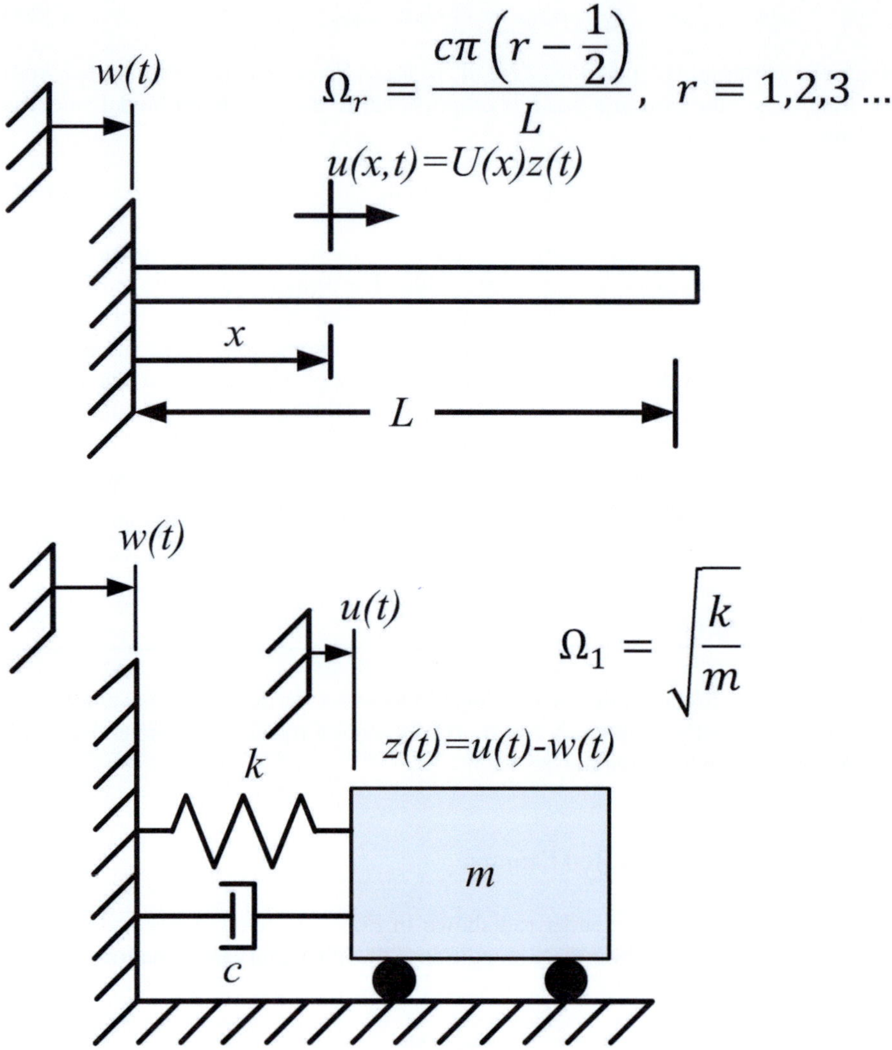

Fig. 7.10 Axially loaded fixed-free slender rod and its SDOF representation

Now, we can write

$$U_r(x) = C\sin\left(\frac{(2r-1)\pi x}{2L}\right),\ r = 1, 2, \ldots, \tag{7.59}$$

where C is a scaling factor and $U_r(x)$ is called a *mode shape*. Mode shapes are basis functions, meaning that the total motion of the rod can be expressed as:

$$u(x,t) = \sum_{r=1}^{r=\infty} U_r(x) z_r(t) \tag{7.60}$$

This is covered in Chap. 8 on multi-degree of freedom systems. Every point on the pseudo-velocity SRS can be considered as the peak of the modal response at a natural frequency Ω_r. In this example, we will consider only the first mode for simplicity. This assumption reduces the problem to that of an SDOF system, shown at the bottom of Fig. 7.10. The difference between the SDOF rod and the SDOF oscillator is that the magnitude of the response is a function of the position on the rod.

The stress in the SDOF model of the rod is given by:

$$\sigma(x,t) = E\frac{\Omega_1}{c}\cos\left(\frac{\Omega_1}{c}x\right)z(t). \tag{7.61}$$

The maximum stress in the rod can be expressed in terms of the pseudo-velocity SRS for a given shock environment.

$$\sigma_{max}(x) = \sqrt{\rho E}\cos\left(\frac{\pi x}{2L}\right)\text{PVSRS}\Big|_{\Omega_1} \tag{7.62}$$

because $\Omega_1 z_{max} = \text{PVSRS}\Big|_{\Omega_1}$. The stress in the rod is independent of the cross-sectional area and the maximum stress occurs at the base, as expected. With the pseudo-velocity SRS of the shock environment, the designer can use Eq. 7.62 to select the material so that the rod will not suffer over-stress damage from the shock.

The continuous model, partial differential equation description, of a structure is not practical for general structures; however, the insights gained with this derivation are important. The axial rod example shows that the peak stress is a function of the pseudo-velocity SRS because the mode shapes are periodic spatial functions of the natural frequencies. This is true for general structures. The natural frequency that multiplies the relative displacement to form the PVSRS comes from the spatial derivative of the mode shape. A similar result can be obtained for a beam in bending.

7.3.3.2 Cantilevered Beam Example

Consider an undamped, Euler–Bernoulli beam cantilevered at the left end as shown in Fig. 7.11. The constants in Eq. 7.39 cannot be found directly for the fixed-free cantilever beam boundary conditions. However, the four boundary conditions produce a set of four homogeneous equations, which yield the characteristic equation:

$$\cos(\lambda L) + \cosh(\lambda L) = 0. \tag{7.63}$$

The roots of this equation must be determined numerically. Blevins [1] has tabulated the natural frequencies and mode shapes for many basic structures, including beams, rods, and plates. The first three values for λL are: 1.8751, 4.6941, and 7.8548. The natural frequencies are then found using:

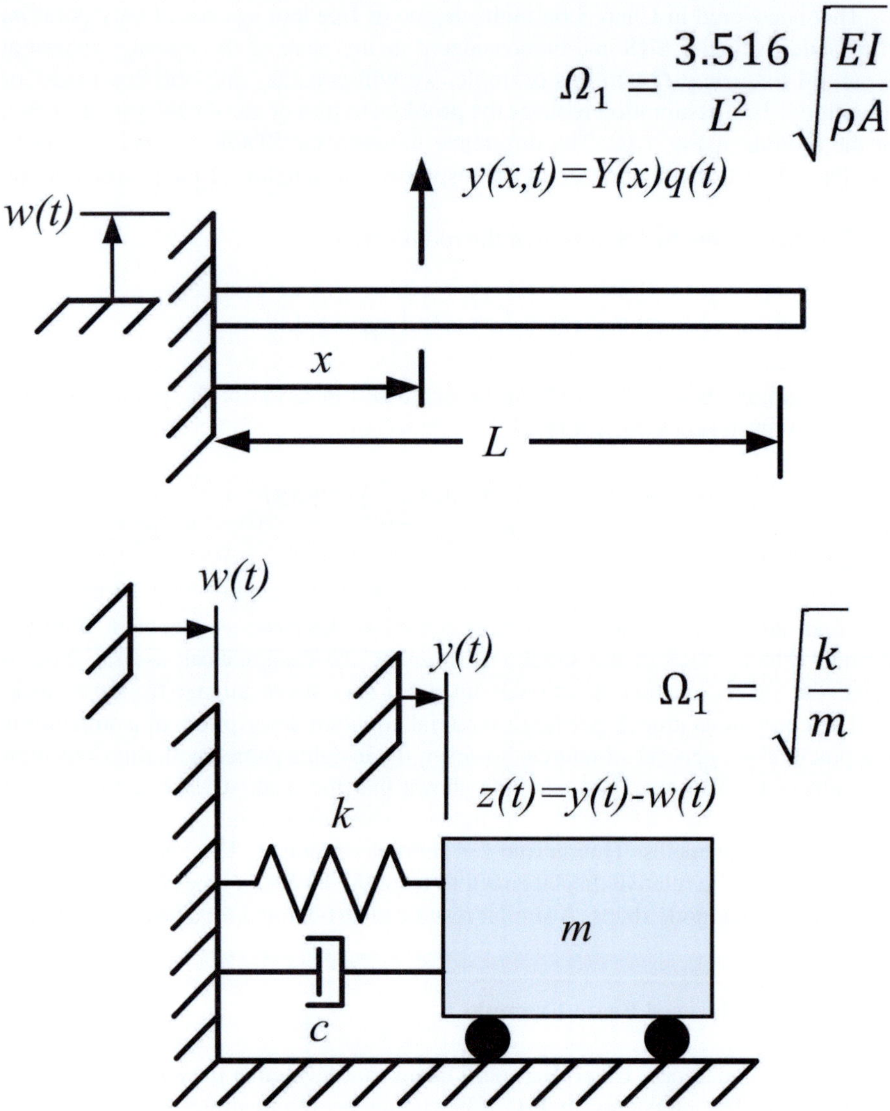

Fig. 7.11 Cantilevered Euler–Bernoulli beam and its SDOF representation

$$\Omega_r = \lambda_r^2\sqrt{\frac{EI}{\rho A}}. \tag{7.64}$$

The expression for the displacement mode shapes is

$$Y_r(x) = C\left[K_r\left(\sin(\lambda_r x) - \sinh(\lambda_r ax)\right) + \cosh(\lambda_r x) - \cos(\lambda_r x)\right], \tag{7.65}$$

where C is a constant scale factor and

$$K_r = \frac{\cos(\lambda_r L) + \cosh(\lambda_r L)}{\sin(\lambda_r L) + \sinh(\lambda_r L)}. \tag{7.66}$$

For the cantilever beam example, the first three values for K_r are: 0.7341, 1.0185, and 0.9992.

Finally, the stress mode shape equation is the second derivative of the displacement mode shape, given by:

$$S_r(x) = C\sqrt{\frac{\rho AE}{I}}\,[K_r\,(\sin(\lambda_r x) + \sinh(\lambda_r x)) - \cosh(\lambda_r x) - \cos(\lambda_r x)] \tag{7.67}$$

To design the beam, we treat the beam as an SDOF system by considering only the fundamental mode. Analogous to the axial rod example, the difference between the SDOF beam model and the SDOF oscillator is that the magnitude of the response is a function of the position on the beam in the beam model. The peak stress in terms of the pseudo-velocity SRS is

$$\sigma_{peak}(x) = \sqrt{\frac{\rho AE}{I}}\Bigg[K_1\left(\sin\left(\frac{\lambda_1 x}{L}\right) + \sinh\left(\frac{\lambda_1 x}{L}\right)\right) - \cosh\left(\frac{\lambda_1 x}{L}\right) - \cos\left(\frac{\lambda_1 x}{L}\right)\Bigg]\mathrm{PVSRS}\Big|_{\Omega_1} \tag{7.68}$$

The design engineer would select the material, ρ and E; the length, L; and the cross-section, A and I, for a given excitation represented by the pseudo-velocity SRS so that the maximum stress in the beam is below the yield stress by some factor of safety.

7.4 Extreme Loading

If the designer has access to the original time history and a good dynamics-capable finite element code, then it would be straightforward enough to simply use the known time history in the finite element model and perform the design. In this case, the only remaining question is to determine the desired factor of safety for the design. Even if the time history of the environment is known precisely and is not subject to random variations, performing a design study with the finite element model directly can be prohibitively time consuming. The assumption here is that the design is to be based on the SRS data.

If the original shock has a substantial velocity change and a relatively flat plateau in the pseudo-velocity SRS, then a simulation can be performed to model that velocity change. This can be done by driving a finite element simulation with a ramp function, increasing the velocity from zero to the SRS value in a time appropriate to the length of the shock pulse. This can also be performed in reverse by starting the simulation with an initial velocity and arresting the motion accordingly.

Modeling a velocity change is relatively straightforward for most finite element codes but is not necessarily for the novice. A much more simplistic approach to analyzing shock is to use a static acceleration load. Designing for shock with a static load case is extremely contradictory but can be made to work. Section 1.4 introduced the need for dynamic analysis of components starting with the equation of motion for an undamped system subjected to an initial velocity, V_0. The derivation in Sect. 1.4 focused on the cantilever beam with a lumped mass at the free end, but is applicable to any simple system. The equation of motion is given by:

$$m\ddot{y}(t) + ky(t) = 0, \tag{7.69}$$

where m is the system mass, k is the system stiffness, and $y(t)$ is the displacement in the shock direction. With initial conditions $y(0) = 0$ and $\dot{y}(0) = V_0$, the homogeneous solution is of the form:

$$y(t) = A\sin(\omega t) + B\cos(\omega t) \tag{7.70}$$

where $\omega = \sqrt{k/m}$ and the constants A and B are determined from the initial conditions. Substituting the initial conditions into Eq. 7.70, the displacement $y(t)$ is given by:

$$y(t) = \frac{V_0}{\omega}\sin(\omega t), \tag{7.71}$$

and the acceleration is given by:

$$\ddot{y}(t) = -\omega V_0 \sin(\omega t). \tag{7.72}$$

The maximum acceleration occurs when the sine term equals -1, thus

$$\ddot{y}(t)_{max} = \omega V_0. \tag{7.73}$$

Substituting the cyclic frequency for the circular frequency, ω, and converting to gravitational units, g, yield

$$A_{max} = \frac{2\pi f V_0}{g}. \tag{7.74}$$

Using Eq. 7.74, an estimate of a static acceleration design load can be made for a dynamic system based on the target system natural frequency. This becomes a somewhat iterative process and the predicted acceleration result here may be low so it is often wise to add a modest safety factor in the range of 1.5–2 when using this methodology. The design static acceleration would then be

$$A_{design} = \frac{2\pi f V_0}{g} SF, \tag{7.75}$$

where SF is the safety factor on the calculated acceleration to ensure the design is appropriately conservative. A less conservative design safety factor can be used if localized yielding is permitted within the component or its attachment points.

For example, if the intent was to design a large naval system with a first natural frequency at about 14 Hz and a pseudo-velocity SRS maximum of about 3 m/s, Eq. 7.74 would suggest a static acceleration design load equal to 27 g. Adding a conservative safety factor of two to this would suggest that a good starting point is to design the system for about 54 g of static acceleration. Likewise, if the component were a very small electronic part with a primary natural frequency of about 1.2 kHz and an assumed velocity change of 2 m/s, the static design acceleration would be 1540 or 3080 g with $SF = 2$. While this sounds like a tremendous load, small parts with extremely high first resonant frequencies will likely weigh much less than 1 kg, so the actual force the component is resisting may still be quite manageable.

The results using this method will not be completely accurate since this analysis methodology uses a static load to approximate a dynamic load; although, it is a reasonable estimate for design decisions. Applying a static acceleration to load a finite element model is a simple task that can be quickly accomplished. Simple analyses also allow for faster and easier design iterations. A more accurate analysis should still be completed, if possible, as the design matures.

Figure 7.12 shows a photograph of one mounting foot of a U.S. Navy Harpoon missile launcher. The mounting foot is a steel weldment bolted to a foundation weldment. When a surface ship is loaded by the explosion of an underwater mine, first motion is always upward as the blast lifts the ship. Looking at the

Fig. 7.12 Photograph of one of the mounting feet for an old U.S. Navy Harpoon missile launcher

photograph, one can easily imagine that failing the launcher foot in compression would be nearly impossible. Even yielding the foot appreciably would be extremely difficult. However, when the shock loading reverses and the ship starts to move back downward, the entire load of the launcher is suddenly transferred from the foot in compression to the bolts in tension. The figure shows four bolts on the visible side, presumably with four matching bolts on the inside. These four bolts have a very small tensile cross-sectional area compared to the compressive cross-section of the launcher foot. This concept was briefly mentioned in Sect. 4.1.5 in the context of the positive and negative SRS. If the direction of motion is confidently known, such as in the surface ship underwater shock case, the differences in the positive and negative SRS can be used to design the stronger interface direction to align with the largest shock loading direction. Regardless, the weakest loading direction for the interface needs to be designed to survive the expected loading condition. If the bolts in tension are less capable than the foot in compression, then the bolts are the limiting case and need to be designed accordingly. Of course, the shear load on the joint must also be considered in addition to the tensile and compressive loads.

One important consideration when designing for shock is an understanding of the allowable damage. Oftentimes, no damage is allowed from a shock event. If the loading is from a transportation environment and the components are expected to arrive in a pristine condition, then transportation shock damage is not allowed. On the other hand, if the shock environment is an abnormal event and the only requirement is to fail in a safe manner, then significant damage might be permissible or even desirable from a shock. Generally, shock requirements fall into three broad categories: no damage permitted; no significant damage or no functional failures; and damage allowed so long as the item remains safe. However, it should be noted that bolt failure is generally not considered acceptable. The reason for this is that in order to ensure a component fails safe or performs without significant degradation, it is necessary that it also remains where it was originally installed. Even if only a small number of bolts fail and the part remains in place, having pieces of debris moving around inside of a system usually has unexpected and often unpleasant consequences.

7.5 Plastic Design of Foundations

In many instances, it is permissible to exceed the allowable yield stress in a component foundation. If a component is not alignment critical, and the foundation is a welded steel structure, yielding in the foundation should occur well before weld cracking. Since plastic deformation can absorb considerably more energy than elastic deformation alone, it is often an acceptable method of shock survivability where repeated exposure is not anticipated. If bolted joints are used in place of welded joints, it will be necessary to design the bolted interface to be stronger than the structural members to achieve the same results. It should also be noted that not all welds are stronger than the parent material and this methodology will not work in those cases.

The energy attributable to plastic work, deforming the structure, can be calculated as:

$$E_{Plastic} = E_{Kinetic} - E_{Elastic}, \tag{7.76}$$

where $E_{Plastic}$ is the energy for plastic work, $E_{Kinetic}$ is the kinetic energy, and $E_{Elastic}$ is the elastic strain energy. The kinetic energy is given by the usual relationship:

$$E_{Kinetic} = \frac{1}{2} m V^2. \tag{7.77}$$

Likewise, the elastic strain energy is given by:

$$E_{Elastic} = \frac{1}{2} k y^2, \tag{7.78}$$

where y is the displacement at yield.

In this way, the amount of energy available to perform plastic work can be estimated and estimates of the resulting displacements can also be made. This methodology should always result in a conservative estimate for the final displacement since the additional energy is required to transition the material from totally elastic to a plastic hinge.

7.6 Fatigue Loading

Shock and fatigue are two concepts that are not usually studied together. A shock event is a short duration extreme loading scenario, whereas fatigue loading is typically a milder, low-level loading that occurs thousands of times, slowly accumulating damage. Many shock events are assumed to be in the one and done category. The building survives the earthquake or the ship survives the underwater mine explosion. However, if a ship survives an underwater mine blast, the ship is not necessarily decommissioned and sent to scrap. The ship will likely be inspected, repaired, and sent back out to rejoin the fleet. Likewise, if the building appears relatively undamaged from an earthquake, it will likely be inspected, repaired, and certified for continued occupancy. Thus, even for severe loadings, the possibility for accumulating fatigue damage exists.

On the other hand, loadings such as transportation shocks can easily accumulate a relatively large number of fatigue cycles. MIL-STD-810G [7] specifies a transportation shock test sequence consisting of 66 shocks for every 5000 km of on-road transportation. Of course, 5000 km is a significant travel distance and 66 shocks is still not the thousands that one would expect for appreciable fatigue accumulation. However, if the component is to be installed on a vehicle, it is easy to see how a typical commercial vehicle lifetime of over 1.5 million km could accumulate a relatively large number of shock events—approximately 20,000 if the MIL-STD-810G methodology is followed.

In general, fatigue is not a serious concern for shock assuming that the expected number of shock events is low. Most components will have an adequate safety margin to cover a few cycles of fatigue exposure regardless. If the number of cycles is high, fatigue design for shock is similar to fatigue design for vibration or other variable loading scenario. An effective yield stress, also known as an endurance limit is derived and then that lower stress number is used for design [8].

7.7 Shock Spectrum and Strain Energy

When a system is excited by a shock event, energy is transferred to the system. That energy is initially in the form of strain energy or kinetic energy and is eventually dissipated through damping mechanisms. Stored strain energy is a function of the stiffness and strain capability of the structural member. For example, a structural member under axial loading has a stiffness, k, given by $k = EA/L$, where E is the modulus of elasticity, A is the cross-sectional area, and L is the length. The strain energy in the member is given by:

$$U = \frac{1}{2}kx^2. \tag{7.79}$$

In this equation, the displacement x is the relative displacement or stretch in the structural member. Since strain $\epsilon = x/L$, the displacement in Eq. 7.79 is given by $x = \epsilon L$. Making this substitution into Eq. 7.79 yields

$$U = \frac{1}{2}\frac{\sigma^2}{E}AL. \tag{7.80}$$

Since AL is the volume of material, Eq. 7.80 can be rewritten in terms of the strain energy per unit volume as:

$$U_v = \frac{1}{2}\frac{\sigma^2}{E}, \tag{7.81}$$

where U_v is the strain energy per unit volume.

Thus, for a mild steel with a yield strength of 350 MPa and a modulus of elasticity of 200 GPa, the elastic strain energy per unit volume is

$$U_v = \frac{1}{2}\frac{\sigma^2}{E} = \frac{1}{2}\frac{(350\,\text{MPa})^2}{200\,\text{GPa}} = 0.3\,\text{MPa}. \tag{7.82}$$

If the system is allowed to yield, then the amount of energy absorbed per unit volume can be considerably larger. Material yield is typically defined at 0.2% elongation whereas plastic strain for ductile materials can be far in excess of 10%, a 50-fold increase in energy absorption capability. It is for this reason that minor yielding can be extremely beneficial to absorbing and dissipating severe shock loads.

Returning to the elastic example, the spectrum velocity at any given frequency can be read directly from the maximum pseudo-velocity SRS plot. The spectrum displacement is related to the spectrum velocity by the expression:

$$x_s = \frac{V_0}{2\pi f}. \tag{7.83}$$

The spectrum displacement, x_s, is the maximum relative displacement between the SDOF mass and its foundation. Thus, the maximum strain energy of the system is given by:

$$U = \frac{1}{2}kx_s^2, \tag{7.84}$$

where k is the foundation stiffness. Substituting Eq. 7.83 into Eq. 7.84 gives

$$U = \frac{1}{2}k\left[\frac{V_0}{2\pi f}\right]^2 = \frac{1}{2}k\frac{V_0^2}{\omega^2}. \tag{7.85}$$

Substituting the $\omega^2 = k/m$ and rearranging give

$$U = \frac{1}{2}mV_0^2. \tag{7.86}$$

Thus, the maximum strain energy in the SDOF system is determined by the maximum pseudo-velocity SRS and the system mass, m. This maximum strain energy can be compared to the strain energy per unit volume calculated in Eq. 7.81 to see if the cross-sectional area of the foundation is adequate to support the expected shock load. If it is too small, then the section needs to be increased accordingly. A similar calculation could be performed assuming a foundation in bending. For the bending case, the stiffness is defined by the end conditions, whether it is a cantilever foundation, a clamped foundation, simply supported, or something different altogether.

One additional point for consideration, once a system is loaded into the plastic regime, the system and material are no longer linear. Therefore, the shock spectrum theory and the normal mode methods are no longer completely valid. However, in practice, a reasonable upper bound for the energy being transferred to a structure can be estimated using these linear techniques and the resulting strain energy is a reasonable estimate or approximation of plastic deformation.

7.7.1 Example: Machinery Foundation

Consider the above spectral method for sizing a machinery foundation. How large must the structural members be to support and restrain an arbitrary piece of machinery under shock load? For this example, we will assume a 1000 kg equipment mass. The foundation supporting the machinery is a commonly available mild steel with a 350 MPa yield strength. According to Eq. 7.82, the elastic strain energy per unit volume that can be absorbed is 0.3 MPa. If we assume that the design shock spectra shows a maximum pseudo-velocity of 4 m/s, a relatively harsh shock for large equipment, then the required energy absorption is found from Eq. 7.85 to be 8000 N m. The volume of steel required to react the 4 m/s shock load of the 1000 kg machine is then:

$$V = \frac{U}{U_v} = \frac{8000\,\mathrm{N\,m}}{0.3\,\mathrm{MPa}} = 0.0267\,\mathrm{m}^3. \tag{7.87}$$

This volume assumes that the subfoundation is perfectly rigid, which is an over-simplification but often appropriate for design. If we assume that the machine is mounted on four pedestals, then each pedestal needs 0.00668 m^3 of material. This could be accomplished with an 300 mm tall I-beam pedestal approximately 300 mm deep with 300 mm flanges and nominal 25 mm flange and web thickness as an example. Other shapes and sizes could be easily derived using the above data. If the pedestals were allowed to yield, they could be significantly lighter.

With the pedestals designed, how large do the bolts need to be to restrain the machine? Under a pure static 1 g loading, the bolts required to restrain 1000 kg are actually quite small. A single Class 10.9 M8 bolt can react approximately 47 kN, more than enough to hold the machine. However, under shock the effective load is much higher. For the machinery example, it is assumed that the upward shock load and the downward shock load are not equal. This is a common assumption in shipboard shock design where the initial movement is always to compress the foundation. Likewise, the bolts are not required to absorb the full shock load as flexibility in the foundation and subfoundation components can also absorb shock load. For this example, we will assume the that the bolts are required to absorb a velocity change of 1 m/s as opposed to the 4 m/s that the foundation was required to absorb.

If the bolts are required to react a 1 m/s velocity change, then the required strain energy for the 1000 kg machine is 500 N m. Class 10.9 bolts have a nominal yield strength of 940 MPa, significantly higher than the mild steel of the foundation. With this tougher steel, the maximum strain energy per unit volume is 2.21 MPa. Using these values, the volume of bolt required to react the shock load is

$$V = \frac{U}{U_v} = \frac{500\,\mathrm{N\,m}}{2.21\,\mathrm{MPa}} = 0.000226\,\mathrm{m}^3. \tag{7.88}$$

Since bolts are just cylinders, the size can be determined with a few assumptions. The bolt volume is defined as:

$$V = N_B \pi r^2 l, \tag{7.89}$$

where N_B is the number of bolts, l is the bolt grip length, and r is the bolt radius. For this example, we will assume that we need four bolts, each 100 mm long. Substituting N_B and l into Eq. 7.89 yields a required bolt radius of 13.4 mm. Since that is not a standard bolt diameter in Class 10.9, we would select the slightly larger M30 bolts. Of course, there are other considerations, such as applying appropriate factors of safety when designing bolted joints to ensure that bolts are properly loaded and have adequate installation tolerances. This calculation is intended to be a guideline for design, and to be adjusted for individual circumstances.

7.7.2 Example: Small Component Attachment Screws

The previous example works because there is a well-defined velocity change associated with the shock. How does the analysis work when there is no well-defined velocity change? Pyroshock events, for example, are characterized by little or no net velocity change. Using the previously defined methods would imply that we need no screws to restrain the component during the shock event, an obvious fallacy.

As an example, assume that we need to restrain a small 0.5 kg component subjected to a 5000 g 1 kHz shock test with three equally spaced screws. As before, we assume the screws will be Class 10.9 with a nominal yield strength of 940 MPa. A simple static analysis using a 5000 g static load implies that the component can be restrained with three M4 screws (3.32 mm diameter rounded up to the next available size). The problem with this analysis is that we have shown in Chaps. 5 and 6 that the maximum in the SRS is often significantly larger than the defined shock, often two to six times greater for oscillatory type shocks. If we simply repeat the static analysis doubling the load, the screw size increases to M5 screws (4.7 mm diameter rounded up to the next available size). For static analysis, there is no need to specify screw length to solve the equation. This is fundamentally at odds with the need to understand how much strain energy a part can absorb.

If we return to the velocity analysis presented above, the maximum momentary velocity from the 5000 g 1 kHz shock is approximately 7.8 m/s. This can be obtained by application of Eq. 3.21. Using the methods described here, a 7.8 m/s velocity change implies M13 screws (12.08 mm diameter rounded up to the next available size) at 20 mm length are required to restrain the component. In actuality, all three of these answers are likely wrong. While M13 screws would certainly restrain the small component, they are larger than required. In contrast, M4 or M5 screws are likely to be too small because they have not accounted for the dynamics.

The best solution to this problem, if test data are available, is to pull the velocity change from measured test data. Referring back to Fig. 6.4, there was a very steep slope in the SRS near the peak which became more shallow at low-frequencies, and trended to the expected 6 dB/octave slope. If these data are available for the system in question, it is likely that the actual velocity change is about half of the peak, implying that the screws should be designed for a velocity change of about 4 m/s rather than the peak of 7.8 m/s. This implies that the component could be restrained with three M8 screws (6.2 mm diameter rounded up to the next available size). M8 screws would actually restrain the part up to a velocity change of 5 m/s implying some additional margin.

An M8 screw is actually fairly substantial for a 0.5 kg component and may not be desirable. Here again, more knowledge of the actual environment and performance requirements are useful. Since the actual calculation called for a 6.2 mm screw, an M6 screw might be acceptable if minor permanent deformation is allowed. It may also be acceptable if the anticipated velocity change is lowered due to some external attenuation of the shock load. An alternate option is to step up to the stronger Class 12.9 screws.

7.8 Shock Design Guidance

Many times shock design guidance is simple and straightforward. Understanding that the system is going to experience considerable loading applied and removed very rapidly does not agree with loose tolerances, sloppy manufacturing, brittle parts, or substandard materials. Much of this is simply good engineering practice. Other aspects of shock design are not as obvious. The following list is intended to provide some pointers to the solution of shock design problems. In addition to the list provided here, MIL-DTL-901E [9] also provides similar design guidance.

- It is important to consider the complete dynamic environment during the design phase. Shock, vibration, and acoustic requirements must be considered together for the best design.
- All components will have motion under shock. Nothing is absolutely rigid although some parts may be significantly more stiff than other parts.
- Clearance and sway space must be accounted for under shock loads. Components near one another must have sufficient clearance to accommodate out-of-phase motion.
- Be sure to provide sufficient cable length to accommodate relative motion between components under shock loads.
- Component stress and deflection are, to a large extent, determined by the lowest natural frequencies of the part when mounted on its foundation or supports.
- The shock resilience of a component is a function both of its actual static strength and its ability to withstand deflection.
- Components designed for a specific static acceleration are capable of withstanding accelerations with a significantly higher amplitude if the temporal duration of

the load is small. In other words, the maximum dynamic acceleration capability is always higher than the maximum static acceleration capability.

- Use ductile materials whenever possible. Ductile materials are typically defined as materials with greater than 10% elongation, although higher elongations are better.
- Friction cannot be relied upon to maintain alignment of components under shock. Likewise, slotted holes and oversized holes are not appropriate for shock loading.
- Avoid stress concentrations in designs. Likewise, design mounting locations to withstand the expected shock loads and deflections. Components should not be allowed to fail at mounting locations.
- Do not design components that mount on top of foundations or other components where the lowest vibration modes coincide.
- Locate component mountings on locally stiff locations of the next higher assembly or locations of minimal deflection.
- Welds should be made stronger than the individual members. Likewise, do not locate welds in high-stress areas.
- There is an optimal range for system flexibility. Components that are very stiff can develop high forces due to having high natural frequencies. Likewise, components that are too flexible can move and collide with nearby structure. Your design should balance these two competing requirements.
- Structural members should be designed to store the maximum amount of elastic energy. This is accomplished by having a uniform stress distribution through the structure.
- Longer bolts can absorb more strain energy than a shorter bolts. Longer bolts also provide more reliable joints under dynamic loads. More generally, longer parts are capable of absorbing more strain energy than smaller parts because there is more material volume in which to distribute strain energy.
- Use shock isolators where dynamic forces cannot be managed with foundation structure.

In addition to the above general guidance, it is also important to understand that the characteristics of the shock will change as the shock propagates through the structure. The shock profile when it enters the structure may be very different from the shock profile at some location deep within the system. Typically, structural members tend to behave like a low-pass mechanical filter as the shock propagates. Each component or joint that the shock wave crosses will alter the shock wave. Typically reducing the high-frequency excitation and increasing the low-frequency motion. Thus, as the shock wave propagates, the peak acceleration decreases while the excitation frequency decreases (meaning that the motion period increases).

7.9 Summary

This chapter presented basic information regarding the design of structures to withstand shock loading. The first aspect of designing for shock is to understand which parts of an SRS are critical to the design. Using one or more SRS curves to evaluate a design can be complicated by the fact that subtle and often insignificant changes in the waveform will manifest as differences in the SRS. It is important to focus on the overall trends in the SRS and not necessarily on the small differences between two SRS curves. Second, the proof that stress and pseudo-velocity are proportional was presented. This is a critical component of design since many engineers become incorrectly focused on peak acceleration as a measure of damage potential where the real focus should be on the pseudo-velocity or relative displacement. Guidelines for sizing component structure were also provided. These guidelines should be used in conjunction with appropriate safety factors and finite element analysis to ensure the most appropriate design is achieved. Finally, general guidelines for designing systems and components to resist shock are presented.

Problems

7.1 The concept of an equivalent SRS is often fundamental to evaluating the success or failure of a shock test. What is less understood is whether minor deviations between two SRS curves are significant. Perform a trade study with some sample data to demonstrate how apparent SRS differences can be generated while maintaining equivalent peak acceleration.

7.2 In this chapter, it was stated that two shocks with the same velocity change but different peak accelerations will have difference shock bandwidths. Demonstrate this with an example.

7.3 The stress–velocity relationship was derived for an axially loaded rod that is free at both ends. Derive the stress–velocity relationship for a rod fixed at one end and free at the other end. Do the results change? What about for a rod fixed at both ends?

7.4 The stress–velocity relationship was also derived for a simply supported beam in bending. Simply supported boundary conditions were used because the equation has a closed-form solution in this case. Solve the equations for a cantilever beam in bending boundary conditions. Even though a closed form solution does not exist, it should still be possible to understand the relationship between stress and velocity.

7.5 What size, number, and grade or class of screws should be used to restrain a 1 kg component subjected to a 4000 g 0.5 ms haversine shock load? What if the loading was changed to a 4000 g 0.5 ms oscillatory shock load?

7.6 It was stated in this chapter that a static acceleration load could be used as a preliminary design load for shock applications so long at the real shock load was used later to verify the appropriateness of the design. Design a cantilever beam and apply both a dynamic shock load and the approximate equivalent static load and compare the location and magnitude of the resulting strain in both models. How close are the results? Is this appropriate? You may want to use a finite element analysis to solve this problem.

7.7 The equations for strain energy and material volume can be optimistic because they do not account for stress concentration factors. Of course, stress concentrations should be avoided in shock applications where possible. How do stress concentrations alter the energy-volume relationships? Hint: it is not simply a multiplier on the energy since the original assumption involves strain energy distributed through a volume.

7.8 Mechanical shock is a high-speed application of load. In general, material properties vary with the load application speed. For most metals, this tends to occur in one direction. Is this beneficial or detrimental to the shock problem? Research this for a few common materials such as steel, aluminum, and titanium and make a statement about how this alters the design space.

References

1. Blevins, R. D. (2001). *Formulas for natural frequency and mode shape*. Malabar: Krieger Publishing Company.
2. Hunt, F. V. (1960). Stress and strain limits on the attainable velocity in mechanical vibration. *Journal of the Acoustical Society of America, 32*(9), 1123–1128.
3. Kinsler, L. E., & Frey, A. R. (1962). *Fundamentals of acoustics* (2nd ed.). New York: Wiley.
4. Gaberson, H. A. (2012). The pseudo velocity shock analysis stress velocity foundation. In: *Proceedings of the 30th International Modal Analysis Conference*, Jacksonville, FL.
5. Timoshenko, S. (1937). *Vibration problems in engineering* (2nd ed.). New York: D. Van Nostrand Company, Inc.
6. Gaberson, H. A., & Chalmers, R. H. (1969). Modal velocity as a criterion of shock severity. *Shock and Vibration Bulletin, 40*(Part 2), 31–49.
7. United States Department of Defense. (2014). *Department of defense test method standard; environmental engineering considerations and laboratory tests*, MIL-STD-810G (w/Change 1), 15 April 2014.
8. Shigley, J. E., & Mischke, C. R. (1989). *Mechanical engineering design* (5th ed.). New York: McGraw-Hill.
9. U.S. Department of Defense. (2017). *Detail specification, requirements for shock tests, H.I. (high-impact) shipboard machinery, equipment, and systems*, MIL-DTL-901E, Washington, DC, 20 June 2017.

Chapter 8
Multi-Degree-of-Freedom Systems

Chapter 3 introduced the equations of motion and response of the single degree-of-freedom oscillator. The SDOF oscillator provides insight into the structural dynamics of linear systems, but real systems are more complex. The dynamics of structures are governed by partial differential equations but only relatively simple structures, such as beams and plates, can be analyzed by solving the differential equations directly. Evaluating the response of general structures requires multi-degree-of-freedom (MDOF) models. These models can be idealizations like the SDOF model, albeit more complex, or more typically, they are models created with finite element software. The finite element models are often very large, containing perhaps millions of degrees of freedom. However, it is impractical and often unnecessary to work with such large models so we create much smaller, reduced order models that retain and represent the important characteristics of the structure.

Figure 8.1 shows three examples of MDOF models. Figure 8.1a is a representation of a finite element model of a simple supported beam containing five elements with two degrees of freedom at each node point. Figure 8.1b is an idealized representation of a four degree of freedom spring-mass-damper model containing two translational freedoms and two rotational freedoms. Finally, Fig. 8.1c shows a finite element model of a multi-degree of freedom structure previously used by the authors to study the distribution of shock energy in a structure with rich modal content.

The shock response spectrum was introduced in Chap. 4 as a tool to gain insight into the frequency content of a transient excitation. We also described how the SRS may be used to design a structure that can be represented as single degree-of-freedom system to survive shock loads. For MDOF systems the SRS is used primarily for structural design and efficient estimation of the peak response of a structure to transient excitation.

Every structural dynamics textbook (e.g., Craig and Kurdila [1], Thomson [2], Chopra [3]) covers MDOF systems and modeling in detail. In this chapter we review the basic concepts needed to understand the response of complex structures to shock

C. Sisemore, V. Babuška, *The Science and Engineering of Mechanical Shock*,
https://doi.org/10.1007/978-3-030-12103-7_8

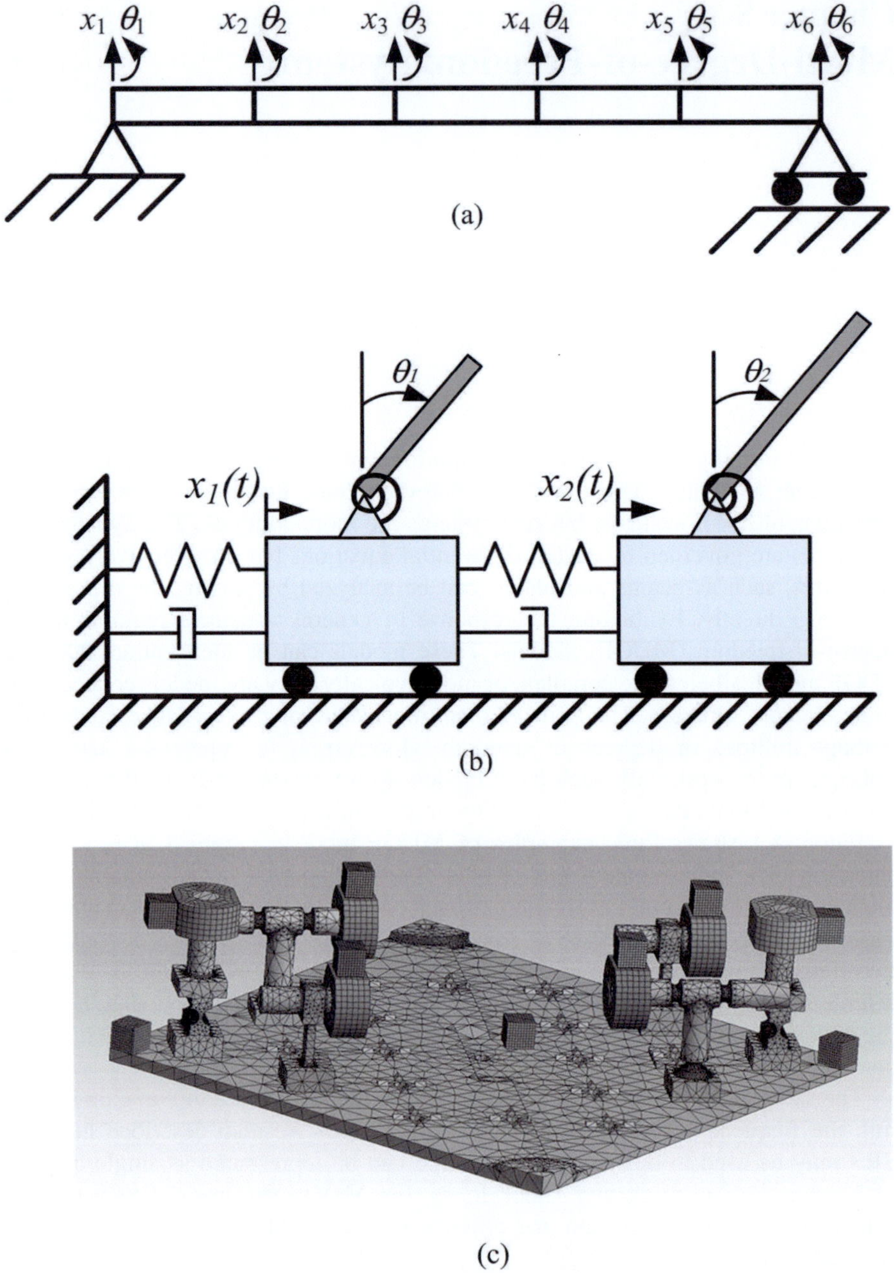

Fig. 8.1 Examples of multi-degree-of-freedom systems. (**a**) Discretized simply supported beam with 12 degrees of freedom, (**b**) Four degree of freedom spring-mass-damper system, (**c**) Finite element model of a plate with mechanical components

(i.e., transient) excitation and how to use the SRS for transient analysis of MDOF systems. Specifically, we explain:

1. the MDOF equations of motion of linear structural dynamic systems;
2. mode shapes, modal coordinates, and the transformation of the equations of motion into modal space;
3. how to estimate the peak response of MDOF systems subjected to shock loading with the SRS.

8.1 Introduction to MDOF Models

The equations of motion for multi-degree-of-freedom systems are very similar to the equation of motion of the single degree-of-freedom oscillator introduced in Chap. 3. The fundamental difference is that instead of the mass, m, spring stiffness, k, and viscous damping c, terms being scalar constants, they are matrices.

We will not discuss the derivation of the matrix equations of motion or how to assemble the mass, stiffness and damping matrices, and the applied forcing, as this can be found in all structural dynamics textbooks. For simple systems one can do this with Newton's laws or with Lagrange's equations. For more complicated structures, finite element methods are best.

8.1.1 *Equations of Motion in Absolute Coordinates*

The matrix equation of motion of an undamped linear, multi-degree-of-freedom system subject to base excitation is:

$$\begin{bmatrix} \mathbf{0} & \mathbf{0} \\ \mathbf{0} & \mathbf{M} \end{bmatrix} \begin{Bmatrix} \ddot{\mathbf{w}} \\ \ddot{\mathbf{x}} \end{Bmatrix} + \begin{bmatrix} \mathbf{K_{ww}} & \mathbf{K_{wx}} \\ \mathbf{K_{xw}} & \mathbf{K} \end{bmatrix} \begin{Bmatrix} \mathbf{w} \\ \mathbf{x} \end{Bmatrix} = \begin{Bmatrix} \mathbf{0} \\ \mathbf{0} \end{Bmatrix} \tag{8.1}$$

where the bold font signifies a vector or matrix quantity, and specifically, $\mathbf{x}(t), \dot{\mathbf{x}}(t) \in \mathbb{R}^{m\times 1}$. The initial conditions are: $\mathbf{x}(t_0) = \mathbf{x}_0$ and $\dot{\mathbf{x}}(t_0) = \dot{\mathbf{x}}_0$. The excitation is the prescribed motion, which is fully characterized by $\mathbf{w}(t), \forall\ t \geq 0$, where $\mathbf{w}(t) \in \mathbb{R}^{l\times 1}, l \leq 6$.

8.1.2 *Equations of Motion in Relative Coordinates*

Equation 8.1 can be written in terms of relative coordinates referenced to the degrees of freedom at the base whose motion is prescribed, by defining:

$$\mathbf{x}(t) = \mathbf{z}(t) + \mathbf{L}\mathbf{w}(t) \tag{8.2}$$

where $\mathbf{L}$ is the rigid body transformation matrix. Substituting Eq. 8.2 into Eq. 8.1 gives:

$$\mathbf{M}\ddot{\mathbf{z}}(t) + \mathbf{K}\mathbf{z} = -\mathbf{M}\mathbf{L}\ddot{\mathbf{w}}(t) - (\mathbf{K_{xw}} + \mathbf{K}\mathbf{L})\,\mathbf{w}(t) \tag{8.3}$$

Base motion will not induce any static forces. For this to be true,

$$(\mathbf{K_{xw}} + \mathbf{K}\mathbf{L}) = \mathbf{0} \tag{8.4}$$

which means that the rigid body transformation matrix is:

$$\mathbf{L} = -\mathbf{K}^{-1}\mathbf{K_{xw}} \tag{8.5}$$

With the static forces due to base motion eliminated Eq. 8.1 becomes:

$$\mathbf{M}\ddot{\mathbf{z}}(t) + \mathbf{K}\mathbf{z}(t) = -\mathbf{M}\mathbf{L}\ddot{\mathbf{w}}(t) \tag{8.6}$$

This is the fundamental matrix equation of motion for MDOF shock problems. The structure is excited by enforced accelerations at its base. These accelerations can be in any or all six axes. Since it is a linear model, the total response is the sum of the responses to individual excitation, so without loss of generality, we assume that the input is a scalar, $\ddot{\mathbf{w}}(t) = \ddot{w}(t)$.

The equation of motion in Eq. 8.6 looks very much like Eq. 3.19 in Chap. 3 without the damping term. This is a result of defining coordinates relative to the moving base. This may not seem like the best choice for structural analysis where internal stresses are functions of the relative displacement of adjacent components. However, if the quantity of interest is a function of the relative motion between structural degrees of freedom, this will be captured in the output equation, so nothing is lost and much is gained by the transformation of the equations of motion to base relative coordinates. This is discussed in more depth in Sect. 8.1.5.

8.1.3 Equations of Motion in Modal Coordinates

For the SDOF case, we divided by the mass to get Eq. 3.11. We could do the same thing with Eq. 8.6, but there is something more effective that we can do for MDOF systems. We can transform the equations of motion so that they decouple into parallel SDOF systems.

Every system represented by Eq. 8.6 can be written in terms of generalized coordinates, $\mathbf{q}(t) \in \mathbb{R}^{N\times 1}$ through a similarity transformation. Let

$$\mathbf{z}(t) = \Phi\mathbf{q}(t) \tag{8.7}$$

where $\Phi \in \mathbb{R}^{m\times N}$, $N \leq m$. Substituting Eq. 8.7 into Eq. 8.6, and premultiplying by Φ^{T}, gives

$$\hat{\mathbf{M}}\ddot{\mathbf{q}}(t) + \hat{\mathbf{K}}\mathbf{q} = -\Phi^{\mathrm{T}}\mathbf{M}\mathbf{L}\ddot{\mathbf{w}}(t) \tag{8.8}$$

where $\hat{\mathbf{M}} = \Phi^{\mathrm{T}}\mathbf{M}\Phi$, $\hat{\mathbf{K}} = \Phi^{\mathrm{T}}\mathbf{K}\Phi$, and $\hat{\mathbf{M}},\ \hat{\mathbf{K}} \in \mathbb{R}^{N\times N}$.

The columns of matrix Φ are called basis vectors. They are linearly independent so Φ has full column rank. Because Φ is an $N \times m$ matrix, where $N \leq m$, the matrices in Eq. 8.8 are usually much smaller than the matrices in Eq. 8.6. An appropriately chosen transformation reduces the number of differential equations of motion but represents the dominant dynamic behavior.

There is a particular transformation matrix, Φ, that diagonalizes the matrices, $\hat{\mathbf{M}}$, $\hat{\mathbf{K}}$, and decouples the equations of motion so that structure can be represented as N parallel SDOF systems.

In Chap. 3 the circular natural frequency was identified as one of the fundamental parameters of vibrating SDOF systems, along with the damping ratio. This is also the case for MDOF systems except that an n-dimensional MDOF system has n natural frequencies. Associated with each natural frequency is a unique deformation of the structure called a normal mode or mode shape.

The natural frequencies and their corresponding mode shapes are found by solving an algebraic eigenvalue problem. Consider the homogeneous form of Eq. 8.6

$$\mathbf{M}\ddot{\mathbf{z}}(t) + \mathbf{K}\mathbf{z}(t) = \mathbf{0} \tag{8.9}$$

The free vibration response has the form

$$\mathbf{z}(t) = \mathbf{Z}\cos(\Omega t - \alpha). \tag{8.10}$$

Substituting this expression into Eq. 8.9 yields the algebraic eigenvalue problem

$$\left(\mathbf{K} - \mathbf{M}\Omega^2\right)\mathbf{Z} = \mathbf{0}. \tag{8.11}$$

Equation 8.11 has a non-trivial solution if and only if

$$\det\left(\mathbf{K} - \mathbf{M}\Omega^2\right) = 0. \tag{8.12}$$

The roots of Eq. 8.12 are the eigenvalues, which are the squares of the circular natural frequencies. Associated with each eigenvalue, Ω_i^2, is an eigenvector, $\mathbf{Z}_i$. The eigenvectors are unique only to within a multiplicative constant, so they can be normalized, i.e., scaled. The eigenvectors can be scaled such that $\Phi^{\mathrm{T}}\mathbf{M}\Phi = \mathbf{I}$, and $\Phi^{\mathrm{T}}\mathbf{K}\Phi = \Lambda = \mathrm{diag}\left(\Omega_i^2\right)$. The normalized eigenvectors are the normal mode shapes, or just mode shapes, and the matrix of normalized eigenvectors is called the mode shape matrix. The mode shapes are linearly independent and orthogonal basis functions. The matrix of eigenvectors is the transformation matrix that decouples the equations of motion.

A more complete explanation of the eigenvalue problem and its role in structural dynamics can be found in any structural dynamics textbook. The important points are:

1. the N natural frequencies, mode shapes, and damping ratios (which are addressed in the next section) are the fundamental parameters of N-dimensional MDOF structural dynamic systems;

2. the mode shape matrix decouples and reduces the dimension of the matrix equation of motion while retaining the dominant dynamic characteristics.

When the transformation matrix, Φ, is the mode shape matrix, the generalized coordinates, $\mathbf{q}(t)$ are called modal coordinates and the transformed equations of motion are in modal space. The equations of motion of the undamped system in modal coordinates are

$$\mathbf{I}\ddot{\mathbf{q}}(t) + \Lambda^2\mathbf{q}(t) = -\Psi\ddot{\mathbf{w}}(t) \tag{8.13}$$

where $\mathbf{I} \in \mathbb{R}^{N\times N}$ is the $N \times N$ identity matrix, and $\Lambda^2 = \text{diag}\left(\Omega_i^2\right) \in \mathbb{R}^{N\times N}$, $k = 1: N$ is the diagonal modal stiffness matrix. On the right-hand side of Eq. 8.13

$$\Psi = \Phi^{\mathrm{T}}\mathbf{ML}, \ \ \Psi \in \mathbb{R}^{N\times 1} \tag{8.14}$$

is the vector of modal participation factors. The modal participation factors are coefficients that distribute the excitation to the modal degrees of freedom. Since all of the matrices on the left-hand side of Eq. 8.13 are diagonal, Eq. 8.13 is a system of independent, parallel SDOF oscillators. This means that all of the theory and machinery of SDOF oscillators discussed in the previous chapters can be used on MDOF systems.

8.1.4 Damping

Equation 8.6 is the matrix equation of motion for an undamped system, but all structures have some energy dissipation mechanisms. The matrix equation of a damped, linear elastic MDOF system is:

$$\mathbf{M}\ddot{\mathbf{z}}(t) + \mathbf{C}\dot{\mathbf{z}}(t) + \mathbf{K}\mathbf{z}(t) = -\mathbf{ML}\ddot{\mathbf{w}}(t) \tag{8.15}$$

where $\mathbf{C}$ is the viscous damping matrix. For the transformation by the mode shape matrix, Φ, to decouple Eq. 8.15, the physical viscous damping matrix cannot be arbitrary; it must be proportional to the mass and/or stiffness matrices, i.e., $\mathbf{C} \propto \mathbf{K}$, or $\mathbf{C} \propto \mathbf{M}$. This is called proportional damping. One particular form of proportional damping is

$$\mathbf{C} = \alpha\mathbf{M} + \beta\mathbf{K} \tag{8.16}$$

which is called Rayleigh damping, named after Lord Rayleigh. Rayleigh damping permits us to specify the damping ratios for two modes by selecting appropriate values of α and β. The damping ratios in the remaining modes are given by:

$$\zeta_k = \frac{1}{2}\left(\frac{\alpha}{\Omega_k} + \beta\Omega_k\right). \tag{8.17}$$

While mass and stiffness matrices depend on the geometry and physical properties of the structure, the mechanisms of energy dissipation are usually more complex and not directly tied to the geometry. This makes the Rayleigh damping model more restrictive than we would like. The solution is to define the damping matrix in modal space with mode specific damping ratios. The modal damping matrix, **D**, is a diagonal matrix with the on-diagonal terms:

$$D_k = 2\zeta_k \Omega_k, \ k = 1 \ldots N_d, \ N_d \leq N \tag{8.18}$$

where N_d is the number of modes of interest for which damping ratios are specified. When $N_d < N$, damping in the remaining modes can be specified using stiffness proportional damping. For this, the modal damping matrix is transformed back into physical space, and a stiffness proportional term is added:

$$\mathbf{C} = \Phi^{-\mathrm{T}} \mathbf{D} \Phi^{-1} + \beta \mathbf{K}. \tag{8.19}$$

The result is called generalized proportional damping. We do not have to explicitly invert Φ because it can be shown that

$$\Phi^{-1} = \Phi^{\mathrm{T}} \mathbf{M} \tag{8.20}$$

which is important since Φ is usually tall, not square.

This damping matrix is then transformed back into modal space and the damping ratios in this matrix are:

$$\zeta_k = \begin{cases} \text{specified } \zeta, & k = 1 \ldots N_d \\ \frac{\beta \Omega_k}{2}, & k > N_d. \end{cases} \tag{8.21}$$

For shock analysis with the SRS, constant modal damping is the most appropriate and most often, $N_d = N$. However, if the physical damping matrix is needed, such as when the equations of motion must be integrated directly, the generalized proportional damping matrix given by Eq. 8.19 should be used. An in-depth treatment can be found in most structural dynamics textbooks.

The modal equations of motion for a linear, proportionally damped system subjected to a prescribed base acceleration are:

$$\mathbf{I}\ddot{\mathbf{q}}(t) + \mathbf{D}\dot{\mathbf{q}}(t) + \Lambda^2 \mathbf{q}(t) = -\Psi \ddot{w}(t) \tag{8.22}$$

The matrices on the left-hand side are diagonal so the matrix equation of motion is that of N parallel SDOF oscillators.

In this section we only consider viscous damping, which is a form of elastic energy dissipation. Viscous damping is convenient but it is only an approximate representation of physical energy dissipation mechanisms. Most energy dissipation is inelastic, usually a function of a hysteretic phenomenon such as internal friction, fluid resistance, elastoplastic stiffness, or backlash in mechanisms. All hysteretic mechanisms are non-linear and therefore problem specific. All of these mechanisms remove energy from the system so we can define equivalent viscous damping that removes the same amount of energy as the inelastic mechanism. A detailed explanation of equivalent viscous damping can be found in most structural dynamics textbooks.

8.1.5 The Output Equation

Just as with the SDOF case, the equations of motion in Eq. 8.22 are only part of the problem statement. We need an output equation to relate quantities of interest to the states in the equations of motion. While the modal states are independent, the modal responses get combined in the output equation to give the physical output quantities of interest. Generally, the quantities of interest are the same physical quantities introduced in Chap. 3: pseudo-velocity, relative velocity, and absolute acceleration. The main difference is that we can recover the quantities at one or more points on the structure.

The output equation combines the modal contributions at each instance of time, so all temporal effects such as the relative phase of the modal responses are included. The general form of a linear output equation is:

$$\mathbf{y}(t) = \left[\mathbf{C_d} \; \mathbf{C_v} \right] \begin{Bmatrix} \mathbf{q}(t) \\ \dot{\mathbf{q}}(t) \end{Bmatrix} + \mathbf{C_r}\ddot{\mathbf{w}}(t). \tag{8.23}$$

The system response vector $\mathbf{y}(t)$ is the weighted sum of the modal time histories, where the weighting coefficients are the elements of output matrices, $\mathbf{C_d}$, $\mathbf{C_v}$, etc.

The output equations for relative displacement, pseudo-velocity, relative velocity, and absolute acceleration are

$$\begin{aligned} \mathbf{y}_{RD}(t) &= \mathbf{B_D}\Phi\mathbf{q}(t) \\ \mathbf{y}_{PV}(t) &= \mathbf{B_{PV}}\Lambda\Phi\mathbf{q}(t) \\ \mathbf{y}_{RV}(t) &= \mathbf{B_{RV}}\Phi\dot{\mathbf{q}}(t) \\ \mathbf{y}_{A}(t) &= \mathbf{B_A}\left[\Phi\ddot{\mathbf{q}}(t) \; \mathbf{L}\ddot{\mathbf{w}}(t) \right]. \end{aligned} \tag{8.24}$$

where $\Lambda = \text{diag}(\Omega_k) \in \mathbb{R}^{n\times n}$, $k = 1 \ldots n$, and matrices $\mathbf{B_{RD}}$, $\mathbf{B_{PV}}$, $\mathbf{B_{RV}}$, and $\mathbf{B_A}$ are selector matrices with n columns and as many rows as there are outputs. For example, if we are interested in the relative displacement at p points on a structure, then $\mathbf{y}_{RD}(t) \in \mathbb{R}^{p\times 1}$ and $\mathbf{B_D}$ will have p rows, and N columns, $\mathbf{B_D} \in \mathbb{R}^{p\times N}$. Comparing Eq. 8.24 with Eq. 8.23 $\mathbf{C_d} = \mathbf{B_D}\Phi$ for relative displacement, and so on. The absolute acceleration may be approximated using the relationship between relative displacement and absolute acceleration of an undamped system:

$$\mathbf{y}_A(t) \approx -\mathbf{B_D}\Phi\Lambda^2\mathbf{q}(t) \tag{8.25}$$

There is another, more subtle, difference when referring to relative quantities of interest, such as relative velocity, as we have defined them here. The term *relative* means with respect to the moving base. For structural analysis, where internal stresses are functions of the relative displacement of adjacent components, we would want to define relative quantities between degrees of freedom. We refer to these quantities as *proximate* or *q-relative* to distinguish them from the quantities relative to the moving base. These concepts are illustrated with two examples at the end of this chapter.

Output equations for other quantities of interest, such as proximate (q-relative) displacement, can be formulated as well, and the equations need not be linear. For example, strain energy is a quantity of interest for which the output equation is non-linear. Energy response spectra are discussed in Chap. 12.

8.2 MDOF Shock Response Spectra

For complex, (i.e., MDOF) systems, the shock response spectrum is used to understand the peak response of the structure to transient excitation and through that, the damage potential of the excitation. Usually we use a finite element model of the structure to compute the transient response but this is often computationally intense. Instead, to quickly estimate the structure's responses to a variety of shock loading cases the shock response spectrum can be used. This approach does not require explicitly computing the responses for each loading case through a system model. The approximate response of an MDOF system with the SRS is very fast because the computational burden is very low. We only need the SRS of the excitation and the modal properties of the system which come from solving an eigenvalue problem. Eigenvalue analysis is very computationally economical.

The analysis of a shock on an MDOF system with an SRS is very similar to the analysis of shock on an SDOF structure with the SRS. There are five differences for the MDOF case:

1. The SRS is in modal space, where each SDOF oscillator represents a potential mode of the structure;
2. Shock response spectra with various damping ratios are calculated because damping varies with modal frequency;
3. For multi-axis loading, SRS are computed for each input direction;
4. The SRS responses are scaled by structure specific weighting factors;
5. The scaled SRS responses at selected natural frequencies are combined to estimate the peak value of the physical quantity of interest.

8.2.1 Relationship to the SDOF Equation of Motion

Recall from Chap. 4 that the shock response spectrum consists of extremal responses of a series of SDOF oscillators. Each row in Eq. 8.22 is the equation of motion of a single degree-of-freedom oscillator. This means that each modal equation of motion can be viewed as one of the SDOF oscillators in the SRS. Unlike the SDOF case, the base excitation applied to each SDOF system is scaled by a modal participation factor, Ψ_k defined in Eq. 8.14. The equation of motion of a single degree-of-freedom oscillator introduced in Chap. 3 is:

$$\ddot{z}(t) + 2\zeta\omega_n\dot{z}(t) + \omega_n^2 z(t) = -\ddot{w}(t). \tag{8.26}$$

The kth row of Eq. 8.22 is:

$$\ddot{q}_k(t) + 2\zeta_k\Omega_k\dot{q}_k(t) + \Omega_k^2 q(t) = -\Psi_k\ddot{w}(t). \tag{8.27}$$

The left-hand sides of the equations are the same if the parameters are the same, i.e., $\zeta = \zeta_k$ and $\omega_n = \Omega_k$. The difference between these equations is on the right-hand side. In Eq. 8.27 the base acceleration is scaled by the modal participation factor Ψ_k. Because both equations are equations of motion of a linear system, the states of Eq. 8.27, $(q_k(t),\ \dot{q}_k(t))$ are related to the states of Eq. 8.26, $(z(t),\ \dot{z}(t))$ by Ψ_k:

$$\begin{aligned} q_k(t) &= \Psi_k z(t) \\ \dot{q}_k(t) &= \Psi_k\dot{z}(t). \end{aligned} \tag{8.28}$$

This equation states that the modal response is just the response of a single degree-of-freedom oscillator scaled by the modal participation factor. This is important because it permits us to use the SRS to estimate the response of an MDOF system simply by selecting and scaling the SRS values at specific natural frequencies.

8.2.2 Damping in MDOF Systems

In Sect. 8.1.4 we showed that the modal damping ratio, ζ_k, usually varies with modal frequency so the SRS in the MDOF case should be parameterized by the modal damping ratio. When we use the SRS to understand the transient excitation the choice of damping ratio is arbitrary and typically 3–5% is used (see Chap. 4), but when we use the SRS to assess the structural response of an MDOF system to shock loading system specific modal damping ratios should be used. This means we should compute response spectra with different values of ζ, and use the SRS based on the specific modal damping ratio associated with each natural frequency.

8.2.3 *The Role of the Output Equation*

The output equation is the most important part of the MDOF SRS calculation because it is through the output equation that we combine modal responses to estimate the peak response of the structure. We can write the output equation in terms of the SDOF oscillator states, $z(t)$ and $\dot{z}(t)$, by substituting Eq. 8.28 into Eq. 8.24. The physical response is the sum of the contributions from each vibration mode. For example, the relative displacement $\mathbf{y}_{RD}(t)$ is:

$$\mathbf{y}_{RD}(t) = \mathbf{B_D}\Phi\mathbf{q}(t) = \mathbf{B_D}\Phi\Psi\mathbf{z}(t) = \sum_{k=1}^{N} \mathbf{B_D}\Phi_{:,k}\Psi_k z_k(t) \tag{8.29}$$

where the subscript ":" indicates all rows. The term $\Phi_{:,k}\Psi_k z_k(t)$ is the contribution of the kth mode to the relative displacement vector. Equation 8.29 is a modal superposition equation.

To use Eq. 8.29, we must know the modal time histories, $z_k(t)$. If two modes are "close," perhaps less than an octave apart, then it is likely that the peak response will depend on the temporal interaction of the individual modal responses. On the other hand, if two modes are far apart in frequency, then their interaction may only minimally affect the peak response. The SRS only contains the peak values of the time histories, not when they occur, so all temporal information is lost. Returning to the relative displacement example, we could simply substitute the value of the SRS at frequency Ω_k for the SDOF state variable, $z_k(t)$. Then the output equation becomes

$$\mathbf{y}_{RD}(\text{SRS}) = \sum_{k=1}^{N} \mathbf{B_D}\Phi_{:,k}\Psi_k \text{SRS}_{RD}(k). \tag{8.30}$$

In this case, the straight substitution of the SRS, $\text{SRS}_{RD}(k)$ for the kth SDOF oscillator response time history, $z_k(t)$, means that the output, $\mathbf{y}_{RD}(\text{SRS})$, is a weighted sum of the SRS values at the natural frequencies of the structure. This is called the modal summation (MS) method. In this method, there is no guarantee that the output will be positive. There are other modal combination functions, which are discussed next. These combination methods treat the absence of temporal information in response spectra in different ways.

8.2.4 *Modal Response Combination*

A variety of modal summation methods exist to combine the independent modal SRS values to approximate the peak response. Five of them are described here:

1. Absolute Method (ABS)

2. Square Root of the Sum of Squares Method (SRSS)
3. Naval Research Laboratory Method (NRL)
4. Close Method (CL)
5. Combined Quadratic Combination Method (CQC)

The methods described can be used with subsets of modes and can be used together. The NRL and the CL methods are part SRSS method and part ABS method. We can also pair with the CQC method with the ABS method. All of these methods, as well as others, have been built into commercial finite element software, such as ANSYS and NASTRAN. Gupta's book [4] includes an in-depth description of modal combination methods from an earthquake engineering perspective.

Let $\mathbf{G}(k)$ be the vector of SRS quantities of interest from mode k defined as:

$$\mathbf{G}(k) = \mathbf{B}\Phi_{:,k}\Psi_k \mathrm{SRS}(k, \Omega_k, \zeta_k). \tag{8.31}$$

For example, if relative displacement is the quantity of interest, then we use the relative displacement SRS and $\mathrm{SRS}(k, \Omega_k, \zeta_k) = \mathrm{SRS}_{RD}(k, \Omega_k, \zeta_k)$ as in Eq. 8.30.

For simplicity of exposition, consider just a scalar quantity of interest, i.e., the ith element $g_i(k)$ in vector $\mathbf{G}(k)$:

$$g_i(k) = B_{i,k}\Phi_{i,k}\Psi_k \mathrm{SRS}(k, \Omega_k, \zeta_k). \tag{8.32}$$

8.2.4.1 Absolute Method (ABS)

With the ABS method, the absolute values of the SRS modal quantities of interest are simply summed together:

$$R_i^{\mathrm{ABS}} = \sum_{k=1}^{N} |g_i(k)|. \tag{8.33}$$

This is the most conservative method because it assumes the peak values occur at the same time and in phase. It provides an upper bound on the combined extremal response, and can be overly conservative. The ABS method has a linear algebra interpretation. Consider each modal SRS value, $g_i(k)$, as an element of a vector of length N. The ABS value is just the 1-norm (i.e., the sum of the modal SRS magnitudes) of this vector.

8.2.4.2 Square Root of the Sum of Squares Method (SRSS)

The SRSS method was first introduced in 1952 [5]. It estimates the total response as the most probable value by taking the square root of the sum of the squares of the modal SRS:

$$R_i^{\mathrm{SRSS}} = \sqrt{\sum_{k=1}^{N} g_i(k)^2}. \tag{8.34}$$

This method assumes that the modal responses are independent and the peak responses combine randomly. The SRSS method is less conservative than the ABS method and is recommended when the modes are well separated. It may underestimate the actual response and represents a lower bound on the response. The SRSS also has a geometric interpretation. The SRSS is the Euclidian norm or 2-norm of this vector of modal SRS.

8.2.4.3 Naval Research Laboratory Method (NRL)

The NRL method is a hybrid of the ABS method and the SRSS method. It uses the absolute value of the modal SRS at a dominant mode (usually the fundamental mode) and augments this with the SRSS of the other modes.

$$R_i^{\mathrm{NRL}} = |g_i(j)| + \sqrt{\sum_{k=1,k\neq j}^{N} g_i(k)^2}. \tag{8.35}$$

This method assumes that the peak response from the dominant mode occurs at the same time and in phase with the sum of the squares of the other modal responses, which combine randomly.

8.2.4.4 Close Method (CL)

The CL method is a generalization of the NRL method. Rather than identifying only one dominant mode, the CL method assumes that there is more than one.

$$R_i^{\mathrm{CL}} = \sum_{j=1}^{N_J} |g_i(j)| + \sqrt{\sum_{k=1,k\neq j}^{N} g_i(k)^2} \tag{8.36}$$

This method assumes that the peak responses from the dominant modes occur at the same time and in phase with each other like in the ABS method, and with the sum of the other modal SRS values, which combine randomly.

8.2.4.5 Complete Quadratic Combination Method (CQC)

The first four methods were conceived in the 1950s and 1960s. The CQC method [6, 7] is a relative newcomer dating only to 1980. It is another generalization of the SRSS method and has become very popular, particularly in the civil and earthquake engineering communities. Reference [6] has been cited more than 173 times since its publication.

The CQC method has been shown to be better than the other methods, like SRSS and NRL, which were shown to give poor results in certain cases when the system modes are closely spaced. The method was developed originally with quantities of interest based on relative displacement; however, it has been used with other response spectra, such as absolute acceleration spectra. The CQC method uses covariance weighting factors for pairs of modes. The expression for CQC estimate of the shock response is:

$$R_i^{\mathrm{CQC}} = \sqrt{\sum_{j=1}^{N} \sum_{k=1}^{N} g_i(j) \rho_{jk} g_i(k)} \tag{8.37}$$

where

$$\rho_{jk} = \frac{8\sqrt{\zeta_j \zeta_k} \left(\zeta_j + r\zeta_k\right) r^{3/2}}{\left(1 - r^2\right)^2 + 4\zeta_j \zeta_k r \left(1 + r^2\right) + 4r^2 \left(\zeta_j^2 + \zeta_k^2\right)} \tag{8.38}$$

and

$$r = \frac{\Omega_k}{\Omega_j}. \tag{8.39}$$

If the constant modal damping is assumed, Eq. 8.38 reduces to

$$\rho_{jk} = \frac{8\zeta^2 (1 + r) r^{3/2}}{\left(1 - r^2\right)^2 + 4\zeta^2 r (1 + r)^2} \tag{8.40}$$

The cross-frequency weighting terms capture, in a probabilistic sense, the interactions of the modes. When the modes are well separated, the off-diagonal terms, ρ_{jk}, approach zero. If $\rho_{jk} = 0$ for $j \neq k$, the CQC method reduces to the SRSS method.

8.2.5 *Multi-Axis Shocks*

Structures often are exposed to multi-axis shocks. During an earthquake, the ground moves in all three axes. A pyroshock on a launch vehicle will be distributed by the load path into all three axes at the shock-sensitive part. The shocks in each axis may be correlated or uncorrelated. If the shocks are correlated, then there is a deterministic relationship at each time between the different axis shocks. If the shocks are uncorrelated, then there is no relationship between them.

One of the properties of linear systems is superposition, which states that the total response is the sum of responses to individual loadings. However, superposition does not apply to the modal combination methods in MDOF SRS analyses. Each modal combination method treats simultaneous excitation in multiple directions differently. To understand how each modal combination method handles multi-axis loading, let the modal SRS at output point i in direction r be

$$g_{ir}(k) = B_{ik}\Phi_{i,k}\Psi_{i,r}\mathrm{SRS}_r(k, \Omega_k, \zeta_k). \tag{8.41}$$

This is a directional generalization of Eq. 8.31.

8.2.5.1 Absolute Method (ABS)

The ABS method for excitation in multiple directions is a simple extension of the single axis equation:

$$R_i^{\mathrm{ABS}} = \sum_{r=1}^{l}\sum_{k=1}^{N} |g_{ir}(k)|. \tag{8.42}$$

8.2.5.2 Square Root of the Sum of Squares Method (SRSS)

In the multi-axis SRSS method, the modal responses are first combined over the excitation directions and then over the modes with the RSS

$$R_i^{\mathrm{SRSS}} = \sqrt{\sum_{k=1}^{N}(B_{ik}\Phi_{i,k}S_k)^2} \tag{8.43}$$

where

$$S_k = \sqrt{\sum_{r=1}^{l}\Psi_{i,r}^2\mathrm{SRS}_r(k, \Omega_k, \zeta_k)^2}. \tag{8.44}$$

8.2.5.3 Naval Research Laboratory Method (NRL)

For the multi-axis excitation case, the NRL method peak response is computed as

$$R_i^{\mathrm{NRL}} = |B_{ij}\Phi_{i,j}S_j| + \sqrt{\sum_{k=1,k\neq j}^{N} (B_{ik}\Phi_{i,k}S_k)^2} \tag{8.45}$$

where S_k is defined in Eq. 8.43.

8.2.5.4 Close Method (CL)

The CL method for multi-axis excitation follows the same idea as the SRSS method. The equation for the CL method applied to multi-axis loading is a generalization of the NRL method.

$$R_{\mathrm{CL}} = \sum_{j=1}^{N_J} |B_{ij}\Phi_{i,j}S_j| + \sqrt{\sum_{k=1,k\neq j}^{N} (B_{ik}\Phi_{i,k}S_k)^2}. \tag{8.46}$$

8.2.5.5 Complete Quadratic Combination Method

The CQC method cannot be used directly for multi-axis excitation because the frequency ratio does not exist so the cross-modal coefficients cannot be computed. However, the CQC result in each direction can be combined using the ABS or SRSS method.

8.2.6 Examples

In this section we illustrate how shock response spectra approximate peak response to a base acceleration with two examples. The first is a 2-degree-of-freedom spring-mass-damper model subject to a haversine shock input. The second is a four-degree-of-freedom spring-mass-damper model, representing a building subjected to a random transient base acceleration.

8.2.6.1 Example 1: Two-Degree-of-Freedom System

Consider the two-degree-of-freedom system shown in Fig. 8.2. The physical parameters of the system are shown in Table 8.1. Modal damping ratios were defined and the damping matrix in physical space was created with Eq. 8.19. The modal effective masses (MEM) are the squares of the modal participation factors Ψ_k and they sum to the total mass of the system. Modal effective masses are useful for determining how many modal degrees of freedom should be retained in a reduced order model. In this example, since there are only two modes, we use both in the modal representation.

The rigid body vector is:

$$\mathbf{L} = \begin{Bmatrix} 1 \\ 1 \end{Bmatrix} \tag{8.47}$$

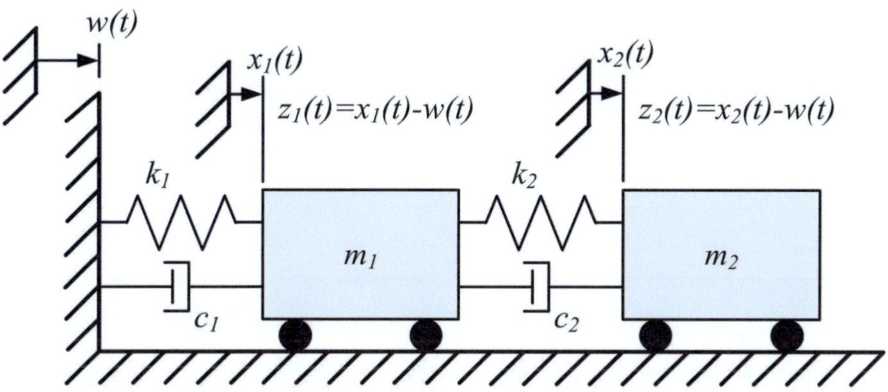

Fig. 8.2 Two-degree-of-freedom system

Table 8.1 2 DOF example parameters

Parameter	Value
m_1	1 kg
m_2	2 kg
k_1	3,00,000 N/m
k_2	1,00,000 N/m
Ω_1	30.3 Hz
Ω_2	102.4 Hz
ζ_1	0.035
ζ_2	0.02
Φ_1	$[-0.1908, -0.6941]^T$
Φ_2	$[-0.9816, 0.1349]^T$
Ψ	$[-1.5791 - 0.7118]^T$
MEM	$[2.49340.5066]^T$ kg

We assume that the system is at rest initially and is subjected to a prescribed based acceleration. The equations of motion in terms of base relative coordinates are:

$$\begin{bmatrix} 1 & 0 \\ 0 & 2 \end{bmatrix} \begin{Bmatrix} \ddot{z}_1 \\ \ddot{z}_2 \end{Bmatrix} + \begin{bmatrix} 25.2777 & -3.2851 \\ -3.2851 & 27.5598 \end{bmatrix} \begin{Bmatrix} \dot{z}_1 \\ \dot{z}_2 \end{Bmatrix} + \begin{bmatrix} 4 & -1 \\ -1 & 1 \end{bmatrix} \times 10^5 \begin{Bmatrix} z_1 \\ z_2 \end{Bmatrix} = \begin{Bmatrix} -1 \\ -2 \end{Bmatrix} w(t) \tag{8.48}$$

The natural frequencies and mode shapes are given in Table 8.1. The output quantities of interest are the absolute accelerations of the two masses and the force in the spring between mass 1 and mass 2, denoted as spring 2. The output equations are:

$$\begin{Bmatrix} y_{A_{m1}} \\ y_{A_{m2}} \\ y_F \end{Bmatrix} = \begin{bmatrix} -400000 & 100000 \\ 50000 & -50000 \\ -100000 & 100000 \end{bmatrix} \begin{Bmatrix} z_1 \\ z_2 \end{Bmatrix} + \begin{bmatrix} -25.2777 & 3.2851 \\ 1.6425 & -13.7799 \\ 0 & 0 \end{bmatrix} \begin{Bmatrix} \dot{z}_1 \\ \dot{z}_2 \end{Bmatrix} \tag{8.49}$$

The magnitudes of the transmissibility response functions between the absolute accelerations at the two masses and the base acceleration are shown in Fig. 8.3. Because Fig. 8.3 is a plot of absolute acceleration transmissibility the low-frequency magnitude must be unity. Figure 8.4 shows the magnitude of the frequency response function between the force in spring 2 and the acceleration applied at the base.

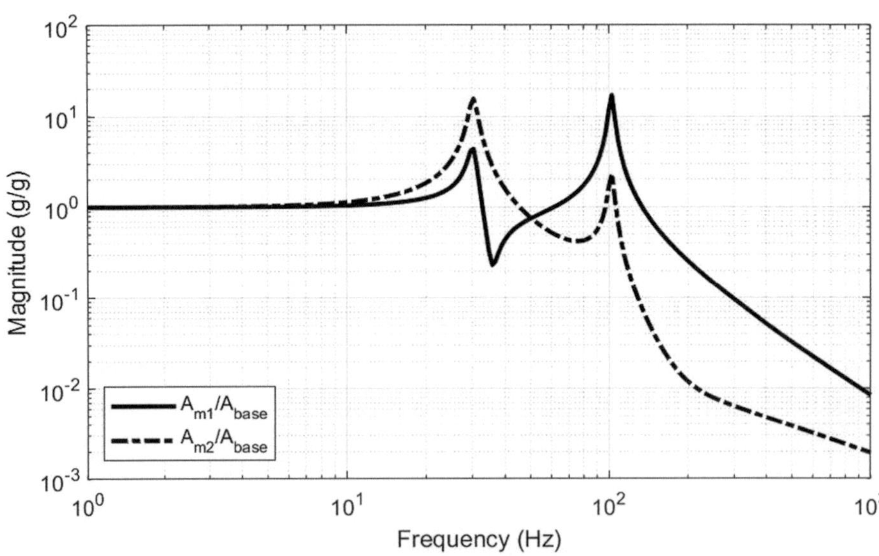

Fig. 8.3 FRF Magnitude of base acceleration to absolute acceleration of mass 1 and mass 2

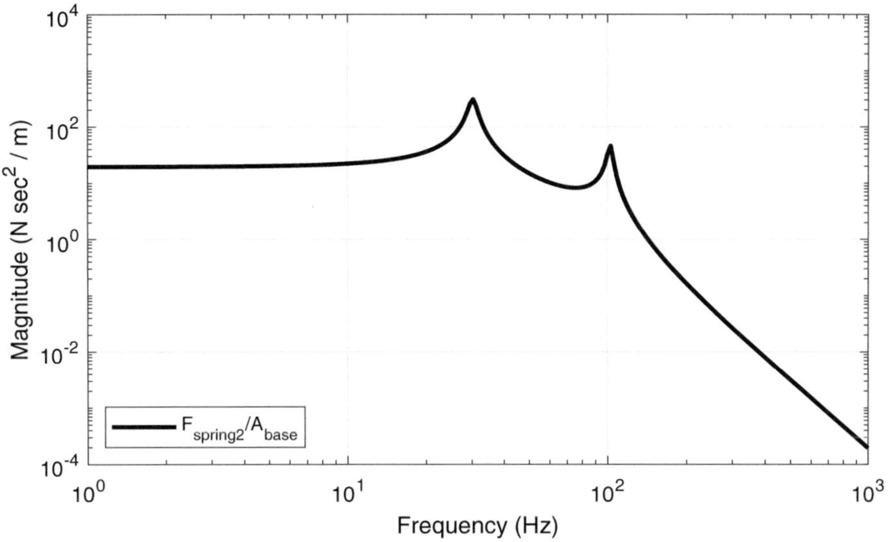

Fig. 8.4 FRF Magnitude of base acceleration to force in spring 2

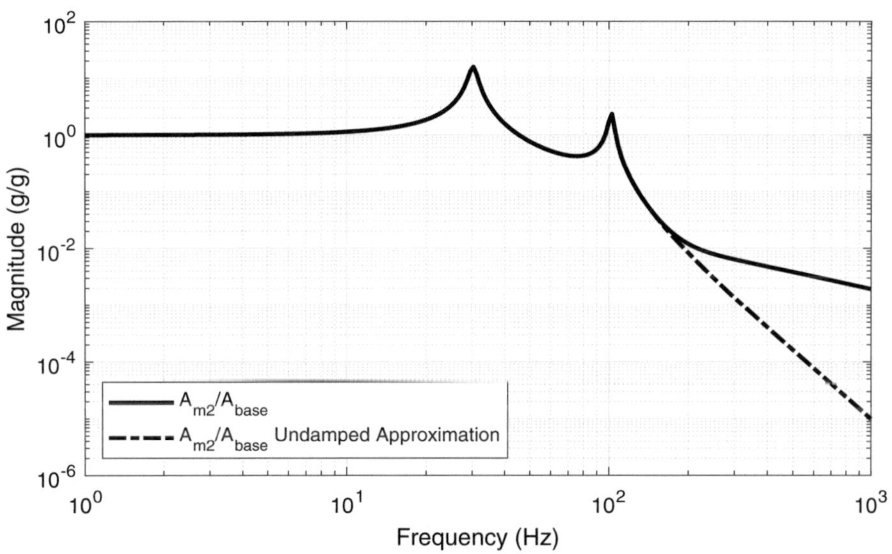

Fig. 8.5 TRF magnitude of base acceleration to absolute acceleration of mass 2—exact and undamped approximation

Figure 8.5 shows the effect of the undamped approximation of absolute acceleration in Eq. 8.25 on the transmissibility between the absolute acceleration of mass 2 and the base. In this example, the approximate expression is accurate in the frequency range of the modes but rolls off much faster at high frequencies.

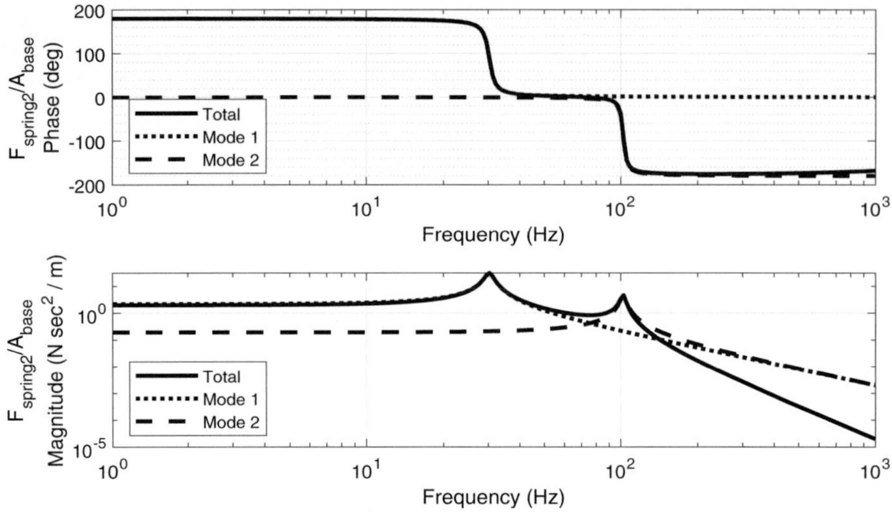

Fig. 8.6 TRF magnitude and phase of base acceleration to force in spring 2 with modal contributions

The contributions of the individual modes to the total transmissibility between the spring force in spring 2 and the base acceleration is shown in Fig. 8.6. The magnitude plot shows that at low frequency, the total response is due to the first mode and the phase angle plot shows that the spring force is out of phase with the applied acceleration. This means that a positive applied base acceleration will produce a compressive force in the spring. At high frequency, the total response magnitude is less than the magnitudes of the individual modal contributions and the modes are in phase. The spring force is small which means that the two masses are moving together, essentially as a rigid body.

The base acceleration is a 12 ms, 1000 g haversine shown in Fig. 8.7. The maxi-max absolute acceleration response spectrum and relative displacement response spectra are shown in Figs. 8.8 and 8.9, respectively.

The exact acceleration response time histories of the two masses are shown in Fig. 8.10. A modal approximation of the accelerations is shown in Fig. 8.11. The responses are slightly different. There are two reasons for the difference. The modal solution was computed with a constant damping ratio of 3% for both modes and the undamped approximation (Eq. 8.25) of the absolute acceleration.

The spring force is shown in Fig. 8.12. The first mode is the largest contributor to the total spring force. The first spring force cycle induces a compressive force in the spring. This is what we expect from the transmissibility response function, Fig. 8.6.

The peak values of the response quantities of interest are summarized in Table 8.2. The peak values were computed assuming a constant, 3% damping ratio, rather than the modal damping ratios of each mode. The ABS method overpredicts the acceleration of mass 1 and is quite close on the acceleration of mass 2. This

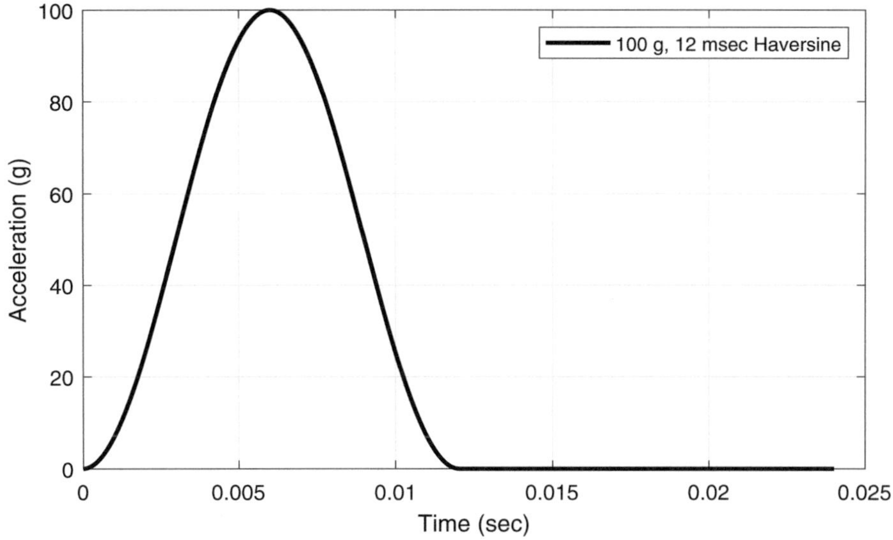

Fig. 8.7 Base acceleration haversine time history

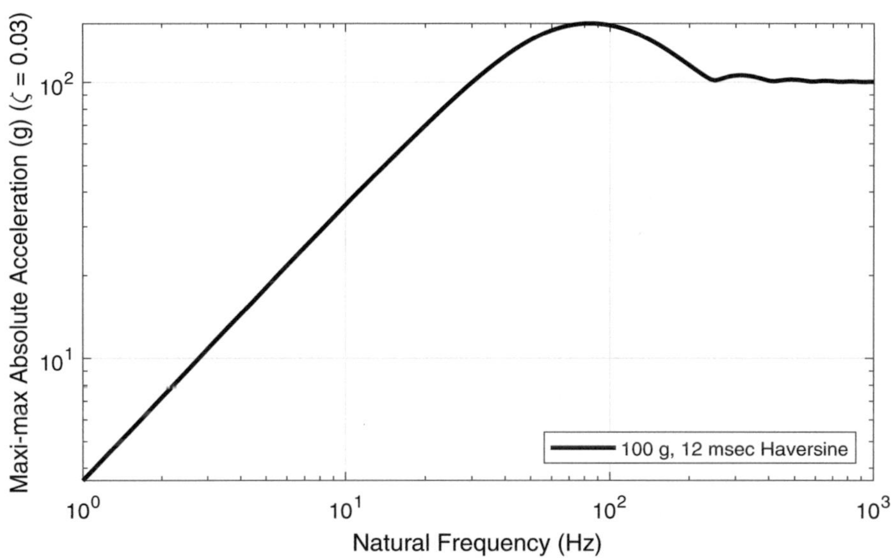

Fig. 8.8 Base acceleration haversine maxi-max absolute acceleration SRS

is expected because the ABS estimate is conservative by its nature. The SRSS and CQC methods slightly underpredict the peak accelerations of both masses. The SRSS and the CQC results are almost the same indicating that there is little interaction between the modes.

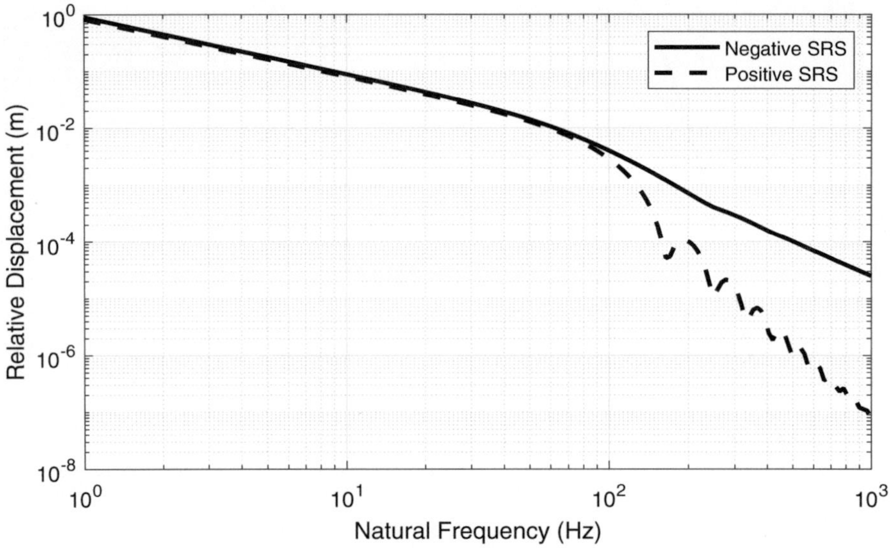

Fig. 8.9 Base acceleration haversine positive and negative relative displacement SRS

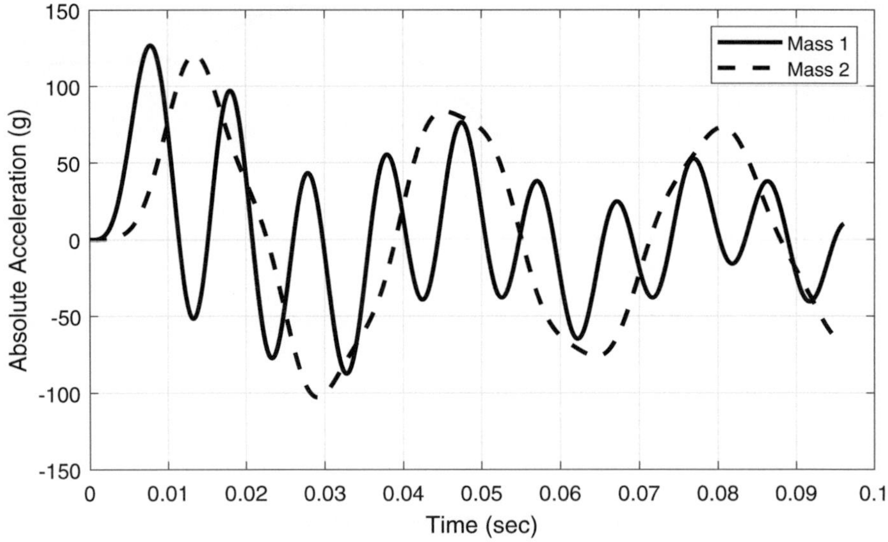

Fig. 8.10 Absolute acceleration of masses 1 and 2—exact solution

Response spectrum methods cannot estimate separate compressive and tensile internal spring forces so the last two rows show the same values except for the sign. The MS method underpredicts the spring force. The ABS, SRSS, and CQC methods

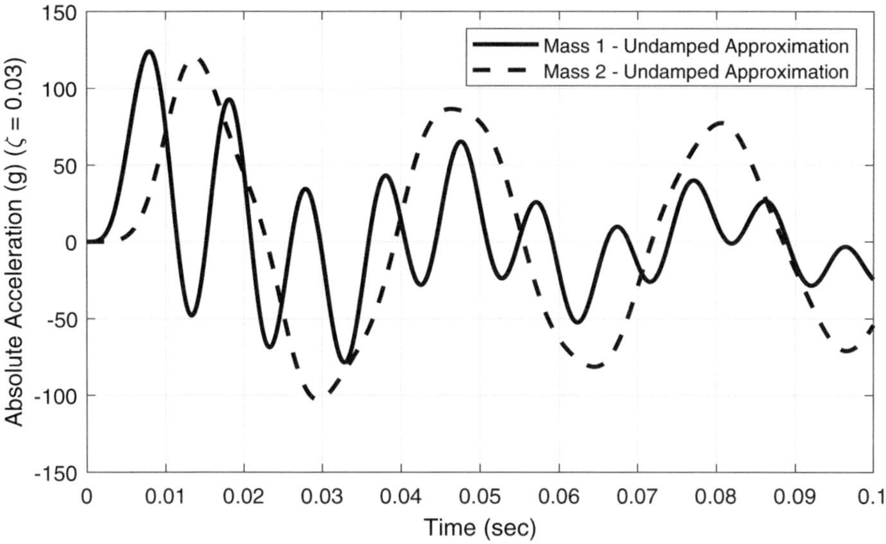

Fig. 8.11 Absolute acceleration of masses 1 and 2—approximate modal solution

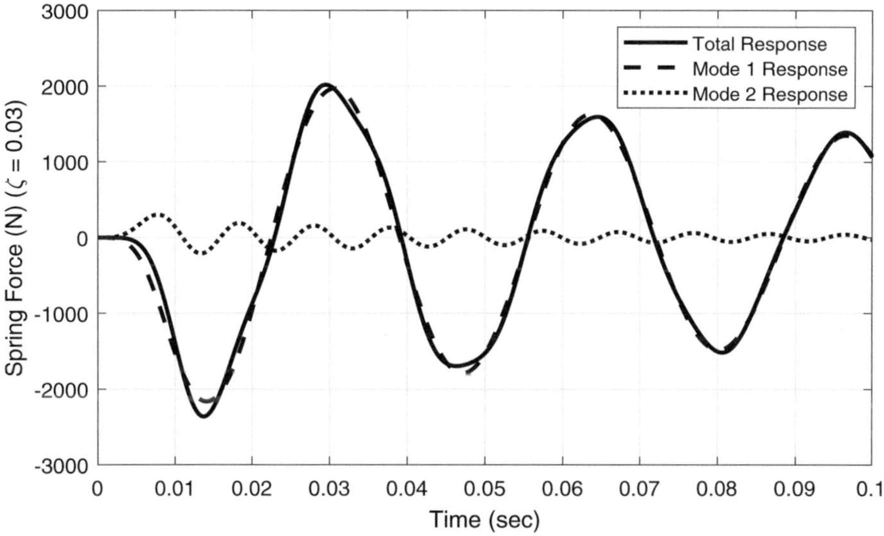

Fig. 8.12 Modal spring force response

all overpredict the tensile force. The ABS method overpredicts the compressive force while the SRSS and CQC methods underpredict the compressive force.

This example problem illustrates the variability of modal combination methods. This type of analysis is useful for preliminary design and design studies but should not be used for detailed analyses.

Table 8.2 Peak response quantities of the 2 DOF model

Peak response quantity	Exact value	Modal approximation	MS	ABS	SRSS	CQC
Maxi-max acceleration mass 1 (g)	126.8	124.1	142.3	142.3	116.0	116.1
Maxi-max acceleration mass 2 (g)	120.6	120.3	94.8	125.6	111.3	111.2
Tensile force spring 2 (N)	2030	2056	1896	2512	2225	2224
Compressive force spring 2 (N)	−2400	−2401	−1896	−2512	−2225	−2224

8.2.6.2 Example 2: Four-Degree-of-Freedom System

Figure 8.13 shows a four-degree-of-freedom system. It represents a structure (e.g., a two-story building) with internal mechanical systems. The internal mechanical systems are represented by masses $M3$ and $M4$. As in the previous example, the system is initially at rest and the excitation is a prescribed base acceleration. The output quantities of interest are the absolute accelerations of mass 1 and mass 2 and the pseudo-velocities of mass 3 and mass 4. The model parameters are shown in Table 8.3.

In this example, the base acceleration is the 400 ms, random transient shown in Fig. 8.14. The maxi-max absolute acceleration SRS is shown in Fig. 8.15.

The transmissibility response function magnitudes for the absolute acceleration of the two primary masses (m1 and m2) are shown in Fig. 8.16. Only the first two modes are observable from these two masses. The magnitudes of the pseudo-velocity transmissibility response functions for the interior masses are shown in Fig. 8.17. All four modes are observable in these TRFs but modes 3 and 4 are so close together that their contributions are overlapping and impossible to distinguish. This is typical of very closely spaced modes.

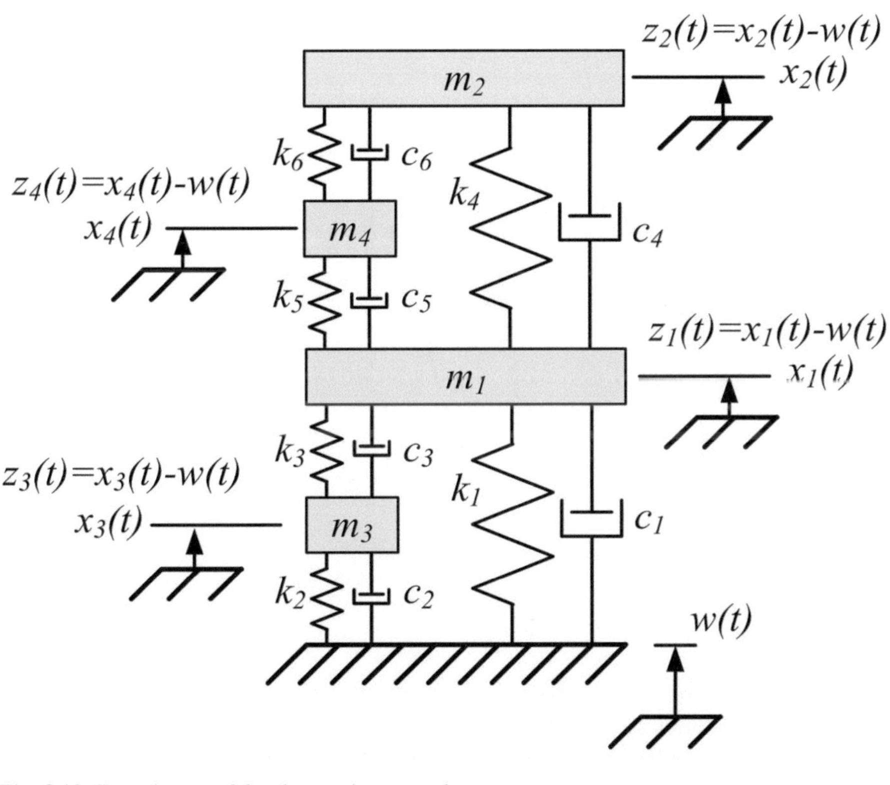

Fig. 8.13 Four-degree-of-freedom spring mass damper system

Table 8.3 4 DOF example parameters

Parameter	Value
m_1, m_2	5 kg
m_3, m_4	1 kg
k_1	5,00,000 N/m
$k_2 - k_6$	2,00,000 N/m
Ω_1	28.4 Hz
Ω_2	68.8 Hz
Ω_3	102.1 Hz
Ω_4	109.1 Hz
ζ_1	0.05
ζ_2	0.03
ζ_3	0.03
ζ_4	0.02
Φ_1	$[0.1666, 0.3903, 0.0905, 0.3026]^T$
Φ_2	$[0.3649, -0.1961, 0.3421, 0.1582]^T$
Φ_3	$[0.0441, -0.0805, -0.7543, 0.6236]^T$
Φ_4	$[-0.1928, -0.0524, 0.5330, 0.7032]^T$
Ψ	$[3.1773, 1.3438, -0.3129, 0.0302]^T$
MEM	$[10.10, 1.81, 0.09, 0.001]^T$ kg

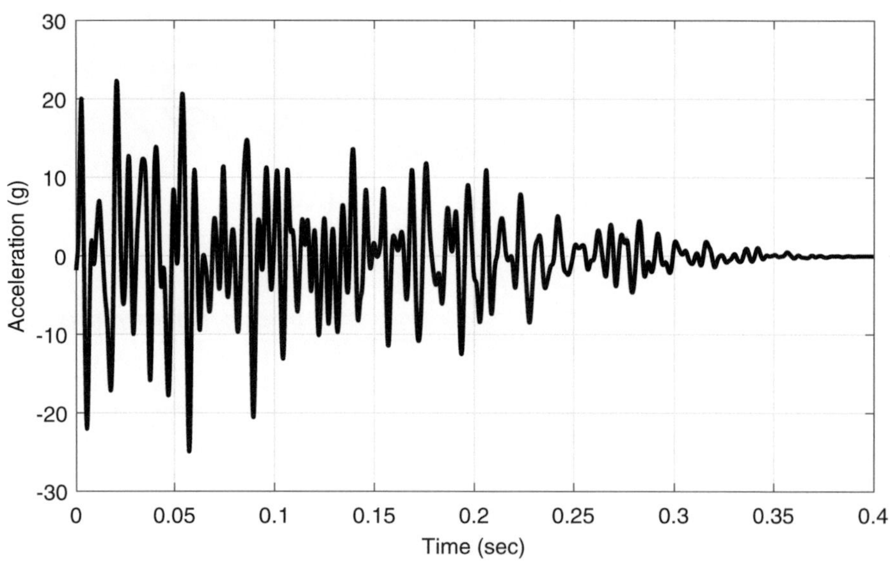

Fig. 8.14 Transient base transient acceleration input

The peak responses to the base transient acceleration estimated with the mode combination methods are listed in Table 8.4. In this example the peak value estimated with the NRL method is shown as well. The first mode was assumed to be the dominant one. This is a good assumption since the modal participation factor

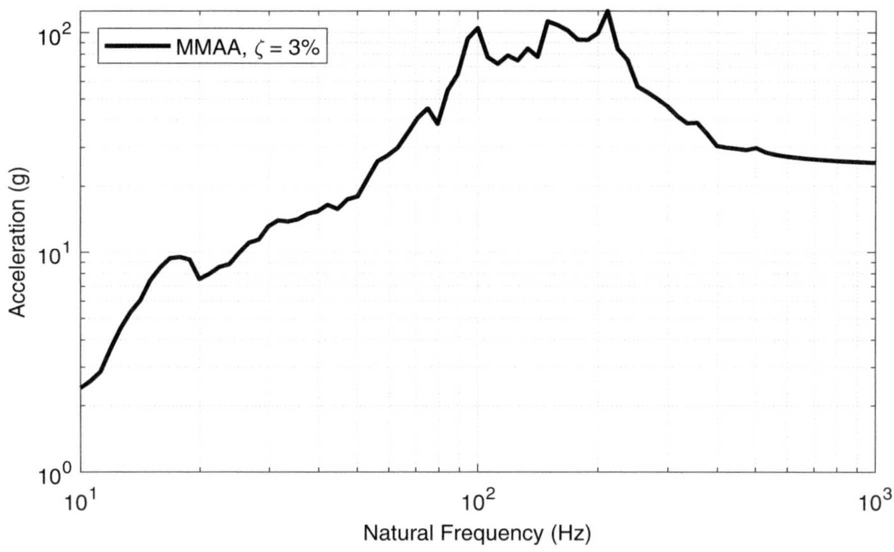

Fig. 8.15 Transient base acceleration maxi-max absolute acceleration SRS

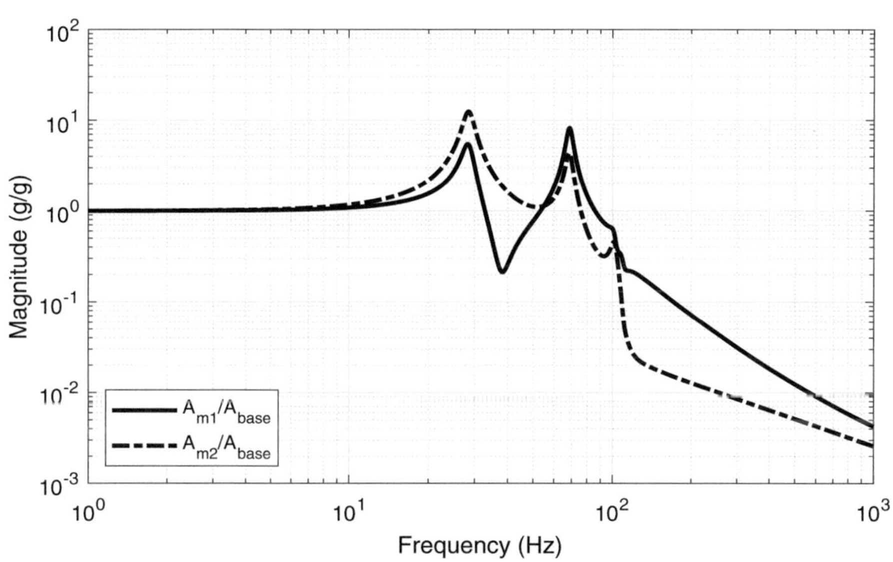

Fig. 8.16 TRF magnitude of base acceleration to absolute acceleration of mass 1 and mass 2

of this mode is the largest. As with the 2DOF example, each modal combination method gives slightly different estimates. The CQC method estimate is closest to the exact peak values of the peak absolute accelerations of mass 1 and mass 2, but the estimated peak values are not conservative. The ABS and NRL methods estimate higher peak accelerations so they are conservative. The NRL estimate is

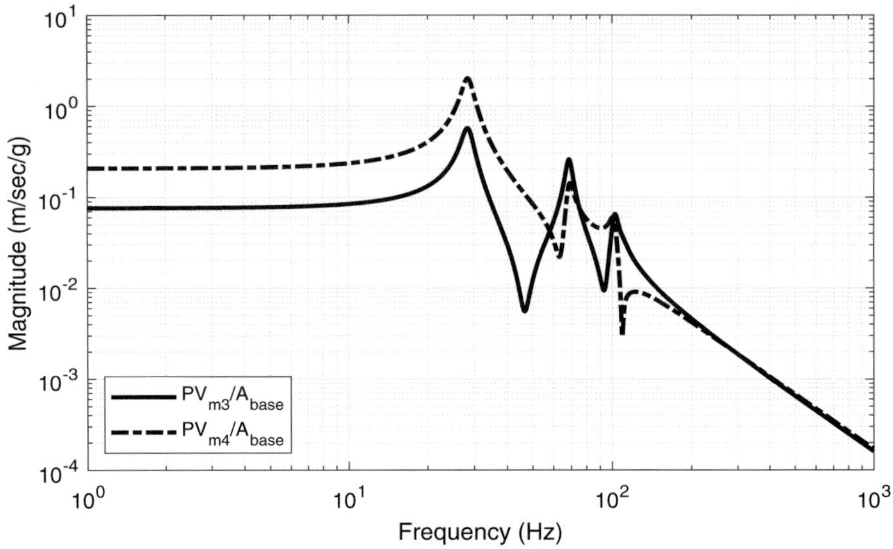

Fig. 8.17 TRF magnitude of base acceleration to pseudo-velocity of mass 3 and mass 4

Table 8.4 Peak response quantities of the 4 DOF model

Peak response quantity	Exact value	MS	ABS	SRSS	NRL	CQC
Maxi-max acceleration mass 1 (g)	19.37	22.97	26.46	18.54	24.77	19.57
Maxi-max acceleration mass 2 (g)	19.36	6.93	27.05	10.22	24.8	17.74
Maxi-max pseudo velocity mass 3 (m/s)	0.85	1.61	1.61	0.68	1.34	0.96
Maxi-max pseudo velocity mass 4 (m/s)	2.3	2.39	3.00	0.42	2.80	2.42

less conservative than the ABS estimate. This shows that the NRL method is a good choice when the total response is dominated by a single mode.

The CQC method produces the best estimate of the peak pseudo-velocities. The estimates are close to the true values and are conservative. The NRL and ABS methods both overestimate the true peak pseudo-velocities by more than the CQC method, and the SRSS method underestimates the peak pseudo-velocities.

This more complex example problem again illustrates the variability of modal combination methods. The results reinforce that this type of analysis is useful for preliminary design and design studies but it is not appropriate for detailed analyses.

8.3 Summary

In this chapter we explained the equations of motion for multi-degree-of-freedom systems. The equations of motion should be transformed into modal space, which provides insight into the response characteristics and also reduces the dimension of the matrix equations.

We showed that the response of an MDOF system to shock excitation can be approximated with the SRS. The SRS of a shock is scaled by structure specific weighting factors and these scaled values are combined to produce an approximate extremal response.

We presented five modal combination methods and showed that the estimated peak values differ from the true value with two examples. While the peak responses in an MDOF can be efficiently estimated with the SRS and modal combination, the results should be used for preliminary design and design analyses, not detailed analyses.

Problems

8.1 Show that Eq. 8.20 is true.

8.2 Compute the cross-modal coefficient matrix in the CQC method (Eq. 8.38) for the 4-degree-of-freedom example problem using the parameters in Table 8.3. How do the weighting factors for the closely spaced modes differ from the weighting factors of the other modes?

Compute the cross-modal coefficient matrix with a constant damping ratio. Comment on how much the damping ratio affects the coefficients and the amount of modal coupling.

8.3 Create a model of the 4 DOF system in Example 2. Use the base relative velocity of mass 2 as the output quantities of interest. Apply the haversine base excitation in Example 1 to the base of the 4 DOF system. Compute the exact peak responses from the time histories and compare with the estimated peak values obtained with the modal combination methods.

References

1. Craig, R. R., & Kurdila, A. J. (2006). *Fundamentals of structural dynamics* (2nd ed.). ISBN-13: 978-0471430445.
2. Thomson, W. T., & Dahleh, M. D. (1998). *Theory of vibrations with applications* (5th ed.). ISBN 0-13-651068-X.
3. Chopra, A. K. (2016). *Dynamics of structures* (5th ed.). ISBN-13: 978-0134555126.
4. Gupta, A. K. (1992). *Response spectrum method in seismic analysis and design of structures*. ISBN 978-0849386282.
5. Goodman, L. E., Rosenbluth, E., & Newmark, N. M. (1952, June). *Asiesmic design of elastic structures founded on firm ground*. Technical Report to the Office of Naval Research, Project NR-064-183.
6. Wilson, E. L., der Kiureghian, A., & Bayo, E. P. (1981). A replacement for the SRSS method in seismic analysis. *Earthquake Engineering and Structural Dynamics, 9*, 187–194.
7. der Kiureghian, A. (1980). *A response spectrum method for random vibrations*. Report No. UCB/EERC-80/15. Berkeley, CA: Earthquake Engineering Research Center, University of California.

Chapter 9
Shock Testing

Shock testing is a critical part of shock engineering. Analysis and predictions are important to the field of mechanical shock, but shock testing provides the proof that the engineering was done correctly. Shock testing is performed using numerous different methods ranging from shock machines, shaker systems, actual system drops or crashes, and even live-fire tests. This chapter discusses some of the more popular shock testing methods and the strengths and weaknesses of the different methods. Shock testing is a specialized field requiring specialized equipment. As such, it is imperative that the equipment be matched to the test requirements.

9.1 Purpose of Shock Testing

Understanding the purpose of shock testing is important to ensuring that the tests are properly conducted. While the purpose of shock testing, and environmental testing in general, may seem obvious, there are in fact two similar but distinctly different goals for testing. One approach uses environmental testing to reproduce a particular environment and to evaluate the system's performance to that environment. The other approach uses environmental testing to establish a minimum level of system robustness. The difference between these two approaches is subtle but important.

If the goal is to mimic a particular environment, then the test input needs to match the environmental loading and match it well. In addition, the test fixture needs to represent the test article's installation configuration, including the dynamic characteristics of the next level of assembly. If the goal is to establish robustness, the test inputs can be combinations or envelopes of similar environments. The test fixture can also be less representative of the next level of assembly. If the input is only an approximation or an amalgamation of different inputs, then it seems less important to match the boundary impedance to the next higher assembly.

C. Sisemore, V. Babuška, *The Science and Engineering of Mechanical Shock*,
https://doi.org/10.1007/978-3-030-12103-7_9

Typically, the goal of environmental testing, and shock testing in particular is to establish robustness and not necessarily to mimic a specific environment, although this is frequently misunderstood. The reason for this confusion is that most test specifications are combinations of environments or worst-case environments that are somewhat removed from the nominal or expected environment. As a result, most test specifications only bear a passing resemblance to any particular exact environment from which they were derived. If the inputs do not match a specific environment, expending significant effort to make the test fixture exactly match the next level of assembly may not have the desired effect. As such, the goal inherently becomes to establish a level of robustness.

9.2 Drop Tables

There are two basic styles of drop tables in common use: free fall drop tables and accelerated fall drop tables. Free fall tables operate by lifting a table or carriage to a predetermined height above a reaction mass and dropping the table, allowing it to fall under gravitational acceleration. Accelerated fall tables typically use some form of spring-assisted drop mechanism such that the carriage is accelerated toward the reaction mass faster than the acceleration of gravity alone. There are several advantages of an accelerated fall drop table: much higher impact velocities can be obtained, the machine height is proportionately shorter, and the spring-assisted fall helps to overcome any friction in the guide mechanisms. The free fall drop table also has several advantages: the operating mechanism is simpler and lower velocity levels can be more easily achieved. The higher impact velocities of the accelerated fall tables in turn make it much more difficult to perform low acceleration shocks. For this reason, companies that are largely focused on consumer products may choose to use free fall drop tables since the average drop height is approximately equal to the height of a person's waist.

Most drop tables are configured to provide a haversine shock or a half-sine shock pulse. Drop tables can also be configured to generate the triangular saw-tooth style shocks. Figure 9.1 shows a drawing of the basic components of a drop table shock test machine. The machine has a large, heavy reaction mass at the base. The reaction mass is typically mounted on some form of hydraulic shocks or snubbers to isolate and protect the building floor from repeated shock exposure. On top of the reaction mass, some form of pulse shaper material is used to control the shock pulse width and shape. In many cases, the pulse shaper is comprised of various layers of stiff felt although other materials such as paper, plastic, or composites can be used. The carriage rides on guide rods to ensure that it impacts the pulse shaper squarely and carries the article under test. The carriage itself is typically fairly robust so that its fundamental resonant mode is beyond the shock excitation frequency. The carriage also contains fast-acting braking mechanisms to hold the carriage in place prior to the shock and release the carriage cleanly. The test fixture is usually a customer-designed component and interfaces the test article with the

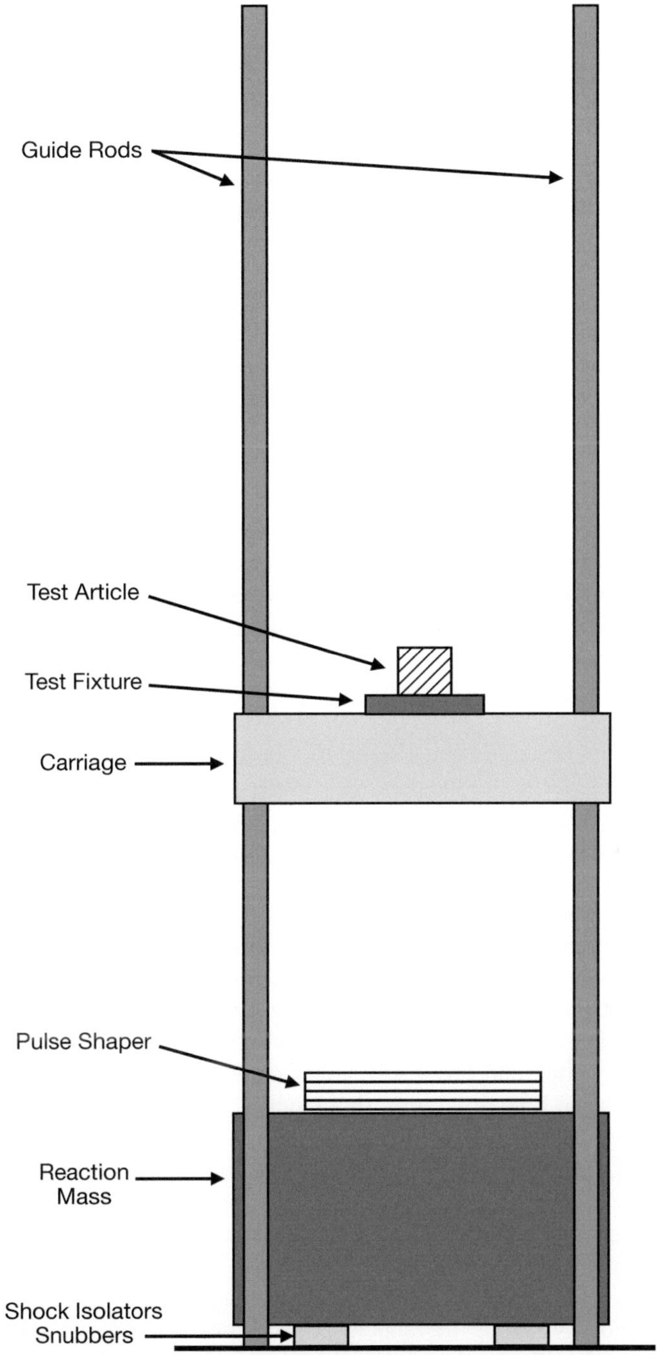

Fig. 9.1 Schematic of a simple drop table shock test machine

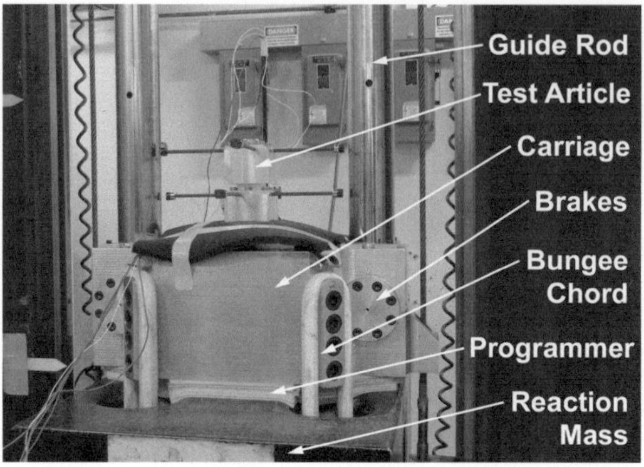

Fig. 9.2 Photograph of an accelerated fall drop table carriage

carriage. The test article is then installed on the test fixture and rides the carriage during the shock event. Sometimes, springs or bungee chords are used to accelerate the carriage downward into the reaction mass and at other times the carriage is allowed to free fall depending on the test levels required.

Figure 9.2 shows a photograph of an accelerated fall drop table shock machine focused on the carriage area with the individual components that are labeled. The relatively heavy carriage in this example is accelerated downward by four bungee chords, one on each corner, onto a reaction mass (only partially visible in this photo) that is topped with several layers of felt programmer that act as a pulse shaper. The unit under test is attached to the carriage's upper side between the two guide rods. In this example, failure of the unit under test was expected so foam sheets were used to cover the carriage top to protect the machine and instrumentation when test parts were ejected as a result of the shock.

Drop shock machines also come in several sizes depending on the shock parameters sought. A drop table with a smaller carriage can typically be accelerated faster resulting in higher velocity impacts. Smaller carriages also have higher natural frequencies though the pulse duration can be much smaller. Of course, smaller carriages can only be used for smaller, lighter weight components. Large drop tables can carry more weight and physically larger components but are limited to lower velocity impacts and longer duration shock pulses to prevent resonating the larger, lower-frequency carriages. Common drop table carriage sizes are 150 mm (6 in), 300 mm (12 in), and 600 mm (24 in) square although any size could be designed and built.

With the desire to test ever-increasing shock levels, an interesting innovation for the traditional drop shock machine has recently been developed. Several drop shock machines have recently been built in which the reaction mass, rather than remaining stationary during the shock, is accelerated upward as the carriage is pulled downward with the two meeting at a predetermined point. Due to the excessive weight of the reaction mass, pneumatics are used to propel it upwards, while bungee chords are used to accelerate the carriage downward. The resulting impact can generate velocity changes on the order of 55 m/s (180 ft/s) for small parts. An impressive shock test to be sure.

Drop tables produce classical shock pulses, typically a haversine shock, although they can produce the other classical shock pulses. The classical shock pulses all have a nonzero velocity change in common. Obviously, the actual test must start and end in a stationary, zero-velocity state. Rather, what is meant by the nonzero velocity change is that immediately prior to the impact the component is traveling with some known velocity and immediately after the shock, the velocity is zero. This type of shock is fundamentally different from some of the other shock machines in which the components have zero velocity immediately prior to the shock and zero velocity immediately after the applied shock. This initial velocity condition also means that the parts under test have comparable momentum. Thus, if a part is cracked or damaged prior to or during the test, the momentum of the damaged part will almost always propagate the crack or grow the damage, frequently with great rapidity. Where cracks may form in other tests, drop table tests have a tendency to convert damaged parts into projectiles.

9.2.1 Measurable Duration on a Drop Table

Up to this point, the duration of a classical shock pulse has always referred to the full theoretical pulse width. However, the full theoretical pulse width is often quite difficult to determine from experimentally collected data. For example, the classical haversine shock that has been discussed at length is defined by the well-known equation:

$$\ddot{w}(t) = \frac{A}{2}\left[1 - \cos\left(\frac{2\pi t}{T}\right)\right]. \tag{9.1}$$

The derivative of this equation represents the slope of the acceleration versus time curve and is given by:

$$\frac{d}{dt}\ddot{w}(t) = \frac{A\pi}{T}\sin\left(\frac{2\pi t}{T}\right). \tag{9.2}$$

From Eq. 9.2, it is apparent that the slope at $t = 0$ and $t = T$ is zero. With an analytically produced waveform, it is possible to determine the time at which the slope transitions from zero to a nonzero number, this is not possible with experimental data. The reason is that we do not measure zero experimentally; rather the measured signal is the instrumentation's noise floor before and after the shock event. As a result, there is no clear transition in the measured signal from the preshock data to the shock pulse initiation. It is impossible to definitively determine that the shock has begun until the signal rises decisively above the noise floor.

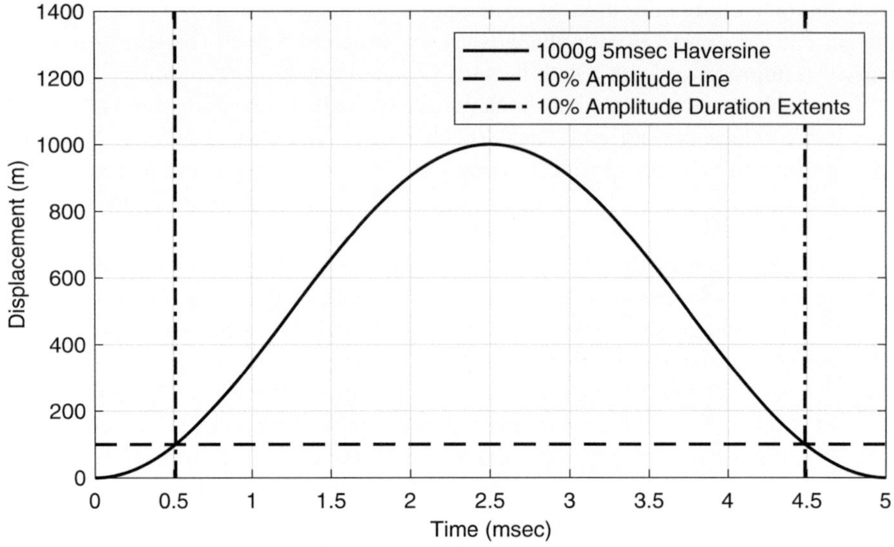

Fig. 9.3 Plot of a classical haversine pulse showing the 10% amplitude duration

Since it is relatively trivial to determine the maximum acceleration amplitude from an experimentally measured pulse, the shock laboratory will typically measure durations at a fraction of the maximum pulse amplitude. This amplitude fraction could be any reasonable value; however, 10% is a popular amplitude level for this measurement. Ten percent is often selected because it is sufficiently above the noise floor for consistent measurements but still low enough to capture the bulk of the recorded pulse. This is shown conceptually in Fig. 9.3. Figure 9.3 shows a classical 1000 g haversine acceleration shock pulse with a 5 ms duration. Ten percent of the maximum amplitude is 100 g, shown by the horizontal dashed line. The intersection point can be readily solved by substituting $\ddot{w}(t) = 0.1A$, in the 10% amplitude case, and rearranging Eq. 9.1 to solve for t as:

$$0.1A = \frac{A}{2}\left[1 - \cos\left(\frac{2\pi t}{T}\right)\right]. \tag{9.3}$$

Rearranging gives

$$\frac{2(0.1A)}{A} = 1 - \cos\left(\frac{2\pi t}{T}\right), \tag{9.4}$$

and clearing the amplitude, A, gives

$$\cos\left(\frac{2\pi t}{T}\right) = 1 - 0.2 = 0.8. \tag{9.5}$$

Taking the inverse cosine of both sides and rearranging give the result

$$t = \frac{T}{2\pi} \arccos(0.8). \tag{9.6}$$

The solution to Eq. 9.6 is $t = 0.1024T$ for the 10% amplitude case described here. Thus, the signal first crosses the 10% amplitude line at approximately 10% of the theoretical duration. Likewise, the signal recrosses the 10% amplitude line at $t = 0.8976T$. Thus, the measured duration between the two 10% amplitude line crossings is 79.52% of the theoretical duration, or about 80% when rounded. As a result, the 10% amplitude duration is approximately 0.8 times the theoretically duration. Or from the other viewpoint, the theoretical duration is about 1.25 times the 10% amplitude duration measured in the laboratory.

This is a somewhat subtle difference in the definition of the classical haversine but it is significant. It is frequently assumed that all parties involved in test specification development and execution are using the same variable definitions; but, it is not necessarily true. As a test engineer it is important to understand whether or not the laboratory's common measurement practices are understood by the test specification developers. Likewise, it is imperative that the test specification developers understand what the laboratory personnel are measuring and calibrating against. For this reason, it is always preferential to clearly state whether the defined pulse duration is the full duration or some smaller duration, such as the 10% amplitude duration.

Returning to the example shown in Fig. 9.3, the pulse shown is a 1000 g 5 ms haversine; however, it may also be described as a 1000 g 4 ms 10% amplitude duration pulse. While the later definition may seem somewhat cumbersome, it is clear where the first definition could be misinterpreted. For example, if your 5 ms duration pulse specification was interpreted to be the full duration when you intended it to be a 10% amplitude duration, then you many have gotten a 200 Hz shock pulse instead of the 160 Hz pulse you intended. Likewise, if the 5 ms duration pulse specification was interpreted to be the 10% amplitude duration when you thought it was the full duration, then you would have received a 170 Hz shock instead of the 200 Hz pulse you intended.

9.3 Resonant Fixtures

Resonant plate testing is a less common test method for simulating high energy mechanical shock in the laboratory although it has been performed for many years. The resonant plate test methodology approximates a pyroshock event and its simulation in the laboratory has many practical benefits as compared to an actual pyroshock field test, not the least of which is not having to work with live explosives for the test. Other benefits include testing in a controlled environment and significantly less time between tests. However, when the pyroshock is simulated in

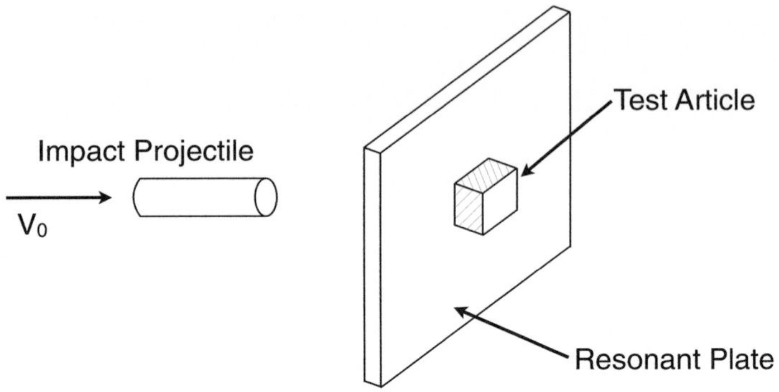

Fig. 9.4 Schematic of a resonant plate shock test

the laboratory with resonant beam or resonant plate test there will be a small velocity change.

Figure 9.4 shows a drawing of the basic components of a resonant plate test. The article under test is attached to the front of a tuned resonant plate with free-free boundary conditions and a gas gun is used to fire a projectile at the back center of the plate. A pendulum can also be used to excite the plate with a hammer strike. The projectile, or hammer impacting the plate excites the plate's bending modes, which excites the test article. The vibration environment is very modally rich. Pulse shaping material can also be used between the impact projectile and the resonant plate to help shape the pulse duration. Impact projectiles with different masses can be used as well as changes in impact velocity to control the momentum transferred to the plate. Resonant plate testing is fundamentally different from drop table testing because the article under test does not have any initial velocity prior to the shock. Additionally, while the plate is technically in a free-free configuration, it must nonetheless be constrained as a practical matter, usually with rope or other compliant material. As such, the plate vibrates post impact but does not have any appreciable translation, quite the opposite from a drop table where the article under test has both initial velocity and substantial displacement from the test's beginning to end.

Figures 9.5 and 9.6 show photographs of a resonant plate shock machine at Sandia National Laboratories [1]. Figure 9.5 shows the front of the plate where the test article would be mounted and Fig. 9.6 shows the back of the plate where the projectile impacts to provide the shock excitation. This particular resonant plate design is suspended with rope inside of a frame to create essentially free-free boundary conditions. A gas gun is used to drive a steel projectile into the back of the plate. The impacting anvil is shown at the plate center in Fig. 9.6. The anvil is typically covered with a thin layer of felt or other pulse shaping material to tailor the energy transfer from the projectile to the plate. The bars bolted to the plate sides are used to increase the system mass but also incorporate constrained layer damping into the plate to shorten the shock excitation time.

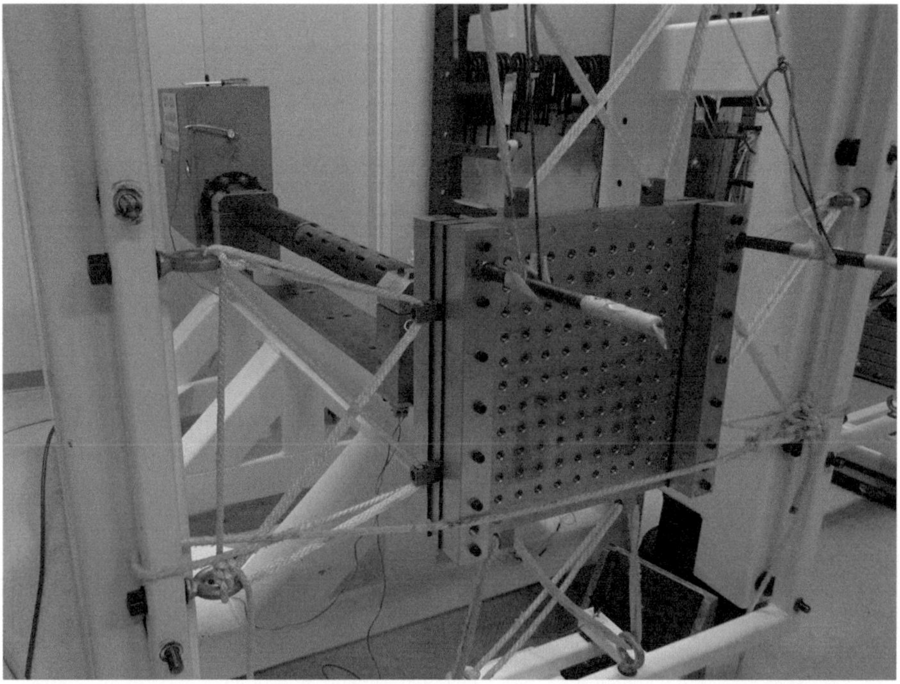

Fig. 9.5 Front view of a resonant plate shock machine with gas gun behind the plate

The front of the resonant plate is built with a grid of threaded inserts to accommodate a wide range of test fixtures. The two pieces of all-thread rod protruding from the top of the plate are used in conjunction with some additional rope to level the plate prior to each shock event. Installing test articles on only one side of a free-free plate has a tendency to tilt the plate forward creating an undesirable impact angle with the gas gun projectile. Figure 9.6 shows the anvil on the back centered on the plate; however, recent testing with off-center impacts and test articles mounted off-center has yielded some interesting shock response results. These off-center impacts are yielding responses that mimic multi-axis inputs. Further research in this area seems promising and is ongoing.

Figure 9.7 shows a close-up view of the corner of the resonant plate from Fig. 9.5 showing the damping bar details. The damping bars in this example are approximately 50 mm square and run the length of the plate. Silicone rubber approximately 3 mm thick is sandwiched between the damping bar and the plate. The bolts shown in Fig. 9.7 pass through the front side damping bar, damping material, resonant plate, rear damping material, and rear damping bar. Other damping materials have been used in place of the silicone rubber, including polyurethane and felt. In addition, different sizes and materials for damping bars could also be used to tune mass or stiffness.

Fig. 9.6 Rear view of a resonant plate shock machine showing impact anvil and gas gun muzzle

One important consideration when incorporating this type of damping is that the damping bars need to be isolated from the resonant plate. The goal of pyroshock testing is to excite high-frequency responses and any metal-on-metal vibration can create additional, and often unwanted, excitation. This design also incorporates a nylon bushing in the resonant plate at each bolt location so that the damping bar bolts do not come into contact with the resonant plate [1]. The large piece of all-thread rod seen in Fig. 9.7 is used for leveling the plate as discussed previously. The all-thread rod is threaded into the damping bar only and does not pass through the rubber damping material or touch the resonant plate.

Resonant plates are designed such that the plate's primary excited frequency corresponds to the primary frequency in the test specification SRS. Gross tuning of the plate is accomplished with the design of the plate's physical dimensions and material. The natural frequencies of a plate can be calculated directly from handbook equations based on closed-form solutions. The natural frequencies of a free-free square plate are given by the equation:

$$f_i = \frac{\lambda_i^2}{2\pi a^2}\sqrt{\frac{Eh^3}{12\gamma(1-\nu^2)}} \tag{9.7}$$

where E is the plate's modulus of elasticity, h is the plate thickness, a is the side length, γ is the mass per unit area, ν is the plate material Poisson's ratio, and finally,

Fig. 9.7 Close-up view of damping bars and damping material attached to a resonant plate

λ_i^2 is a constant and a function of the mode shape indices [2]. Values for λ_i^2 are 13.49, 19.79, and 24.43 for modes one through three, respectively. Given a desired plate material and thickness, the required plate size for a given frequency can be easily calculated by manipulating Eq. 9.7.

One of the most interesting concepts in resonant plate design is that the first mode of the plate is typically not excited. The excitation at the plate center excites the first bending mode of the plate, which is actually the second fundamental plate mode. Figure 9.8 shows plots of the first three mode shapes of a free-free square plate. As can be seen, the first mode is the traditional "flapping" mode of the plate. The second plate mode is the first bending mode. Finally, the third mode is the lowest plate bulging mode. It is frequently and erroneously assumed that the first bulging mode is the one excited by a center impact with a projectile; however, the bulging mode has typically too high frequency to be well excited and hence the motion is dominated by the lower bending mode. The first mode is not excited by the projectile impact because the impact occurs along the plate's node line, that is the line of minimum displacement of the first plate mode. As a result, the first mode is not substantially excited by a center impact.

As was stated earlier, the resonant plate is sized to vibrate at a specific frequency based on the material properties and the plate's physical size. As a result, it is often tempting to design a 1 kHz plate when a 1 kHz test is required. However, it is prudent

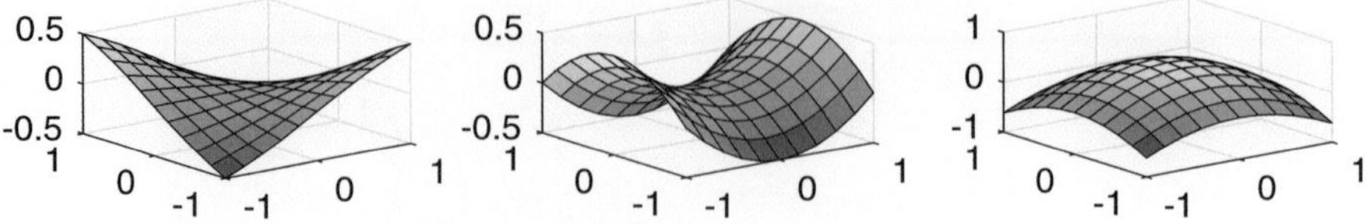

Fig. 9.8 First three mode shapes of a square plate

to design the plate to have a slightly lower resonant frequency than the desired test frequency. There are two compelling reasons for this. First, the addition of a fixture and the unit under test will stiffen the plate and increase the primary response frequency. Second, typical qualification test specifications and test tolerances allow for some slight responses above the test specification levels but look poorly on tests that fall short of the test specifications. For this reason, a slightly lower plate resonant frequency is a more conservative test and typically more acceptable. The amount that the plate frequency is designed below the test specification is not fixed; however, a good practice is to design the plate with a primary bending frequency that is approximately 5% below the desired test specification frequency.

Resonant plates can be built to match any frequency and made of several different materials. Aluminum plates are probably the most common due to their lighter weight and ease of manufacture. For reference, a 1 kHz resonant plate can be made of 50 mm thick aluminum and 500 mm square. From Eq. 9.7, this gives an actual frequency of 963 Hz which is a few percent less than the 1 kHz design frequency. Likewise, a 500 Hz resonant plate could be made of 25 mm thick aluminum still 500 mm square resulting in an actual frequency of 481 Hz. The stiffness of the unit under test and the test fixture typically brings the effective stiffness of the plate back up near the design frequency. Extremely stiff test fixtures can significantly increase the plate's resonant frequency. A few percent is common; however, the author has seen frequency shifts in excess of 20% (over 200 Hz) with an extremely stiff test fixture.

In addition to the test fixture, it is common to add constrained layer damping along the plate's free edges to shorten the system ring-down time as was discussed earlier. Aluminum plates typically have a modest ring-down time although much longer than a pyroshock event. A steel plate would have a substantially longer ring-down time since it has less internal damping. The faster ring-down of aluminum occurs because aluminum is not a tonal material. Tonal metals such as steel and brass have less internal damping and consequently ring for a longer time. For anecdotal evidence of this fact, it only needs to be noted that bells are not made from aluminum.

Another concept that is similar to the resonant plate is a resonant beam. The concept for a resonant beam is similar to that of a resonant plate. A large, heavy beam is designed to have a fundamental resonant frequency, a test article is attached to one side and a projectile impacts the opposite side of the beam. Very few resonant beam test systems are currently in existence. Sandia National Laboratories

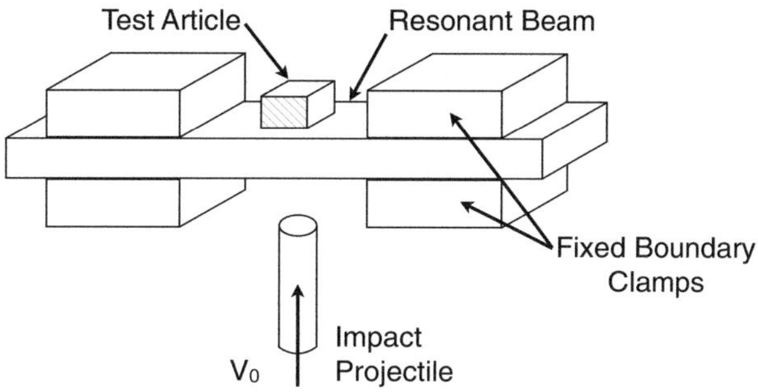

Fig. 9.9 Schematic of a transverse resonant beam shock test

developed and patented this system in the mid-1990s [3]. While it is used extensively at Sandia, like resonant plates, its use is limited.

Figure 9.9 shows a drawing of the basic components of a resonant beam test. The beam at Sandia National Laboratories is oriented horizontally, although this is not a requirement, with the projectile is driven up at the beam from underneath. The test article is attached to the top face of the beam opposite the projectile's impact point. Large clamps are used on each end of the beam to provide clamped-clamped boundary conditions. In this system, the fixed boundary clamps can be moved in or out to change the beam's effective length. The first fundamental natural frequency of a clamped-clamped distributed-parameter beam is given by:

$$\omega = \frac{\lambda_1}{l^2}\sqrt{\frac{EI}{\rho A}}, \tag{9.8}$$

where $\lambda_1 = 22.3733$ for the first bending mode. Thus, increasing the distance between the clamps lowers the resonant frequency, while bringing the clamps closer together raises the beam frequency. As with plate testing, the fundamental beam frequency should be tuned to match the primary frequency in the test specifications. The greatest advantage to this configuration is that the same beam can be tuned to accommodate different test configurations. In contrast, resonant plate tests require different plates for each primary test frequency. The second advantage to this configuration is that the fixed boundary clamps help ensure a zero net velocity change from the shock test—more closely representing the zero velocity pyroshock. The free-free boundary condition of the resonant plate does result in some forward velocity as a result of the projectile's impact on the plate.

One problem with a clamped-clamped resonant system is that the clamped-clamped boundary conditions are an idealization and not actually attainable in the laboratory. This is especially true when the stiffness of the beam is of a similar order of magnitude to the stiffness of the clamps. As such, the effective length of the beam is frequently longer than the distance between the clamps. The actual flexible portion of the beam extends some distance into the clamps, albeit a small distance. Furthermore, the clamps can also deform, as everything deforms under shock to some degree, making the boundary condition somewhat less than mathematically

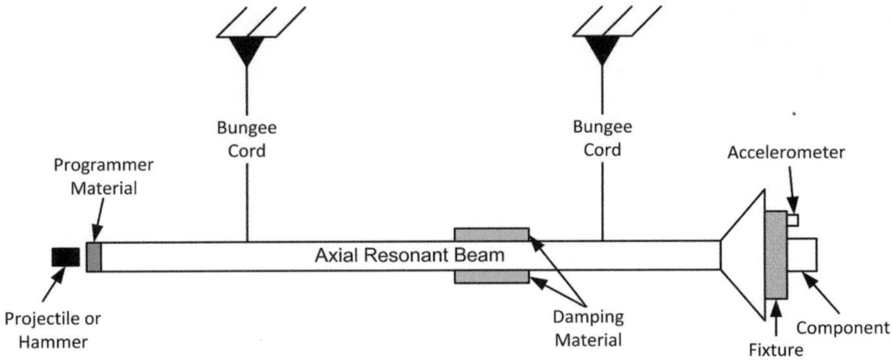

Fig. 9.10 Schematic of an axial resonant beam shock test

fixed. Nevertheless, the design is very useful for simulating pyroshock events with the recognition that its behavior in practice may be slightly different than the theory.

Resonant beams are also used in the axial direction. In this configuration, a projectile or hammer impacts one end of the beam and the unit under test is attached to the other end as shown in Fig. 9.10. The axial bar is often called a Hopkinson bar, after Bertram Hopkinson, a British patent lawyer and Cambridge University professor who came up with the idea in 1914 to measure the pressure produced by explosives. The resonant beam is designed so that axial vibration of the beam imparts an oscillatory shock with the required spectral content to the part. Systems that have discrete attachment points are often tested with a free-free axial resonant beam. As with other impact-induced shocks, the programmer material plays an important role in creating the right spectral content and magnitude of the environment.

In addition to square plates and rectangular beams, other designs have been proposed. Recently, Sandia National Laboratories has studied the use of rectangular resonant plates for some test applications. The rectangular plates are used in conjunction with slightly off-center projectile impact locations with the specific goal being to excite different plate bending frequencies in each axis. Using this configuration, it may be possible to excite a component with slightly different test specifications in two orthogonal axes simultaneously, the intention being to produce a true multi-axis environmental test.

9.3.1 Resonant Plate Acceleration Environment

The shock environment produced in a resonant plate test is a complex, oscillatory environment. Figure 9.11 shows a typical acceleration response of a resonant plate along with the corresponding velocity change. The response is clearly a complex, oscillatory shock. The acceleration has two phases. The first phase, shown in gray,

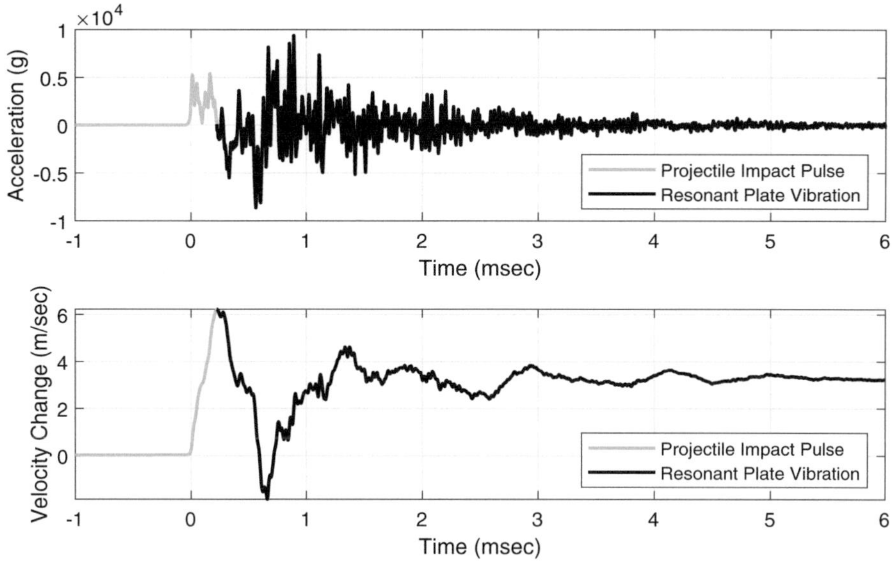

Fig. 9.11 Typical resonant plate environment: top: acceleration; bottom: velocity change

is the impact pulse. The second phase, shown with the black line, is the transverse vibration of the plate. In this example, the projectile imparted an initial velocity change of about 6 m/s. The velocity settled out at about 3 m/s for most of the shock event, before eventually dying out.

The maximum positive (MPAA) and negative (MNAA) absolute acceleration SRS are shown in Fig. 9.12. The MNAA SRS shows a strong "dip" between 200 and 600 Hz, with the minimum occurring at 350 Hz. The response of the SDOF oscillator to the positive acceleration from the impact pulse in the first phase gets canceled out by its response to the plate vibration in the second phase of the complex shock. This characteristic is clearly seen in the response of a 350 Hz SDOF oscillator to the resonant plate environment, shown in Fig. 9.13. The principle of superposition says that the total response of the SDOF oscillator is the sum of its responses to the two phases in the complex shock:

$$y(t) = y_1(t) + y_2(t). \tag{9.9}$$

In this case, the SDOF oscillator response to the plate vibration is 180° out of phase with response to the impact pulse and the magnitudes of the second cycles are similar. This combination causes the destructive interference that reduces the peak negative absolute acceleration.

Every resonant plate test will have such a "dip" in the MNAA SRS. The depth and frequency band of the MNAA dip depend on the pulse duration and the relative

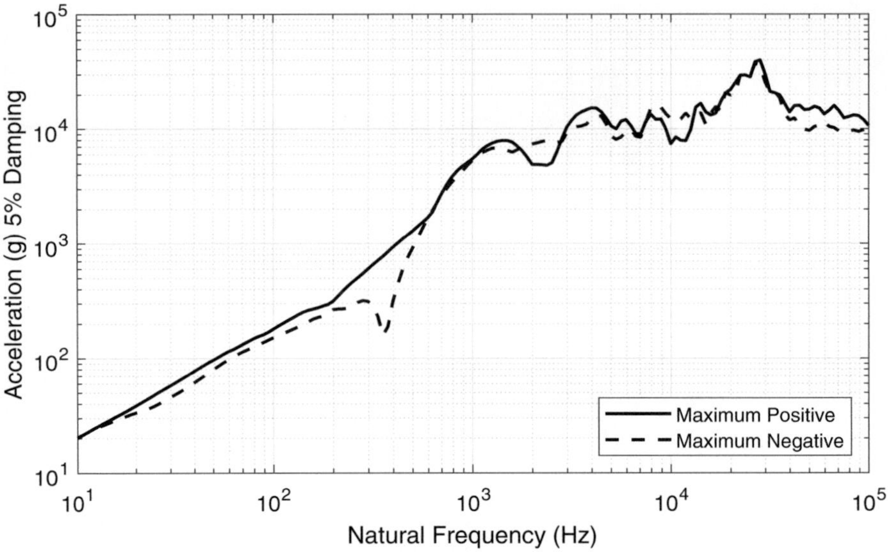

Fig. 9.12 Maximum positive and negative shock response spectra of the acceleration shown in Fig. 9.11

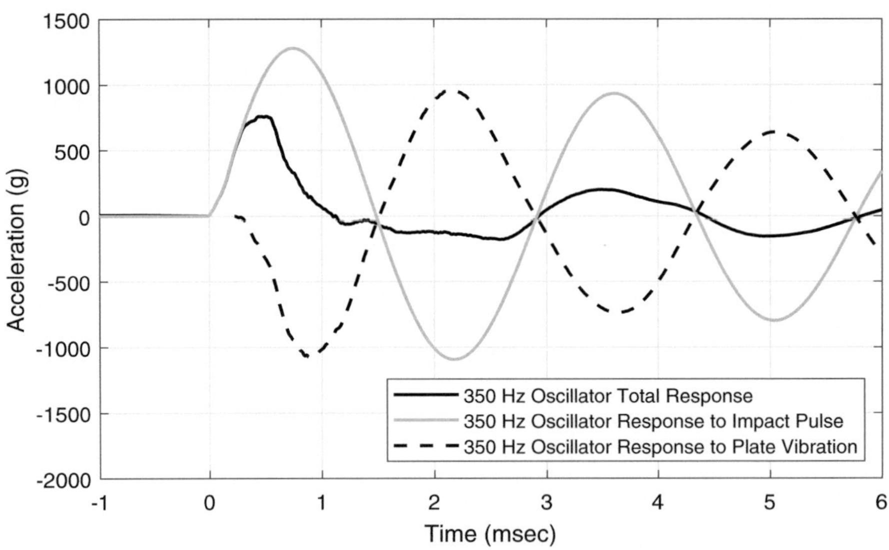

Fig. 9.13 Response of a 350 Hz SDOF oscillator to the acceleration shown in Fig. 9.11

magnitude of the plate vibration to the impulse magnitude. This has practical implications if a component has both positive and negative shock requirements and the "dip" is as large as it is in this example. Because of the "dip," a single resonant

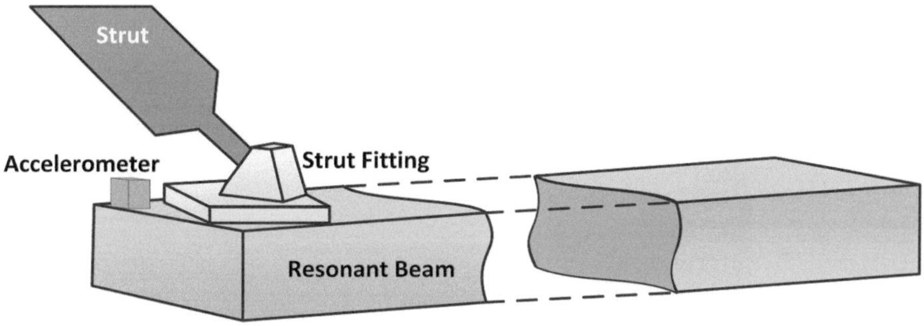

Fig. 9.14 Axial resonant beam with subsystem strut and accelerometer

plate test may be insufficient because the negative shock may be out of tolerance in the frequency band of the "dip." Another test would be required in this case.

9.3.2 Axial Resonant Beam Acceleration Environment

A shock environment that can be generated by an axial resonant beam test is illustrated in this section with an example from a spacecraft subsystem test. The test objective was to simulate a mid-field pyroshock environment. An axial resonant beam test was selected because the subsystem is attached to the spacecraft by struts. In the flight configuration, shocks are transmitted from the strut bases throughout the payload. The test had to introduce shocks into the struts as they would be introduced in flight. Figure 9.14 shows the attachment of a strut to the top of the resonant beam. The beam and subsystem were supported by elastic cords in a free-free boundary condition configuration, similar to that shown previously in Fig. 9.10. An accelerometer mounted at the end of the beam, near the strut mounting plate, measured the environment applied to the strut. A projectile impacted the other end of the beam exciting the axial vibration modes.

Figure 9.15 shows the acceleration time history of one of the shocks applied to one of the struts. The maximum applied acceleration was slightly less than 3000 g and the shock duration was relatively long at about 40–50 ms. The MMAA SRS from the shock is overlaid on the shock test specification in Fig. 9.16 along with required tolerance bounds for this test. The low-frequency portion is out of tolerance because of the velocity change. The SRS above 2 kHz exceeds the upper tolerance bounds indicating some over-test. The subsystem was very robust to shock so the out-of-tolerance test condition was not an issue and this was considered an acceptable test.

This example illustrates the difficulty of replicating a specified pyroshock environment in the laboratory. Even with the generous test tolerance bounds provided

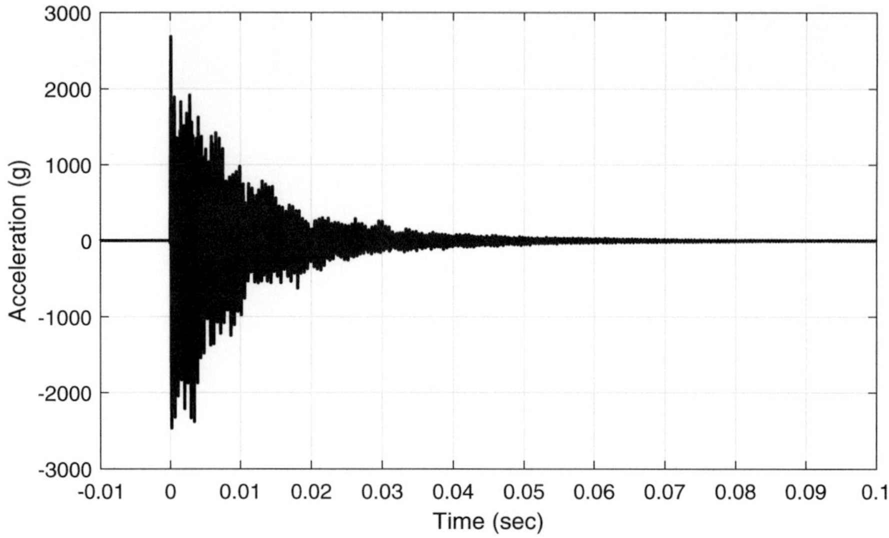

Fig. 9.15 Typical spacecraft component pyroshock test acceleration response

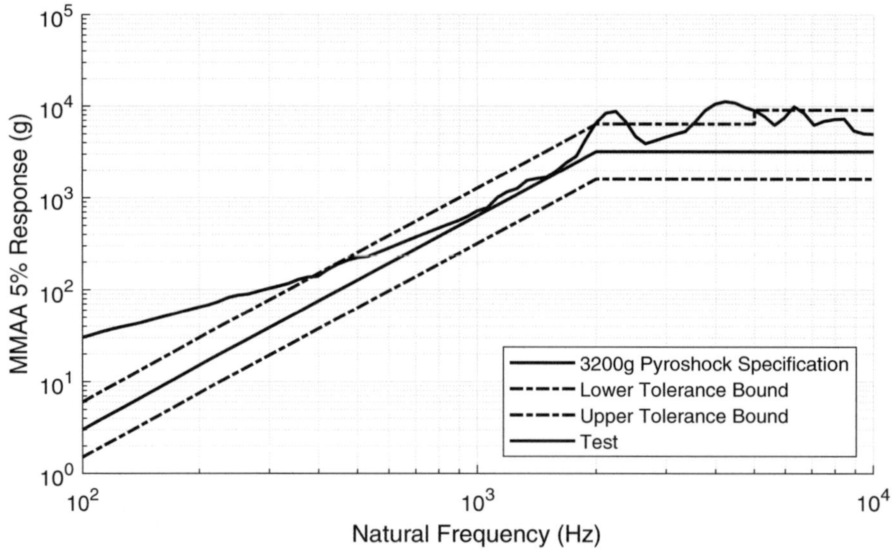

Fig. 9.16 Typical spacecraft component pyroshock test SRS with test measured SRS and test tolerance bounds

by NASA-STD-7003A, it is often difficult to generate an environment that fully lies within the tolerance bands.

9.4 Shaker Shock

The late Dr. Gaberson (1931–2013) was fond of saying "shaker shock is wimpy shock." Certainly, those who operate shaker machines and sell testing services using shakers would disagree. However, shaker shocks are limited to testing lower-level shocks because of equipment limitations on armature motion. Shakers are generally limited to small displacements because the armature has a limited stroke. They are also acceleration and velocity limited by amplifier power. This is not intended to imply that shock machines have no limitations but rather that the shock energy input can be much higher with a shock machine. In contrast, shock profiles from a shaker can be considerably more complex, incorporating drop-outs or notches in specified frequency ranges.

Shaker shocks are severely limited in the low-frequency regime. This limitation is a result of the requirement that the armature returns to rest at its starting point at the end of the shock. This is in contrast to a drop shock machine where the carriage can start high up on the machine and end on the reaction mass. The shaker armature has nowhere to go. As a result, to execute a classical shock on a shaker table actually requires two pulses, the actual shock pulse, and a compensating pulse. From the haversine theory discussed previously, the displacement is open-ended because the final velocity is different from the starting velocity. That cannot be true for a shaker. The shaker must begin and end the shock with zero velocity. To illustrate this, a compensated haversine acceleration time history is shown in Fig. 9.17. In this plot, a sample 100 g amplitude, 10 ms pulse width haversine is plotted at the center and a −2.5 g, 400 ms compensating pulse has been added. This acceleration time history is then integrated to obtain the predicted shaker velocity and displacement. Adding the compensating pulse causes the shaker to pull the armature down prior to the shock, apply the desired shock, and then slowly return the armature to its initial starting point. Thus, the initial and final acceleration, velocity, and displacements are all zero as required. A cursory look suggests that this may be a perfectly reasonable method for executing this shock on a shaker table.

However, Fig. 9.18 provides a different interpretation of the compensated haversine shock. This plot compares the calculated pseudo-velocity SRS from the compensated haversine shock to the SRS from a true haversine shock with the same amplitude and duration. Figure 9.18 shows that the two shocks are very similar above about 5 Hz but drastically different at lower frequencies. In addition to the 400 ms compensating pulse, Fig. 9.18 also shows the effects of shorter compensating pulses on the overall SRS. The SRS from the same 100 g 10 ms haversine shock with a 200 ms and a 100 ms compensating pulse is also plotted for comparison. As can be seen, the overall SRS amplitude decreases slightly with the shorter compensating pulses; however, the greatest effect is the significantly shortened plateau in the SRS. The 100 ms compensating pulse has almost removed the plateau completely.

The above comparison is very elucidating; however, a compensated haversine shock of this magnitude is actually difficult to perform on a shaker due to its high velocity component. The maximum velocity shown in Fig. 9.17 is approximately

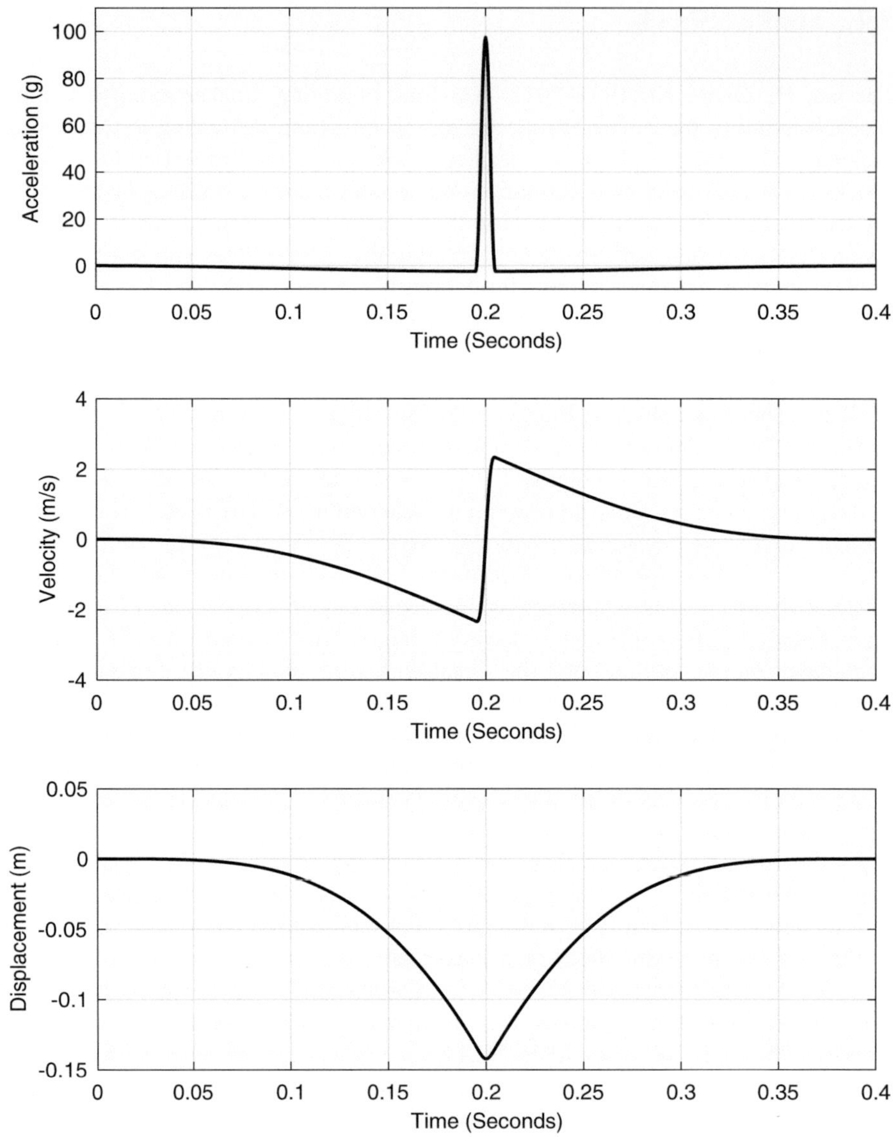

Fig. 9.17 Acceleration, velocity, and displacement time histories from a 100 g 10 ms haversine compensated with a 2.5 g 400 ms inverted haversine

2.34 m/s, more than some shakers are capable of pushing. A much more common method of performing shock testing on a shaker is with the use of decayed sines or some other similar method of developing an oscillatory shock. As was discussed in Chap. 4, there are numerous acceleration time history curves that can generate essentially identical SRS. One popular method for creating shock time histories

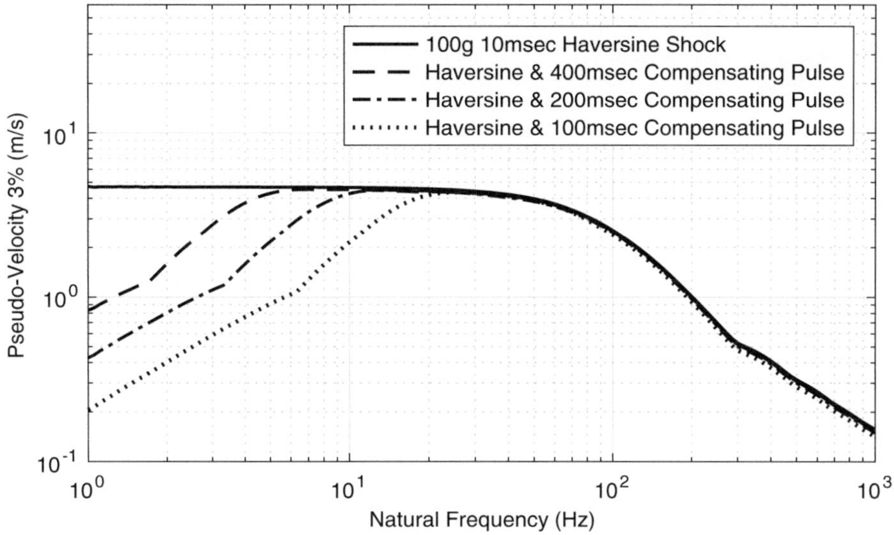

Fig. 9.18 Pseudo-velocity SRS comparison of a 100 g 10 ms haversine shock and the same shock compensated with different length compensating pulses for use on a shaker

is through a summation of decaying sinusoids. An individual decaying sinusoid is described by the equation:

$$x_i(t) = A_i e^{-\zeta_i \omega_i t} \sin(\omega_i t) \tag{9.10}$$

where A_i is the amplitude, ζ_i is the damping coefficient, and ω_i is the frequency. A shock can be comprised of a single decaying sinusoid or a summation of multiple sinusoids with different frequencies, damping, and amplitudes. Thus, a summation of decayed sines is given by:

$$x(t) = \sum_{i=1}^{N} x_i(t) = \sum_{i=1}^{N} A_i e^{-\zeta_i \omega_i t} \sin(\omega_i t). \tag{9.11}$$

An example of a single decayed sinusoid acceleration time history is shown in Fig. 9.19 and a summation of three decayed sinusoids is shown in Fig. 9.20. The three tones used here were selected arbitrarily but it can be easily imagined that if the number of tones were increased and the amplitudes, damping ratios, frequencies, and phases were optimized using some sort of computer algorithm, it would be possible to tailor a shock pulse.

Let us return to the example of the haversine shock performed on the shaker machine. It was noted earlier that the velocity from the compensated haversine shown in Fig. 9.17 was likely too severe for most shakers. An alternative approach would be to derive a similar shock pulse using a summation of decayed sinusoids

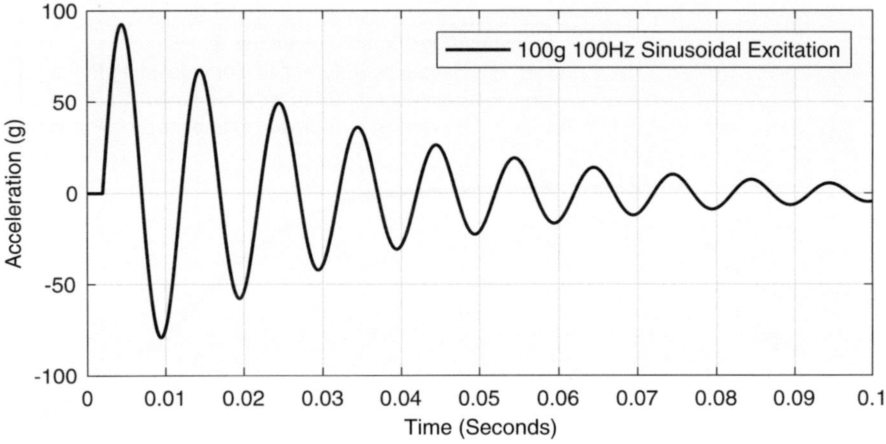

Fig. 9.19 Acceleration time history of a single 100 Hz decayed sinusoid

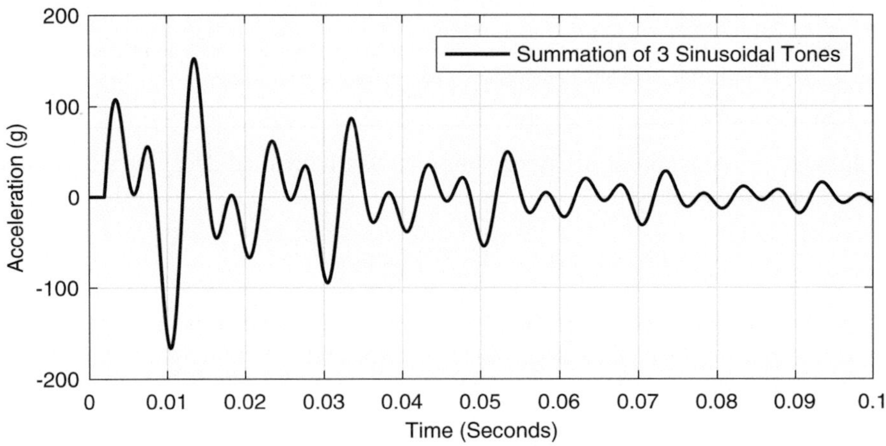

Fig. 9.20 Acceleration time history of a summation of three decayed sinusoids, 100 g 100 Hz, −50 g 150 Hz, and 85 g 200 Hz sine tones

to satisfy the same SRS. This was accomplished using an optimization algorithm designed to match the SRS and minimize the acceleration, velocity, and displacement responses. While this may seem to make this a somewhat unfair comparison, this is traditionally done to bring a high shock into alignment with what can actually be performed on a shaker table. Figure 9.21 shows the resulting time history from a summation of 39 decayed sinusoids of various amplitudes, frequencies, damping coefficients, and phases. The time history responses admittedly look nothing like the original haversine shock. However, the pseudo-velocity SRS comparison of the classical haversine with this decayed sinusoid version is shown in Fig. 9.22. From this plot, the SRS comparison is quite good above 10 Hz. The SRS is actually very

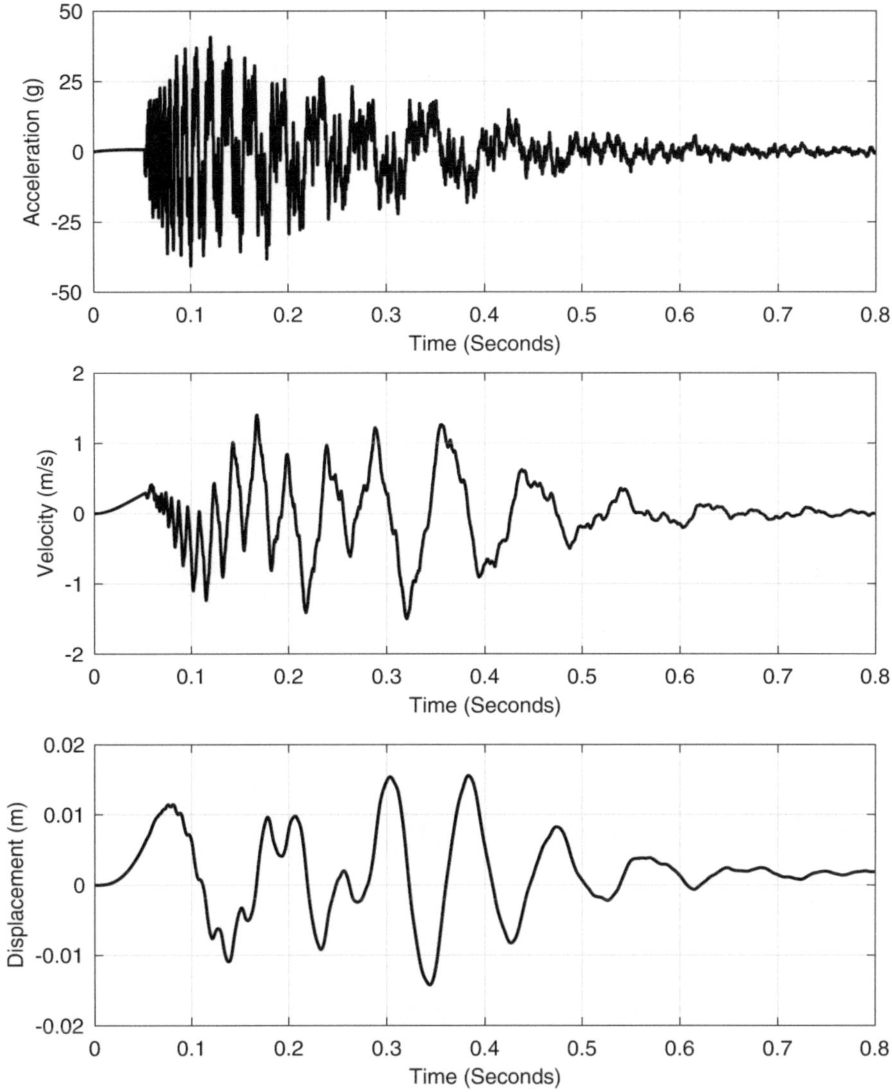

Fig. 9.21 Acceleration, velocity, and displacement time histories from a summation of decayed sinusoids approximating a 100 g 10 ms haversine shock

similar to the single haversine with the 300 ms compensating pulse shown earlier in Fig. 9.18. For this derivation, no decayed sines were used with a frequency below 10 Hz which is why the SRS falls off so dramatically at the low frequencies. This is actually quite common with electrodynamic shakers due to the relatively small displacement limitations of these machines.

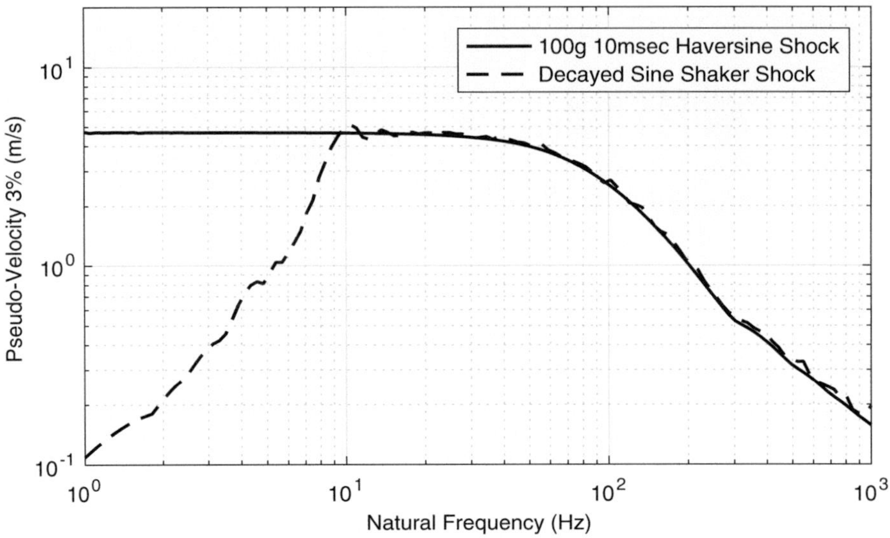

Fig. 9.22 Pseudo-velocity SRS comparison of a summation of decayed sinusoids approximating a 100 g 10 ms haversine shock with the SRS from the same classical haversine

Studying the derived time history plots in Fig. 9.21 highlights some of the sharp differences between shaker shock testing and shock machine testing. The maximum acceleration from the shaker shock is only 40.8 g compared to 100 g from the classical haversine. The maximum velocity is 1.5 m/s compared to 4.9 m/s for the classical shock and 2.3 m/s from the compensated haversine given in Fig. 9.17. Since stress is proportional to velocity, these shocks are clearly not equal. The hope is that they have similar damage potential which is the essence of the SRS. Shocks with similar SRS are supposed to have similar damage potential although this is a theory and is generally true, but it is not necessarily universally true.

How do the time histories vary so dramatically in both form and amplitude and yet the pseudo-velocity SRS peak at approximately the same amplitude? This is a function of the number of oscillatory cycles in the shock. The classical haversine shock has one cycle and must attain the maximum velocity of the SDOF oscillator in one shot. In contrast, the summation of decayed sines consists of multiple oscillatory cycles that work together to increase the amplitude of the SDOF oscillator over time. This is shown graphically in Fig. 9.23. Figure 9.23 shows a 30 Hz SDOF oscillator response to both the 100 g 10 ms classical haversine shock and the summation of decayed sines representation of the same haversine shock shown in Fig. 9.21. The pseudo-velocity response from the classical haversine reaches its maximum absolute value on the first half cycle and decays from there. In contrast, the decayed sine representation requires about ten oscillatory cycles before it drives the SDOF oscillator to the same velocity. This type of oscillatory motion may not actually occur in the real system if the system under test does not behave exactly like an

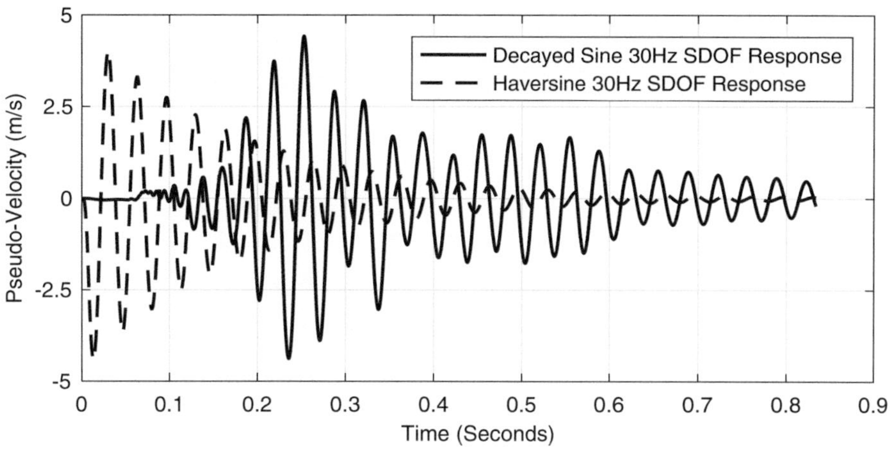

Fig. 9.23 Pseudo-velocity time history comparison of a 30 Hz SDOF oscillator response to the classical haversine shock and the sum of decayed sinusoids representation of the same shock

SDOF oscillator. As a result, the shaker shock representation of a shock machine test is frequently a more benign test and is likely never more severe.

Shaker shock testing is also used to simulate a pyroshock environment in the laboratory. It is not appropriate for near-field or mid-field responses because the excitation levels are too large to be realized with a shaker, but may be considered for components in the far-field region. In the far-field, the shock is a high-frequency oscillatory event with little or no velocity change. Thus, low-frequency limitations of the shaker are not a problem.

9.5 Pyroshock Testing

Because of the short duration and high-frequency nature of a pyroshock, the pyroshock-induced environment that a component experiences will be affected by the distance and structural features in the load path between the pyroshock source and the component. The pyroshock environment is normally characterized by three regions: near field, mid field, or far field [4]. Near-field pyroshock is the most intense as a component in the near field is in close proximity to the detonation. Mid-field and far-field components typically have the advantage of distance and some intervening structure between the component and the detonation to help attenuate the effects of the shock. The definition of near field, mid field, and far field is somewhat subjective but the three regions are generally classified by MIL-STD-810G and are summarized in Table 9.1.

The near field is the region close to the pyrotechnic source, usually within 15 cm. The response in this region is strongly governed by stress wave propagation

Table 9.1 Pyroshock classification regions

Region	Primary frequency	Acceleration amplitude	Response type
Near field	>10 kHz	>10,000 g	Stress wave
Mid field	3–10 kHz	<10,000 g	Stress wave and structural response
Far field	<3 kHz	<1000 g	Structural response

in the material. Pyroshocks typically have frequency content above 10 kHz and acceleration amplitudes greater than 10,000 g. The environment in this region is very difficult to measure. It cannot be confidently measured with accelerometers because they cannot sense the stress waves that dominate the response in this region. Strain gauges or laser vibrometers are options but they have their own unique issues.

The mid field is an area sufficiently far away from the pyrotechnic source so that the stress waves, while still contributing to the response, have also induced structural vibrations. The stress waves transfer some of their energy to lower-frequency waves in the structure creating the structural dynamic response. This is a very nonlinear process. This region is typically 15–60 cm from the source. The structural vibrations typically have their dominant frequency response in the 3–10 kHz range with acceleration amplitudes less than 10,000 g.

Finally, the far field is everything on the structure beyond the mid field. In the far-field region, the structural response is primarily vibration with frequency content below 3 kHz and the maximum acceleration amplitudes less than about 1000 g. The response can be readily measured with accelerometers and simulated with nonexplosive sources, such as shakers and resonant beams or plates. Usually, the intervening structure between the pyrotechnic source and a shock-sensitive component in the far field contains structural features such as joints and corners that attenuate the acceleration amplitudes significantly.

The mid-field and far-field regions are of the most interest to spacecraft designers. This is why pyroshock can be characterized by the MMAA SRS, but with the frequency range extending beyond 10 kHz and out to approximately 50 kHz. A pyroshock SRS is characterized by a low-frequency ramp portion with a slope of approximately 12 dB/octave and a flat portion above the first resonant frequency in the SRS, the knee frequency. The SRS does not contain a duration, but it is understood that the duration of the pyroshock transient is less than 20 ms. MIL-HDBK-340A[5] and NASA-STD-7003A[6] recommend no more than 1/6 octave natural frequency spacing and a dynamic amplification factor, $Q = 10$, for an SRS calculated from pyroshock data.

Figure 9.24 is an example of a mid-field stage separation pyroshock environment. The SRS is shown in Fig. 9.25. This pyroshock has a 1.4 kHz knee frequency and a 4500 g plateau. The dotted line shows a typical pyroshock environment SRS specification. The natural frequency range is typically 100 Hz to 10 kHz.

Because of the high-frequency character of a pyroshock, small parts, brittle parts, and circuit board elements are most at risk of damage. However, it is often not possible to define a pyroshock environment at specific shock-sensitive parts inside

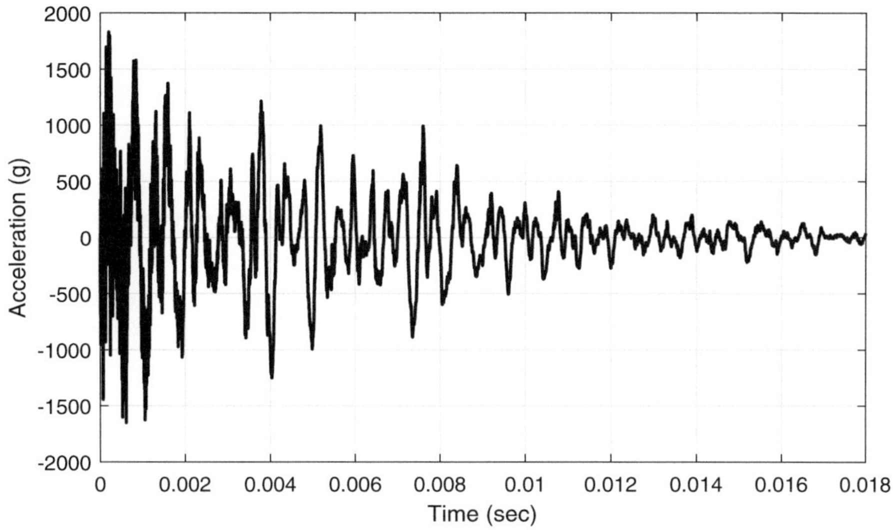

Fig. 9.24 Mid-field stage separation pyroshock acceleration environment

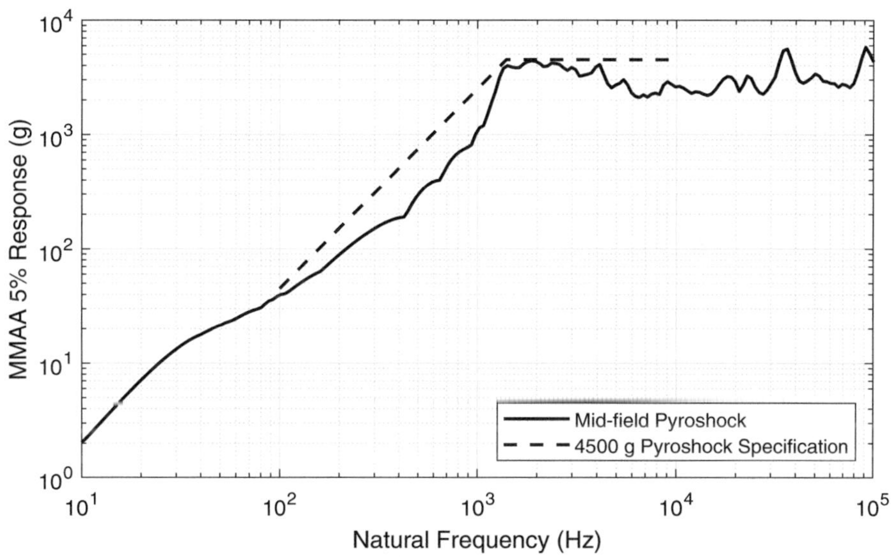

Fig. 9.25 Absolute acceleration SRS of the mid-field stage separation pyroshock environment in Fig. 9.24

a component; they are simply too small. The pyroshock SRS are used primarily to establish test environments for components (i.e., electronic boxes) mounted on the spacecraft structure. These environments are used to design tests to verify the component's robustness to pyroshock transients, or to exempt a component from shock testing requirements. For example, MIL-HDBK-340A [5] exempts units from shock qualification testing if:

1. The qualification random vibration test spectrum when converted to an equivalent shock response spectrum (3σ response for $Q = 10$) exceeds the qualification shock spectrum requirement at all frequencies below 2 kHz.
2. The maximum expected shock spectrum above 2 kHz does not exceed acceleration, g, values equal to 0.8 times the frequency in Hz at all frequencies above 2 kHz, corresponding to a velocity of 1.27 m/s (50 in/s).

Both criteria must be satisfied, so understanding the attenuation of the pyroshock transient between the source and the component containing shock-sensitive parts is critical. The dominant contributor to pyroshock attenuation is the distance between the source and the component of interest. Distance within the system is synonymous with structure geometry, but this is design specific and therefore impossible to generalize.

9.5.1 Pyroshock Test Methods

Pyroshock testing is typically performed using pyrotechnics for components in the near field and either pyrotechnics or specialized mechanical shock machines for components in the mid-field regimes. A true pyroshock does not have any momentum exchange with the component, so there is no velocity change. However, when the pyroshock is simulated in the laboratory with a resonant beam or resonant plate test there will be a small velocity change, due to the movement of the resonating test machine component.

Resonant bcams or resonant plates are a commonly used method for simulating pyroshocks in the laboratory. In a resonant plate test, the shock-sensitive component is mounted to a plate and the other side of the plate is impacted by a hammer released from a pendulum or a projectile fired from a gas gun, as illustrated in Fig. 9.26. Sometimes, the projectile or hammer impacts cannot impart enough intensity and an explosive device must be used. Explosive devices are also used in large structure tests like launch vehicle stage separation testing. In these tests, the actual separation pyrotechnic is used. NASA and the US Air Force have compiled standards for pyroshock testing and analysis [5–8].

Shaker shock testing is another way to simulate a pyroshock environment in the laboratory. It is not appropriate for near-field or mid-field responses because the excitation levels are too large to be realized with a shaker, but it may be considered for components in the far-field region. In the far field, the shock environment is typically described with an acceleration response between 100 Hz up to about 3 kHz or higher.

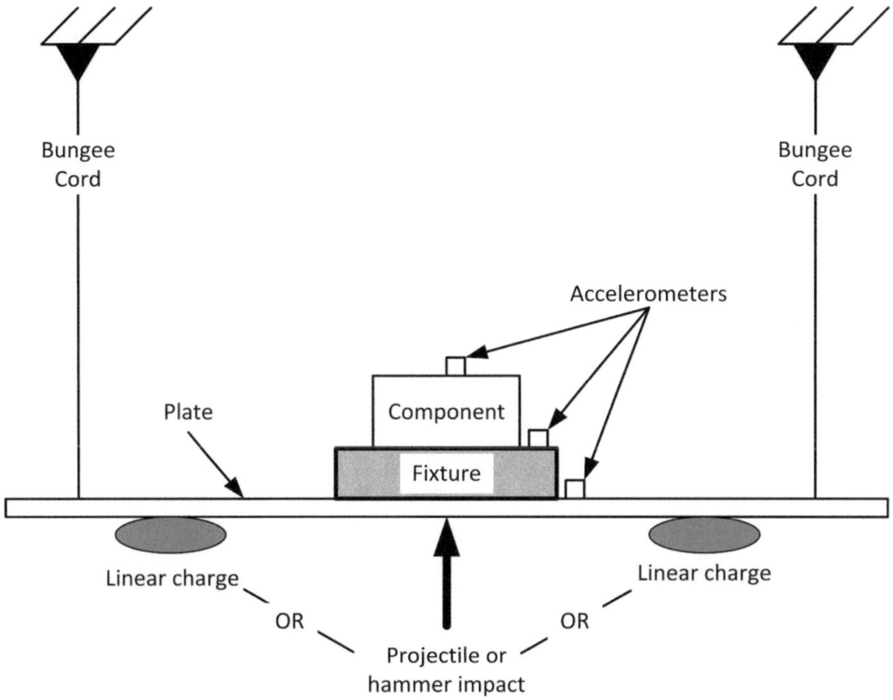

Fig. 9.26 Pyroshock plate test configuration

9.6 U.S. Navy Shock Testing

U.S. Navy research was undoubtedly one of the primary drivers in the field of mechanical shock. Much of the early research and development was performed on shock test machines designed by the U.S. Navy engineers to reproduce damage witnessed in the U.S. Navy wartime engagements. As a result of this, the U.S. Navy has been reluctant to use more generic shock test machines, preferring to continue using their own designs. There is certain justification for this position. The U.S. Navy designed machines have performed well for many years and they work very well for screening shipboard equipment. In addition, shipboard equipment is typically very large and heavy which is not compatible with many of the more generic commercial shock test machines.

MIL-DTL-901E [9] defines three general types of shock tests for the U.S. Navy equipment: light weight, medium weight, and heavy weight. The light-weight test is performed on the light-weight shock machine and is generally used for relatively small components weighing less than about 113 kg (250 lbf) and having limited deflection of resilient mounts. Medium-weight testing is performed on the medium-weight shock machine which has the capability to test reasonably large components weighing less than about 2040 kg (4500 lbf) and having up to 76 mm (3 in) of resilient mount deflection. The weight limits given here are taken from MIL-DTL-901E directly, although, in most circumstances the maximum weight is not actually achieved on each machine. If a component is near the maximum weight, it is usually prudent to move to the next larger shock machine. Heavy-weight testing is performed on a floating barge using live explosives to excite the load. All resilient mounting schemes are allowed on a barge test and the weight limit is generally defined by the barge stability.

The Naval Sea Systems Command (NAVSEA) generally requires all shipboard equipment to be tested on one of the three standard shock test machines described in MIL-DTL-901E. However, MIL-DTL-901E Section 3.1.8 does state that modifications from the basic shock parameters can be approved by NAVSEA.

It is also important to note that the U.S. Navy shock qualification always involves multiple shock tests typically ranging from four to nine hits depending on the size of equipment and the test method used. NAVSEA has approved equipment with fewer shocks but never only one shock. This requirement to test multiple shocks rises from the concern over shock fatigue damage and the desire to propagate damage to failure if possible. A more basic reason for this type of testing is that it is not unusual for a ship to receive multiple near-miss shock events in rapid succession. Typical sea mine warfare doctrine is to deploy multiple mines in an area. Therefore, if a ship encounters one mine, it will very likely encounter more before it can clear the area. In February 1991, the USS *Princeton,* CG-59, triggered an influence mine that detonated under the port side rudder while patrolling the Persian Gulf. The resulting explosion triggered a second mine that detonated forward of the starboard bow. A photograph of some of the hull damage is shown in Fig. 9.27. The hull was significantly damaged just forward of the ship's fantail requiring some creative at-sea patching to strengthen the hull. Despite significant damage to the ship, *Princeton* stayed on station until relieved—a tribute to crew's tenacity and the U.S. Navy shock hardening programs.

9.6.1 U.S. Navy Equipment Designations

MIL-DTL-901E Sec. 1.2.2 defines most shipboard systems and equipment to be either Shock Grade A or Shock Grade B [9]. Shock Grade A equipment is considered essential to the safety and combat capability of the ship. Grade A equipment should not fail under an MIL-DTL-901E type shock event. Shock Grade B equipment is not immediately essential to the ship's safety and combat capability but could become a hazard to personnel or other Shock Grade A equipment. Therefore, Grade B equipment must either not fail during a shock event or fail in a predefined safe manner. The other major requirement for Grade B equipment is that it may not come adrift as a result of a shock event.

In addition to shock grades, MIL-DTL-901E Sec 1.2.4 also defines equipment classes based on the use of resilient mounting schemes. Resilient mounting is defined here as some sort of flexible, shock absorbing coupling between the ship and

Fig. 9.27 Photograph of USS *Princeton's* cracked hull after and the temporary at-sea patch following the 1991 sea mine incident (U.S. Navy photograph)

the equipment. Class I equipment is required to meet its shock requirements without the use of resilient mountings between the equipment and ship structure. In contrast, Class II equipment requires resilient mountings. Finally, Class III equipment is defined as equipment that is sometimes installed in a Class I configuration with resilient mounts and sometimes in a Class II configuration without resilient mounts depending on its location. Therefore, Class III equipment must be proven in both configurations.

MIL-DTL-901E Sec 1.2.6 also defines different requirements for different mounting locations within the ship. Items can be hull mounted, deck mounted, shell mounted, or mounted to the wetted surface. For general surface ship installations (carriers are treated separately), hull mounted means that the item is mounted to the main structural members of the ship. Deck mounted generally defines equipment mounted on the main deck and above, but it is also applicable to items mounted on nonstructural bulkheads below the main deck. Shell mounted items are mounted to the shell plating below the waterline. Wetted surface items are components installed on the ship's exterior.

Mounting orientations are also specified for shipboard equipment. Most common equipment is typically specified as unrestricted orientation. This means that the equipment can be installed in any orientation relative to the ship's principal axes and typically at any location within the ship. However, some equipment can be qualified with a restricted orientation. This condition typically specifies which local coordinate component axis must align with the ship's coordinate system. This type of restricted orientation qualification is usually only approved for very specialized equipment that can only be installed in one orientation or one location in a ship class.

Usually, equipment installation locations on ships are defined by the ship designers. However, in some cases it is possible to define the installation configuration. In those cases, equipment designers typically press for a deck mounted configuration since the deck mounted requirements are less severe than hull or shell mounting. The reason for this is that the ship's deck is a flexible foundation and tends to filter the mechanical shock loads as they travel from the wetted surface to the equipment in question. The deck acts like a mechanical filter trading reduced peak acceleration at the expense of increased deflection.

9.6.2 Light-Weight Shock Machine

The predecessor to the light-weight shock machine was built in 1939 in Britain, and is rumored to have been assembled from parts obtained from the local scrap yard [10]. However, its success at predicting shock performance soon attracted the attention of the U.S. Navy and a modified version was built in 1940 by the General Electric Corporation.

The light-weight shock machine (frequently abbreviated as LWSM) was the first shock machine developed for general naval equipment qualification. The machine has a standard structural steel welded framework with two 181 kg (400 lbf) hammers, one used for vertical excitation and the other used for lateral excitations. The machine is installed on a foundation embedded in a 27 t (60,000 lbf) reinforced concrete reaction mass. The equipment under test is attached to a fixture that is in turn attached to an anvil plate. Figure 9.28 shows a drawing of the light-weight shock machine. The motor on the frame top is used to raise either of the two hammers. The hammer choice depends on the anvil plate orientation. Equipment is typically subjected to nine shock events, three shocks in each orthogonal direction with hammer drops of one, three, and five feet. This not only imposes a significant shock load in each direction but also exposes the equipment to a few fatigue cycles.

The light-weight shock machine is designed for a maximum load of 250 kg (550 lbf) on the anvil plate for surface navy equipment and 136 kg (300 lbf) for submarine components. The most common fixtures used on the light-weight shock machine have weights ranging from 75 kg up to 103 kg, effectively limiting the equipment weight to about 136 kg (300 lbf) for surface ship equipment and much

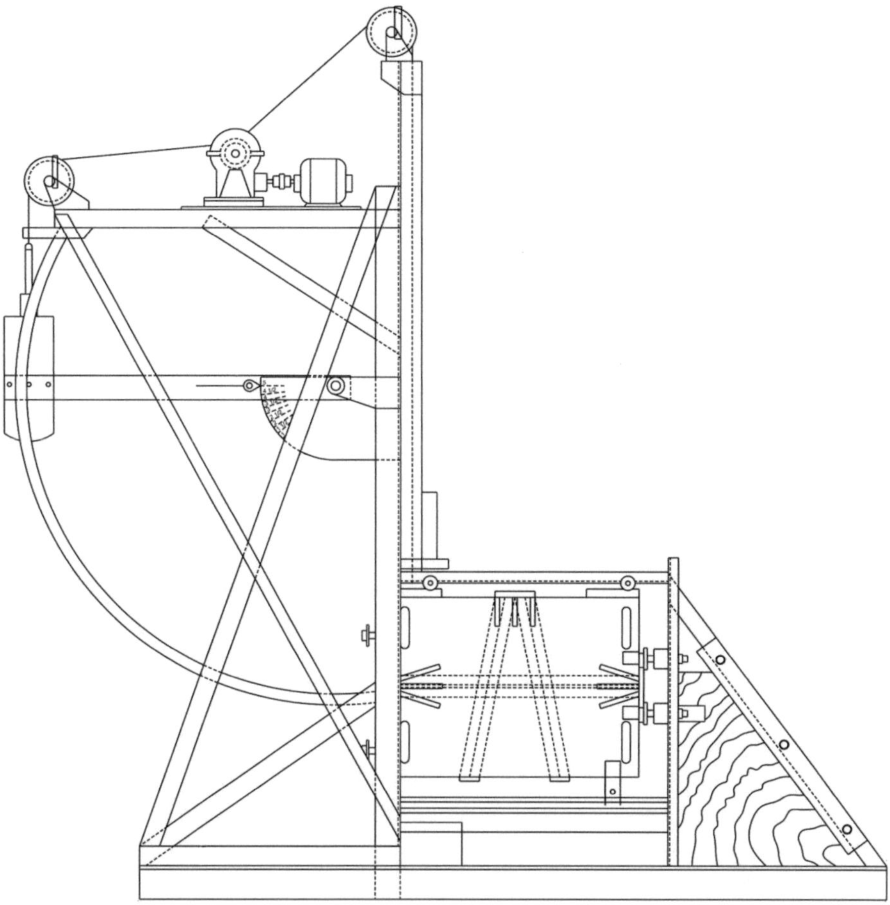

Fig. 9.28 Drawing of the U.S. Navy light-weight shock machine, U.S. Navy drawing [9]

lighter for submarine equipment. Light-weight shock machine test fixtures are intended to represent hull mounted shipboard locations.

The anvil plate has stops installed to arrest the anvil's forward motion at 38 mm (1.5 in). As a result, a normal shock transient on this shock machine has the initial acceleration from the hammer impact on the anvil; the anvil travels forward for 38 mm; the forward motion is stopped and the anvil rebounds back toward the hammer; the anvil plate can actually travel back past its original starting point approximately 9.5 mm (3/8 in) where it again impacts stops and rebounds forward toward its starting location. The maximum shock impacts on this machine are designed to produce an anvil velocity change of about 2.1 m/s (7 ft/s) which amplifies through the anvil-to-fixture interface resulting in a fixture peak velocity of approximately 3 m/s (10 ft/s).

The light-weight shock machine is traditionally known for its severity. In fact, it is often somewhat erroneously stated that the light-weight shock machine is too severe, or even an over-test. While it is true that some equipment that fails a light-weight shock machine test can pass shock testing when the test is repeated on the medium-weight shock machine or the floating shock platform; it is unreasonable to completely dismiss the light-weight shock machine. Two common arguments are typically levied against the light-weight shock machine: the first is that it was designed to reproduce damage seen on equipment designed and built in the 1930s, and the second is that it does not replicate data measured from floating shock platform tests or ship shock trials.

The first argument that the light-weight shock machine was designed to replicate damage seen in 1930s equipment and hence not applicable to modern equipment misses the fundamental point of the machine. The machine was indeed designed to replicate battle damage on certain components; however, the damage was from components that the navy felt should have survived the shock event in order for the ship's crew to continue the fight. In a round-about-way, the light-weight shock machine was designed for a keel shock factor. The U.S. Navy has a design-level keel shock factor at which it is assumed that most of the ship's crew would not only survive the shock event but be physically and mentally capable of continuing the fight. The navy is not interested in reproducing shock damage to parts collected from sunken or destroyed vessels, but rather reproducing damage in scenarios where the equipment failed while the crew was still capable. As such, the shock levels imparted by the light-weight shock machine are still very much relevant today because they indirectly correlate to crew capability.

The second point is true that the measured acceleration and velocity time histories from a light-weight shock machine will not match with those measured on a floating shock platform or in a ship shock trial. However, this is not a limitation of the machine. As was shown in Sect. 7.3, the peak stress is proportional to the maximum velocity and not necessarily the individual shock time history. It is also readily apparent that the actual war-time threat could be any number of practically infinite shock insult possibilities. Therefore, it is entirely unrealistic to dismiss a shock machine on the basis of the fact that its induced time history does not agree with another method's time history since the real-life insult will not agree with either.

9.6.3 Medium-Weight Shock Machine

The first medium-weight shock machine (often abbreviated as MWSM) was built by Westinghouse Electric Corporation in 1942 and was installed at the Naval Engineering Experimental Station in Annapolis, Maryland in 1943. The design intent was to generate a shock intensity roughly equivalent to the light-weight shock machine already in use. The goal was to impart a 2.1 m/s (7 ft/s) velocity change to a 113 kg (250 lbf) test item and a 6 ft/s velocity change to a 2040 kg (4500 lbf) test

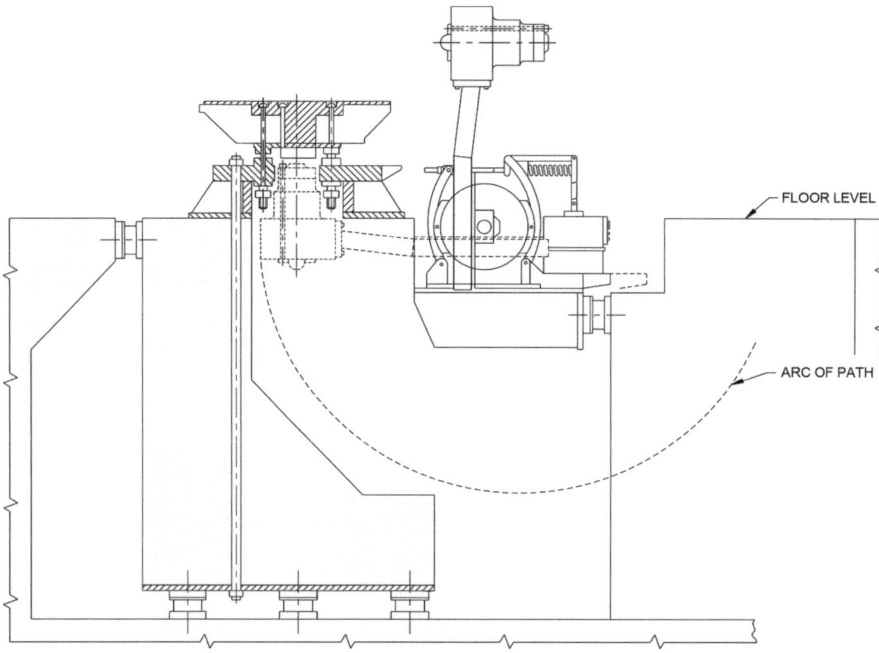

Fig. 9.29 Drawing of the U.S. Navy medium-weight shock machine, U.S. Navy drawing [9]

article. The lower velocity change at the higher weights was considered acceptable since heavier equipment is generally more difficult to accelerate in the shipboard configuration due to increased deflection of the surrounding ship structure.

The medium-weight shock machine uses a 1360 kg (3000 lbf) hammer swinging on a 680 kg (1500 lbf) arm that strikes the underside of a 2040 kg (4500 lbf) pound anvil plate. The entire medium-weight shock machine is installed in a 41 t (90,000 lbf) reaction mass that is isolated on coil springs from the surrounding building structure. The medium-weight shock machine also maintains the correct equipment orientation during the shock event. By having the hammer swing underneath and impact the anvil table from below, the upward vertical shock is oriented the same as the underwater shock event that the machine was designed to replicate. Figure 9.29 shows a drawing of the medium-weight shock machine. Medium-weight shock machine testing requires a minimum of six shocks to be applied to the equipment under test. Here again, the goal is to not only qualify the equipment to a specified shock level but to also expose the equipment to a few fatigue cycles.

Standard fixtures for the medium-weight shock machine are intended to represent hull mounted locations onboard ships. The frequency response of the standard fixtures is generally intended to fall in the 55–70 Hz range although frequencies have been measured up to approximately 90 Hz.

The medium-weight shock machine was originally used for equipment and fixtures with a total weight up to about 3350 kg (7400 lbf); however, it was quickly recognized that the velocity inputs were not reached with the heavier weights. For this reason, the practical upper weight limit on the medium-weight shock machine is about 2040 kg (4500 lbf).

9.6.4 Heavy-Weight Shock Test

Heavy-weight shock testing is typically performed on a floating barge with 27.2 kg (60 lbf) High Blast Explosive (HBX) depth charges to provide the excitation. The first Floating Shock Platform (FSP) was built in 1959 by the Underwater Explosion Research Division (UERD) at Norfolk Naval Shipyard. Four tests are usually required for qualification, the first with the depth charge located off the forward end of the FSP followed by three successively closer depth charges off the side of the FSP. The final shock is performed with the depth charge only 6.1 m (20 ft) abeam of the barge.

The standard Floating Shock Platform (FSP) measures approximately 4.9 m (16 ft) by 8.5 m (28 ft). The FSP can handle equipment weights up to about 27 t (60,000 lbf) provided the combined center of mass of the FSP and equipment is low enough to ensure stability. The navy has, in at least one circumstance, tested equipment heavy enough and with a high center of mass such that an outrigger was added to the back side of the FSP to increase stability although this is by no means common. Typically, equipment is installed on the FSP in the same manner as it is installed on the ship. Mounting locations are duplicated, and for deck mounted equipment a frequency tuned deck is installed in the FSP. There are currently several locations around the USA capable of performing FSP tests.

An alternate barge known as the Large Floating Shock Platform (LFSP) is also defined in MIL-DTL-901E. The LFSP is approximately 9.1 m (30 ft) by 15.2 m (50 ft) and uses 136 kg (300 lbf) HBX depth charges for the shock excitation. The LFSP is capable of testing equipment weighing up to 181 t (4,00,000 lbf) assuming the combined center of mass of the barge and equipment is adequate for stability. This is large enough to test many of the ship's generators and propulsion equipment. The LFSP is not nearly as common of a test as the FSP and there are limited testing locations for this platform.

A third size of barge was built some years ago and designated the Intermediate Floating Shock Platform (IFSP). As the name implies, it is sized between the FSP and LFSP with a deck area of approximately 6.1 m (20 ft) by 12.2 m (40 ft) and a 113 t (2,50,000 lbf) weight capacity. At this time, the only IFSP capable test facility is near Arvonia, Virginia.

Another somewhat similar heavy-weight test platform is the Submarine Shock Test Vehicle (SSTV). This is a submersible section of a submarine hull to which equipment can be attached. Testing is then performed using depth charges with the test vehicle submerged.

One of the advantages to testing equipment on any of these barge platforms is the ability to achieve multi-axis inputs with a single test. All heavy-weight barge tests are two-axis tests at a minimum. The primary excitation shock direction from a surface barge test is vertical as the barge is lifted from the water but there is also a substantial horizontal excitation component.

9.6.5 *Alternative Test Methods*

MIL-DTL-901E does permit custom test methods and test equipment, although this is not particularly common. One of the more interesting alternative test machines was built at the Naval Ordnance Laboratory in White Oak, Maryland. This heavy-weight shock machine was originally designed and built for shock testing torpedoes. It consists of a section of a 16-in gun barrel liner mounted in a pendulum fashion such that it strikes the underside of a pivoting table mounted at a 45° angle. The angle of the impact is used to provide two-axis input to the torpedo and the hammer weight is approximately 4990 kg (11,000 lbf). The machine was relocated to the Naval Surface Warfare Center at Dahlgren, Virginia when the naval laboratory at White Oak was closed. The machine has been used in recent years for shock qualification of various weapon systems and components with weights up to about 3630 kg (8000 lbf).

9.7 Live-Fire Testing

Shock testing in general is designed to simulate the effects of a high-energy event on a system. However, in some cases it is necessary or desirable to simply reproduce the actual event. This can be very instructive and informative as well as quite fun and exciting. After all, there should be an element of fun in one's career [11].

Rail impact tests have often been performed with actual rail cars rolling down and inclined track and coupling to other rail cars. Transportation shock has been tested by driving cargo loaded trucks over roads intentionally designed to represent the poorest of roads in the world. Rifle optics are often tested by firing hundreds of rounds of ammunition down-range. Even fully manned naval warships are subjected to depth charges to test their ability to survive and function during and after an underwater shock event.

Live-fire testing also frequently reveals flaws that do not show up in the laboratory. For example, rifle optics are almost always qualified using actual ammunition fired down-range. The reason for this is that a suitable substitute testing method has not been developed. No other loading condition exactly simulates the firing, recoil loading, muzzle blast, and muzzle flip of an actual rifle shot.

A fascinating example of live-fire testing was a man-portable anti-tank system on which the author was privileged work. The system was tested using practice ammunition in a purpose-built test stand at the range with all units passing the test. The system was then given to military personnel for live-fire testing and the sighting system failed on every unit tested. This was quite an embarrassment for the laboratory that designed the sights. After considerable research and investigation, it was discovered that the method used to secure the weapon in the test stand changed the modal characteristics of the weapon system enough that the sights were not being adequately stressed. In contrast, when the system was removed from the stand and fired by the end user, the results were very different because the modal characteristics were also very different.

Another interesting example of live-fire testing uncovering a problem was from a mortar tube upgrade program. The initial design of the mortar had a screw-on base cap. During live-fire testing, the base cap begun to unscrew resulting in propellant blow-by at the threads. The cap was reinstalled and it promptly loosened with firing. The torque was doubled and the loosening continued. The design was finally replaced with a monolithic design eliminating the problem completely. This type of problem never shows up in an analysis as it is beyond the state of the art for system modeling. The failure also did not show up in laboratory testing. The mortar was successfully fatigue tested in the laboratory using a specially designed tester to simulate the rapid application and release of the internal pressure load associated with the weapon firing.

Examples such as these are the primary reason that the US Department of Defense requires live-fire test and evaluation of new systems. Contrary to what is often stated, analysis capabilities can only inform testing, they have not matured to a level necessary to replace testing. Live-Fire Test and Evaluation (LFT&E) is also codified in law. Title 10, Section 2366 of the United States Code requires major systems and munition programs to undergo survivability and lethality testing before full-scale production and preferably early enough in the program that deficiencies can be cost effectively corrected.

9.7.1 Ship Shock Trials

One of the most involved applications of live-fire testing is the U.S. Navy's full ship shock trials. The U.S. Navy generally shock tests one of the first ships of each new class. Ship shock trials are designed as a large-scale system test used to evaluate the performance of the ship system as a whole. The shock trials are also performed on fully manned ships. As a result, the shock levels imparted to the ship are somewhat below the design-level shock. Figure 9.30 shows a photograph of one shock in USS *Arkansas*, CGN-41, ship shock trials. As can be seen in the photograph, the explosive charge used for a full ship shock trial is quite impressive and the distance is relatively close for a fully crewed operational warship.

Ship shock trials typically have significant instrumentation installed throughout the ship during the shocks. Accelerometers, velocity gages, and strain gages are frequently used to collect as much data as possible about the dynamics of the ship and various ship systems. These data are then used to evaluate systems, validate

Fig. 9.30 USS *Arkansas* ship shock trials, March 1982. U.S. Navy photograph

finite element models, perform pre- and post-test data analysis, as well as numerous other applications. The purpose is to learn as much as possible about the ship and ship systems from these highly complex, extremely expensive tests.

The ship shock trials are often mistakenly assumed to represent a true shock environment. While the applied loading is certainly within the realm of possible loads, it is not necessarily the true load. Underwater sea mines come in numerous sizes and could be detonated at an infinite number of ranges and angles with respect to the ship. The large charges used in full ship shock trial tests are not designed to represent specific torpedoes—although they are often assumed to represent a nuclear torpedo. Rather, the large charge at a larger standoff distance is designed to load the ship with a nearly planar shock wave. The plane wave impact loads the ship largely uniformly and lifts the ship somewhat evenly. Small explosions located close to the ship have a much more localized effect on the ship but can also be more damaging due to induced hull whipping.

Recent ship shock trials include: USS *Mesa Verde,* LPD-19; USS *Winston Churchill,* DDG-81; USS *John Paul Jones,* DDG-53; USS *Theodore Roosevelt,* CVN-71; USS *Mobile Bay,* CG-53; as well as many others. Not all ships undergo exactly the same set of shock loadings, although most are similar. However, the list does underscore the importance of live-fire testing to the military. Each ship is tested as a fully operational warship with a full crew complement. Sometimes, the testing reveals weaknesses in equipment, sometimes the tests reveal weaknesses in training and crew response. However, the goal is always to improve the entire system, both equipment and crew.

9.8 Test Fixture Design

No discussion of shock testing would be complete without at least a cursory discussion of test fixture design. The test fixture is the interface between the unit under test and the shock test machine. As the interfacing component, the fixture brings significant influence to the equipment being tested. Unfortunately, test fixture design is not always given the appropriate level of attention. This is partly a result of numerous competing requirements. The ideal test fixture would cost nothing, weigh nothing, perfectly represent the component interface in the next level of assembly, work for every conceivable test orientation, and with all forms of test equipment. Add to this the self-imposed complication that fixture design is typically considered a minor task of secondary importance and the result is a less than desirable situation.

Fixture weight is one of the more important factors in the design. Fixture weight directly impacts the force that could be applied to the test article. If the shock machine is capable of testing 50 kg but the fixture weighs 25 kg, then the weight of the unit under test is limited accordingly. Increasing the mass on the shock machine also effects the dynamic response of the machine itself. More mass on the shock machine lowers the system natural frequency and the maximum acceleration. This could be especially problematic if the frequencies are moved too close to the frequency range of the shock test.

Fixture stiffness is another important design factor. There are two general lines of thought regarding test fixture stiffness: fixtures should be as rigid as possible, or fixtures should match the stiffness of the next level of assembly. Rigid fixtures are very popular. Part of their popularity rises from the field of vibration testing where stiff test fixtures are frequently designed to have their lowest fundamental natural frequency higher than the frequency range of the test. This design greatly simplifies shaker control and can help with shock testing. On the other hand, overly stiff test fixtures can load the test unit in ways that differ from the next level of assembly. Whether or not this loading difference is significant depends on the system.

Fixture material is also an important consideration with more limited options. Most fixtures are made from aluminum or steel although other choices are available. Aluminum has an advantage for small test fixtures because of its strength and light weight. Steel is a good option for very large test fixtures. Fixture material choices should also include consideration of fatigue life. Most fixtures will be used repeatedly, sometimes for many years. As a result, most fixtures should be designed to the material endurance limit. In addition, aluminum fixtures should make use of steel threaded inserts to minimize thread damage from repeated use.

Shock fixtures should also be designed with positive restraint for the unit under test. Frictional interfaces will always slip under shock loading and cannot be relied upon to carry load. Likewise, bolted joints need to have adequate strength and relatively close-fitting hole sizes. If significant shear load is expected, then the use of shear pins is advisable. Bolted joints tend to loosen under shock loads and care should be taken to prevent this or ensure that bolt torque is checked during a test sequence. Short, stout bolts are preferred for shock testing. Long bolts and small cross-section bolts both tend to stretch under shock loading allowing for momentary preload loss and unanticipated part movement. Fixture designers should consider the sage advice—you should not be testing the test fixture in conjunction with the unit under test. In other words, there should be no risk of test fixture failure during the shock test.

9.9 Summary

There are many different methods of shock testing systems and components. This chapter has presented a brief overview of several popular shock testing options. When specifying a shock test, it is imperative to understand both the goal of the shock test and the capability of the different test machines. Shaker shocks are good for many types of insults but will not impart significant velocity, and hence momentum, to the component under test. Likewise, pyroshocks are known for being a zero-velocity shock event so a shock machine with a large velocity change would be inappropriate. If the test is supposed to represent a drop event, then a drop table should be used, regardless of whether or not another machine is capable of performing the test.

Shaker shocks are a frequent abuser of shock testing capability because of a desire to perform shock testing in conjunction with vibration testing. This typically saves time in the laboratory and often matches the SRS; however, it is often not a very good representation of the specified test. Sometimes, it is necessary to substitute one shock machine test for another shock machine test when equipment or laboratory time is unavailable. When such substitutions are made, it is important to understand the differences in the resulting test. Not only is it important to understand differences in the SRS, but also the differences in the resulting shock loading.

Problems

9.1 Using some readily available drop shock test data, compare the 10% amplitude duration with the full theoretical duration. How close do the two numbers agree with the theory?

9.2 Derive the relationship between the theoretical full duration and the 10% amplitude duration for the other classical shocks.

9.3 Resonant plates or beams are usually made from aluminum. How do the dimensions change if they are made from steel? Design a 500 Hz resonant plate in both steel and aluminum and compare the sizes and weights. Aluminum is usually used because the internal damping is higher and the shock excitation dies out faster. A steel plate will ring for considerably longer, exciting the component under test to a greater extent.

9.4 Resonant plates can also be made in other shapes besides square. Rectangular plates and circular plates can also be built. Research the equations for natural frequency and mode shape for a circular plate and compare with the equations for a square plate. Remembering that the lower modes are more easily excited than the upper modes, how do you think a circular plate would compare with a square plate in terms of size, weight, and modal content?

9.5 Shaker shock is fundamentally different from a shock machine. The reason, as discussed in this chapter, is that a shaker uses a series of oscillations to build up an SRS, whereas a shock machine hits its maximum velocity change or peak acceleration on the first pulse. Typically, the maximum acceleration from a shaker shock time history will be one-half to one-third of that from a shock machine. Create a shaker shock time history with a similar SRS to a classical shock pulse and compare the damage potential of the two signals.

9.6 The U.S. Navy Lightweight Shock Machine is actually a resonant plate type shock machine. A hammer is used to excite a plate where the unit under test is attached. The LWSM standard mounting plate is a rectangular plate approximately 686 mm wide by 864 mm tall and 13 mm thick. The boundary conditions can be approximated as simply supported on the two long edges and free on the two short edges. What are the lowest natural frequencies of the LWSM plate? Do you have any test data to compare with these calculations?

9.7 Fixture design has been a popular debate topic for many years. What are the properties of a good fixture compared to a bad fixture? Model a fixture-component system as a two-degree-of-freedom system and compare the component responses for various frequencies of the fixture and the component. How does the first cycle response compare? How does the overall response compare?

9.8 The purpose of shock testing is typically to demonstrate a minimum level of robustness for a component or system. In this case, the specifics of the shock may be less important than the shock amplitude. Prove this by analyzing a simple system, such as a cantilever beam, subjected to different shock pulses that have the same velocity change. How close are the results?

References

1. Spletzer, M., & Sisemore, C. (2016). Design of a resonant plate shock fixture to attenuate excessive high-frequency energy inputs. In *Proceedings of the 86th Shock and Vibration Symposium*, New Orleans, LA.
2. Blevins, R. D. (2001). *Formulas for natural frequency and mode shape*. Malabar, FL: Krieger Publishing Company.
3. Davie, N. T. (1996). *Simulation of Pyroshock Environments Using a Tunable Resonant Fixture*. U.S. Patent No. 5,565,626.
4. United States Department of Defense. (2014). Department of Defense Test Method Standard; Environmental Engineering Considerations and Laboratory Tests, MIL-STD-810G (w/Change 1), 15 April 2014.

5. United States Department of Defense. (1999). *Test Requirements for Launch, Upper-Stage and Space Vehicles*. Department of Defense Handbook, MIL-HDBK-340A, 1999.
6. National Aeronautics and Space Administration. (2011). *Pyroshock test criteria*. NASA Technical Standard, NASA-STD-7003A, 2011.
7. National Aeronautics and Space Administration. (2001). *Dynamic environmental criteria* NASA-HDBK-7005, 2001.
8. Air Force Space Command. (2014). *Test Requirements for Launch, Upper-stage and Space Vehicles*. Space and Missile System Standard SMC-S-016, 2014.
9. U.S. Department of Defense. (2017). *Detail specification, requirements for shock tests, H.I. (high-impact) shipboard machinery, equipment, and systems*, Washington, DC: MIL-DTL-901E, 20 June 2017.
10. Bort, R. L., Morris, J. A., Pusey, H. C., & Scavuzzo, R. J. (2001). *Practical shock analysis & design*. Course Notes, Society for Machinery Failure Prevention Technology.
11. Pusey, H. C. (2014). *What a fun business: A look back on the professional life of Henry C. Pusey*. Arvonia, VA: HI-TEST Laboratories.

Chapter 10
Temporal Information

One of the primary limitations with shock response spectra is that all temporal information is lost in the non-linear transformation. Couple this with the fact that many different shock signals of many different time lengths can generate the same SRS and one could use a short time signal in one case to achieve an SRS and a long duration signal in another case to achieve the same SRS. One could readily imagine a case where a high acceleration, short duration shock is measured in the field but the laboratory test makes use of a longer duration, lower acceleration shock pulse that yields the same SRS. Some shock transients have a high initial transient rise followed by a longer decay, such as with a decayed sine excitation. Some of the classical shock pulses are relatively symmetric in that the rise and fall are approximately equal. However, the SRS is relatively unaffected by the symmetry, or lack thereof, of the shock pulse or the shock pulse duration. This is a direct result of the non-unique nature of the SRS transformation as discussed in Chap. 4. Are these different tests equivalent? For this reason, there has long been an interest in retaining some minimal temporal information about the original shock event to supplement the spectral data.

Why is this important? It is important because the SRS only shows the peak response. In a short duration intense shock, the peak response is usually reached on the first cycle, so any damage happens immediately. In a long duration transient shock, the structure may experience many cycles whose amplitudes are just below the maximum. In this case, the failure mode may be a function of more than just the peak response. If we know that damage is caused only by the peak response, then two shocks with the same SRS are equally damaging to a structure that can be represented by single degree of freedom system. However, often damage mechanisms are more complex and we need more information about the shock than just the SRS to evaluate its damage potential. Temporal information is another piece of information about the nature of the shock, like the velocity change, and whether it is a classical shock or complex shock that provides more insight into understanding

C. Sisemore, V. Babuška, *The Science and Engineering of Mechanical Shock*,
https://doi.org/10.1007/978-3-030-12103-7_10

the damage potential of a shock and how to simulate a shock measured in the field in the laboratory.

Shock duration is probably the first and simplest effort to preserve temporal information and provide additional information about the shock and its damage potential. While the shock duration does not offset the non-uniqueness of the SRS, it is helpful. Two shock excitations with nominally the same SRS and similar durations may be completely different in terms of waveform shape and character but they are at least starting to represent similar transients. Two waveforms with similar SRS and pulse duration will also of necessity have similar peak amplitudes.

Temporal moments were developed in the 1990s to characterize the temporal nature of the shock. Temporal moments are a method of estimating shock duration as well as several other statistical measures from the waveform. The idea is that the more closely the temporal moments from two shock pulses match, the greater likelihood that they will have the same type of damage potential. While it is obviously true that two very similar shocks will have the same temporal and spectral parameters and should therefore have similar damage potential, it is not clear that all shocks with similar spectral and temporal parameters are always necessarily similar.

Another area of concern with shock temporal information is that the calculations, like the SRS calculations, can be blindly applied but may not necessarily yield the expected results. The temporal moment calculations are more sensitive to noise-floor measurements than the SRS calculations. The details of the temporal estimates and calculations along with their sensitivities are discussed in Sect. 10.2.2.

10.1 Shock Duration

Shock duration, the effective length of the shock pulse, is an important characteristic for properly translating measured test data into realistic test specifications. Shock durations can vary substantially depending on the source and type of shock. For example, a crash test might have durations in the 15–25 ms range where a pyroshock might have much shorter durations. Since the SRS is a non-unique transform, shocks of various durations can produce the same SRS. As a result, it is desirable to preserve some form of temporal information about the shock transient to supplement the spectral data. The shock duration is usually the most common temporal component and is desirable to ensure that shocks synthesized in the test laboratory reasonably represent the shock environment measured in the field. Shock pulse duration is a vital consideration in that respect because a measured field event lasting tens of milliseconds would not be accurately represented by a laboratory shock event measured in seconds. As a result, a measure of the shock duration along with the SRS is a significant improvement over the SRS alone.

The duration of a transient waveform seems like it should be a straightforward measurement; however, the evidence suggests otherwise. The shock duration for a classical, analytically defined pulse is straightforward. A haversine shock or half-sine pulse is defined by an amplitude and duration without ambiguity. However, an experimentally measured waveform often has no clear beginning or end. This is a result of the instrumentation measuring a random noise floor prior to the shock event and the pulse decaying back, not to zero, but to the original noise floor. At what point does the transient cease and the noise floor begin? The transition is frequently ambiguous. The difficulty can be readily understood by reading the current and past revisions of MIL-STD-810.

The current version of MIL-STD-810G [1] contains an interesting definition for effective shock duration. MIL-STD-810G with Change 1 defines the effective shock duration, T_e, to be the time from the first measurement acceleration zero crossing above the noise floor until the perceived termination of the shock. This is a very subjective measurement and open to interpretation. While subjective interpretations of data are not necessarily bad or inappropriate, they do not lend themselves to incorporation in data processing codes since human judgment is required.

The previous version of MIL-STD-810G [2] contains a different definition for the effective shock duration. The 2008 edition of MIL-STD-810G defines the effective shock duration, T_e, to be the minimum length of continuous time that contains the root mean square (RMS) time history amplitudes exceeding 10% of the peak RMS amplitude. This definition has a significant advantage over the current definition in that it is readily translated into a programmable algorithm and not necessarily subject to personal interpretation—unless the instrumentation noise floor is too high. The difficulty with this definition is that it takes a finite amount of time for the shock pulse to rise above the noise floor and reach the 10% threshold. As a result, T_e does not start at the exact moment of the shock initiation. Figure 10.1 shows a simple oscillatory shock pulse with the shock duration, T_e, overlaid on the waveform. In this example, the duration, calculated from the original MIL-STD-810G [2] definition, is approximately 3.65 ms. The duration is defined by the first and last times the waveform exceeds 300 g since that is a tenth of the 3000 g peak amplitude.

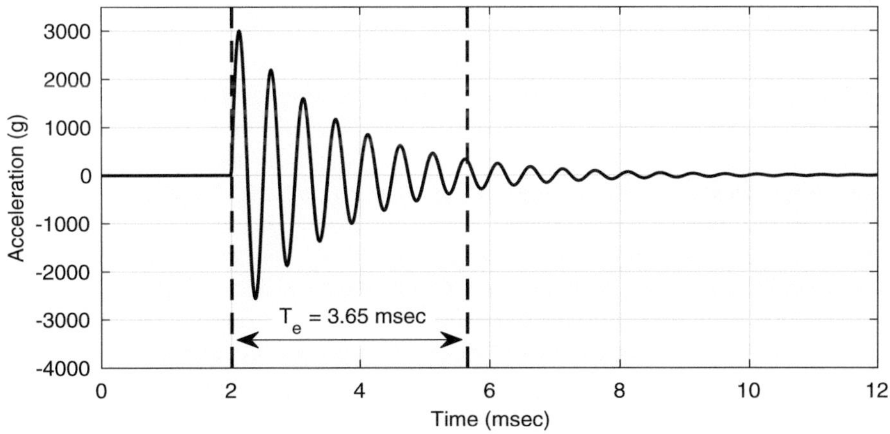

Fig. 10.1 Plot of a 3000 g oscillatory shock response with the shock duration T_e defined from MIL-STD-810G (2008 release)

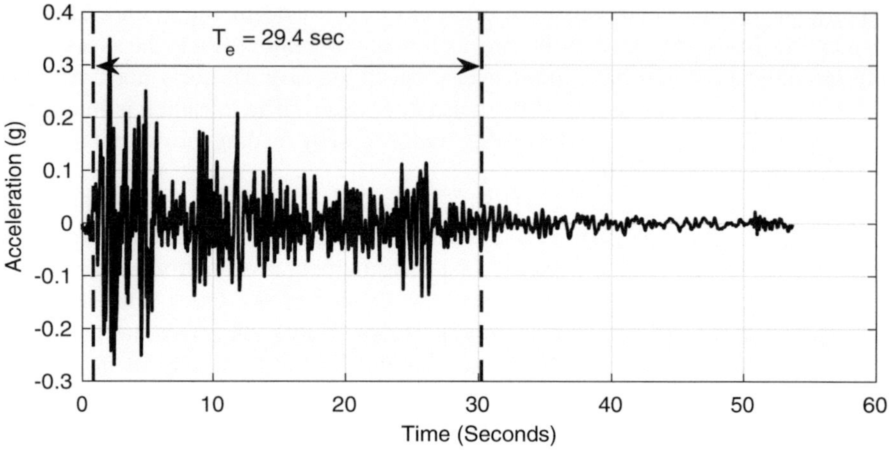

Fig. 10.2 Plot of the May 1940 El Centro earthquake shock time history response with the shock duration T_e defined from MIL-STD-810G (2008 release)

The shock duration example shown in Fig. 10.1 is straightforward but one could readily see how a more interpretive definition such as the one in the newer MIL-STD-810G with Change 1 [1] might be more desirable. With the updated definition that the shock duration extends until the "perceived termination of the shock," it could easily be argued that the shock duration in Fig. 10.1 might more reasonably be in the 6–8 ms range as opposed to the 3.65 ms calculated using the equations. Figure 10.2 shows the acceleration time history response from the May 1940 El Centro earthquake and the resulting shock duration calculated from [2]. While a T_e of 29.4 s is reasonable for this event, reasonable people could choose a different value if one of the more subjective measurement approaches was selected. The intention here is not to definitively state whether the T_e calculation of MIL-STD-810 is more or less accurate than the subjective T_e approach of MIL-STD-810G with Change 1. Rather, the purpose is to highlight the need for consistency when analyzing data and the need to understand the methods used when data are summarized and provided to the analyst without the benefit of insight from the original waveforms. For example, suppose an analyst was asked to synthesize an oscillatory shock waveform similar to Fig. 10.1 with a 7 ms shock duration. With no further information or direction, one could readily imagine several possible waveforms of varying durations that would all equally well fit within the required 7 ms shock duration. This is obviously a less than ideal situation for shock synthesis.

If the two definitions for shock duration in subsequent revisions to MIL-STD-810G are not enough, MIL-STD-810E [3] defines the effective shock duration, T_E (note the subtile change in the subscript) to be the minimum length of time containing all data magnitudes exceeding one-third of the peak absolute magnitude. It then goes on to explain that the response spectrum analysis duration should be $2T_E$ with the start and end-points of the time window situated to encompass the

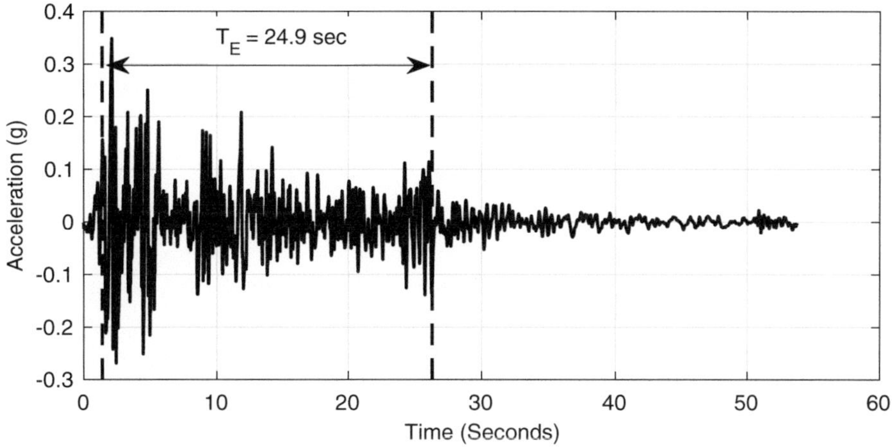

Fig. 10.3 Plot of the May 1940 El Centro earthquake shock time history response with the shock duration T_E defined from MIL-STD-810E

most significant data prior to and following the T_E window. Perhaps this is the worst of both options; a clearly implementable algorithm coupled with required human interpretation.

Figure 10.3 shows the May 1940 El Centro earthquake response shown in Fig. 10.2 and the T_E shock duration calculated from MIL-STD-810E [3]. The duration calculated using the one-third peak method is approximately 24.9 s as opposed to the 29.4 s calculated using the one-tenth RMS method. The MIL-STD-810E method would subsequently use an analysis duration of 49.8 s—calculated as $2T_E$. This is significantly different from the 29.4 s shown in Fig. 10.2 but perhaps closer to what an analyst might subjectively determine is the perceived shock termination point.

MIL-STD-810G [2] implies that the shock durations defined here are generally only applicable to oscillatory type waveforms. The tacit assumption is that the waveform is reasonably symmetric in the positive and negative oscillations. That assumption would preclude the use of these equations for all of the classical shock waveforms since the classical pulses are all one-sided excitations. Nevertheless, it is interesting to apply these methods to a classical pulse to see how the estimates might compare to the known duration. Figure 10.4 shows this comparison for a 5 ms haversine shock. In this case, the pulse duration is exactly 5 ms with no ambiguity. The one-tenth RMS amplitude method of [2] yields a shock duration, T_e equal to 3.98 ms. While this is about 80% of the actual duration, it is probably quite close to what would likely be measured in the laboratory where instrumentation noise is also measured. In contrast, the one-third peak amplitude method of [3] shows a duration, T_E equal to 3.04 ms with a $2T_E$ analysis window of 6.08 ms—longer than the actual event.

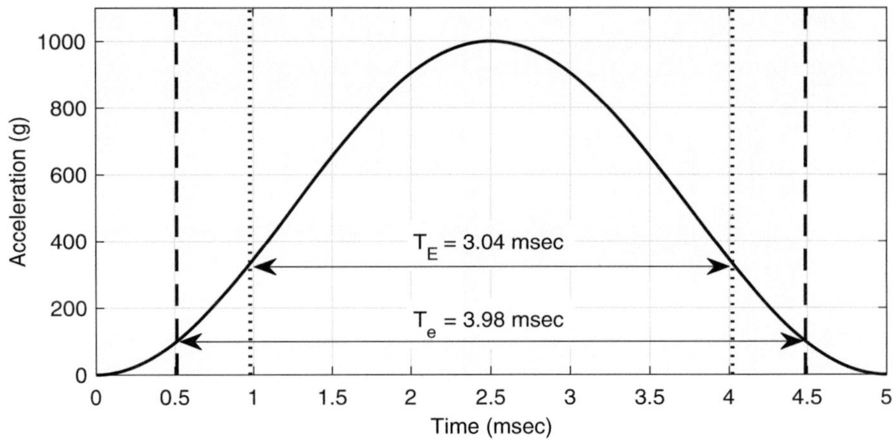

Fig. 10.4 Plot of a 5 ms haversine shock time history with the shock durations T_e and T_E overlaid for comparison

Shock duration seems like it should be a relatively straightforward topic with a clearly defined answer. However, the fact that the definition has changed multiple times over the last several revisions of MIL-STD-810 is an indication that this topic is not as clearly defined as it might at first appear. In reality, shock duration is rather more nuanced. Defining a shock duration is a very important step in understanding the shock event and developing representative test specifications for laboratory experiments. However, it is incumbent upon the engineer to ensure a consistent definition across their work to ensure their intentions are appropriately implemented.

10.2 Temporal Moments

Another method to preserve temporal information is with the use of band limited temporal moments proposed by Smallwood [4, 5]. The concept of temporal moments is analogous to probability density function moments from statistics. Two realizations of a random process will not have the same time histories, because they are random. However, two time histories can be said to be realizations of the same random process if their moments are the same. For example, two realizations of a Gaussian process will have same mean and standard deviation. The idea behind temporal moments is similar. Two shocks that have the same temporal moments can be considered to be equivalent in some sense.

The ith temporal moment $m_i(a)$ of a time history, $x(t)$, about a time location, a, is defined by Smallwood as:

$$m_i(a) = \int_{-\infty}^{\infty} (t-a)^i \, [x(t)]^2 \, dt. \tag{10.1}$$

There are two significant differences between temporal moments and probability density moments. The first is that the definition of temporal moments uses the square of the time history whereas probability moments simply use the probability distribution function as-is. The second significant difference is that temporal moments operate on a time history whereas probability moments are operated on a probability distribution function. Using the square of the time history alleviates the problem that the zeroth temporal moment might not be positive. If the time history in Eq. 10.1 were not squared and $a = 0$, the equation reduces to a calculation of the time history mean. The mean of a transient excitation should be nominally zero, although it could be a small positive or small negative value.

Using the square of the time history also provides a relationship between the time domain and frequency domain through Parseval's identity. Parseval's identity states that the square of the energy in the time history is equal to the square of the energy in the frequency response. This is written as:

$$[x(t)]^2 = [X(\omega)]^2 . \tag{10.2}$$

While Eq. 10.1 defines the actual moments of the time history, when the term *temporal moments* is typically used, it most often refers to the normalized temporal moments about the time history centroid. The reason for this is that the temporal moments as defined in Eq. 10.1 are related to the origin of the time history which is arbitrary, making the definition given in this equation only a portion of the actual working definition. Equation 10.1 puts no restriction on the number of temporal moments that can be calculated although only the first five moments (0–4) are typically used. These first five moments have also been given special names relative to their specific meaning and the details for calculating these respective normalized moments are provided here.

The zeroth order temporal moment is known as the energy and is calculated by assuming $i = 0$ in Eq. 10.1. The zeroth order moment is also independent of a and is typically written as m_0 and is defined as:

$$E = m_0 = \int_{-\infty}^{\infty} [x(t)]^2 \, dt. \tag{10.3}$$

While this is called the energy, it is not actually energy in the pure use of the word. A unit analysis of Eq. 10.1 with $a = 0$ shows that the units are actually power per unit mass. Nevertheless, for historical reasons the zeroth order temporal moment will continue to be called energy.

The first order temporal moment is used to calculate the time history centroid by defining the first moment to be equal to zero. First define the point about which the moment is taken to be $a = \tau$ and assume $i = 0$ in Eq. 10.1. Thus, the first order moment about the point τ is defined as:

$$m_1(\tau) = \int_{-\infty}^{\infty} (t - \tau) \, [x(t)]^2 \, dt. \tag{10.4}$$

To find the signal centroid, set $m_1(\tau) = 0$ and solve for τ.

$$\int_{-\infty}^{\infty} (t - \tau)\,[x(t)]^2\,dt = 0$$

$$\tau \int_{-\infty}^{\infty} [x(t)]^2\,dt = \int_{-\infty}^{\infty} t\,[x(t)]^2\,dt \tag{10.5}$$

Recognize that the integral on the left side of Eq. 10.5 is simply τ times the energy from Eq. 10.3. The integral equation on the right-hand side will be designated m_1' as:

$$m_1' = \int_{-\infty}^{\infty} t\,[x(t)]^2\,dt. \tag{10.6}$$

Solving Eq. 10.5 for τ gives:

$$\tau = \frac{m_1'}{m_0} = \frac{m_1'}{E}. \tag{10.7}$$

As a result of the way τ was defined by setting $m_1 = 0$, τ is naturally normalized by the signal energy. The units of τ are time. All of the higher-order moments will be normalized by the signal energy as well and defined about the centroid. As such, they will be referred to as *normalized central moments*. Defining the temporal moments about the signal centroid alleviates the problem of having the temporal moments be a function of the time origin. Here again the shock origin for an ideal, classical waveform is easy to select whereas the choice for an experimentally recorded signal can sometimes be slightly ambiguous. Regardless, the time history origin is completely arbitrary in relationship to the shock pulse.

RMS duration is the square root of the normalized second central moment with the units also being time. The RMS duration is calculated by using Eq. 10.1 with $i = 2$ and normalizing by the energy, E, as shown:

$$D_T^2 = \frac{m_2(\tau)}{E} = \frac{1}{E}\int_{-\infty}^{\infty} (t - \tau)^2\,[x(t)]^2\,dt \tag{10.8}$$

However, the RMS duration, D_T, does not represent the duration of the shock pulse. Rather, the RMS duration represents the amount of time before and after the shock signal centroid where most of the energy is encompassed. For this reason, the significant portion of the shock duration is always double the RMS duration and the actual shock time history will typically be about three to five times the RMS duration.

Skewness is a measure of the shock pulse symmetry. A symmetric waveform has zero skewness while a pulse with positive skewness has more of the energy content prior to the centroid and a pulse with negative skewness has more energy after the shock centroid. The skewness is calculated using Eq. 10.1 with $i = 3$ and normalizing by the energy, E, as shown:

$$S_T^3 = \frac{m_3(\tau)}{E} = \frac{1}{E}\int_{-\infty}^{\infty} (t-\tau)^3\,[x(t)]^2\,dt. \tag{10.9}$$

The units of the skewness are also time. Most common shocks are going to have either a positive or zero skewness, S_T. The importance of skewness when deriving shock pulses for testing is not so much in the actual value but in matching the intent of the shock. For example, when testing certain shock or crash safety equipment a high positive skewness is desired because it more closely represents the actual environment. An automobile air-bag sensor has no warning that air-bag deployment is needed until the moment of the impact—high positive skewness. On the other hand, if the laboratory test made use of a shock profile with high negative skewness, motion would begin early and build up to the shock at the record end. In effect, this type of testing would provide the sensor with some level of a priori knowledge of the impending shock and very well might alter the test results.

Kurtosis is a shape descriptor of the function being analyzed—the time history in this case. Kurtosis is a measure of the number of extreme outliers in a signal. This description is frequently misinterpreted as a description of peakedness in the signal, a rather ambiguous definition. The kurtosis is calculated using Eq. 10.1 with $i = 4$ normalized by the energy, E, as:

$$K_T^4 = \frac{m_4(\tau)}{E} = \frac{1}{E}\int_{-\infty}^{\infty} (t-\tau)^4\,[x(t)]^2\,dt. \tag{10.10}$$

The units of kurtosis are also time, the same as the centroid, RMS duration, and skewness. While the lower-order temporal moments are relatively easy to understand, the kurtosis is less intuitive. Figure 10.5 shows two time history plots with the same energy and RMS duration but with different values of kurtosis. The plot on the left has a greater kurtosis although not significantly greater. The plot on the left also has slightly more, higher amplitude peaks than the plot on the right. Recall that temporal moments are calculated from the square of the time history. Figure 10.6 shows a more interesting comparison. The two time histories in Fig. 10.6 have the same energy, centroid, and RMS duration but look nothing alike. The curve with the lower kurtosis is a classical haversine while the curve with the higher kurtosis is a single frequency decayed sine shock. In this plot, the decaying sinusoid will have more peaks than the single-peaked haversine, hence the kurtosis should be greater.

While the kurtosis is an interesting discriminator between two or more shocks with differing time histories, it does not yet appear to be poised for widespread application in temporal moments. The numerical differences are too subtle for such widely varying signals for any type of shock synthesis definition. Nevertheless, one will see it reported from time to time and it is appropriate to understand its meaning.

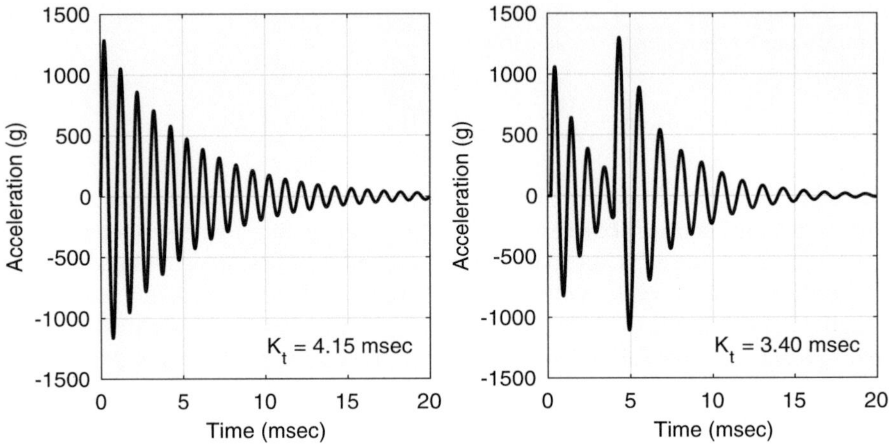

Fig. 10.5 Normalized kurtosis comparison between two shocks with the same energy and RMS duration

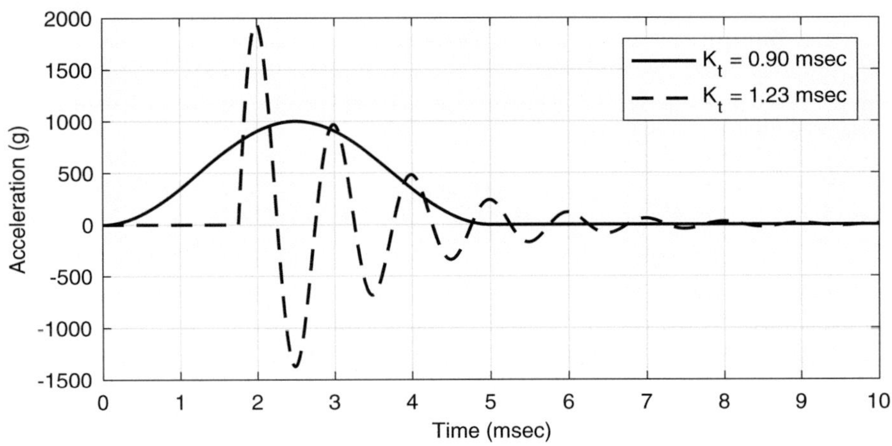

Fig. 10.6 Normalized kurtosis comparison between two different shocks with the same energy, centroid, and RMS duration

10.2.1 Temporal Moment Examples

The use of temporal moments is illustrated with two examples. The first example makes use of the 1000 g 5 ms haversine shock shown in Fig. 4.2. This classical shock pulse has the advantage that the temporal moments can be calculated in closed form using the integral expressions from Eq. 10.1 without the need for numerical integration.

The equation for the classical haversine shock is given by:

$$x(t) = \frac{A}{2}\left[1 - \cos(\alpha t)\right], \tag{10.11}$$

where A is the haversine amplitude, T is the haversine duration, and $\alpha = 2\pi/T$.

The zeroth order temporal moment, energy, is calculated by substituting Eq. 10.11 into Eq. 10.3 and solving.

$$m_0 = \int_0^T \left\{\frac{A}{2}\left[1 - \cos(\alpha t)\right]\right\}^2 dt. \tag{10.12}$$

The limits of integration have been changed here to represent the actual length of the shock pulse. Before and after the shock pulse the time history is zero by definition. The solution to this definite integral is relatively straightforward with a table of integrals. Thus, for the example haversine shock defined here, the temporal moment energy can be shown to be given by:

$$m_0 = \frac{3}{8}A^2 T = 1875\,\mathrm{g}^2\,\mathrm{s} = 1.80 \times 10^5\,\frac{\mathrm{N\,m}}{\mathrm{kg\,s}}. \tag{10.13}$$

As stated earlier, the units are not actually energy but power per unit mass. This result can also be verified by numerically integrating the haversine time history if desired.

The first temporal moment of the haversine shock is given by Eq. 10.4 as:

$$m_1 = \int_0^T (t - \tau)\left\{\frac{A}{2}\left[1 - \cos(\alpha t)\right]\right\}^2 dt. \tag{10.14}$$

However, in this equation, the intent is to set $m_1 = 0$ and solve for τ. Thus the equation becomes

$$0 = \int_0^T t\left\{\frac{A}{2}\left[1 - \cos(\alpha t)\right]\right\}^2 dt - \tau\int_0^T \left\{\frac{A}{2}\left[1 - \cos(\alpha t)\right]\right\}^2 dt. \tag{10.15}$$

The second half of Eq. 10.15 is just the expression for m_0 given in Eq. 10.12. The first half of Eq. 10.15 is m_1' from Eq. 10.6 and the solution to the integral is:

$$m_1' = \int_0^T t\left\{\frac{A}{2}\left[1 - \cos(\alpha t)\right]\right\}^2 dt = \frac{3}{16}A^2 T^2, \tag{10.16}$$

once all simplifications have been performed and the integration constant is added to ensure the integral is zero at $t = 0$. Thus, solving for τ yields:

$$\tau = \frac{\frac{3}{16}A^2T^2}{m_0} = \frac{\frac{3}{16}A^2T^2}{\frac{3}{8}A^2T} = \frac{1}{2}T = 0.0025\,\text{s}. \tag{10.17}$$

The result here is very intuitive since the centroid of the classical pulse is equal to half of the total pulse duration. It is also evident that the signal centroid does not depend on the signal amplitude units.

The second temporal moment about the centroid for the haversine shock is given by Eq. 10.1 as:

$$m_2(\tau) = \int_0^T (t-\tau)^2 \left\{\frac{A}{2}\left[1-\cos(\alpha t)\right]\right\}^2 dt. \tag{10.18}$$

Expanding the polynomial gives:

$$m_2(\tau) = \int_0^T \left[t^2 - 2\tau t + \tau^2\right] \left\{\frac{A}{2}\left[1-\cos(\alpha t)\right]\right\}^2 dt. \tag{10.19}$$

From this equation it is apparent that two of the terms, m_0 and m_1', have already been calculated.

$$m_2(\tau) = \int_0^T t^2 \left\{\frac{A}{2}\left[1-\cos(\alpha t)\right]\right\}^2 dt - 2\tau m_1' + \tau^2 m_0. \tag{10.20}$$

The remaining integral term will be designated m_2' and the solution is:

$$m_2' = \int_0^T t^2 \left\{\frac{A}{2}\left[1-\cos(\alpha t)\right]\right\}^2 dt = \frac{A^2}{4}\left[\frac{1}{2} - \frac{15}{16\pi^2}\right]T^3, \tag{10.21}$$

once all the simplifications have been made. The RMS duration is then given as the square root of the normalized second temporal moment by

$$D_T = \sqrt{\frac{m_2(\tau)}{m_0}} = \sqrt{\frac{m_2' - 2\tau m_1' + \tau^2 m_0}{m_0}}. \tag{10.22}$$

Substituting the results from Eqs. 10.13, 10.15, 10.17, and 10.21 into Eq. 10.22 and simplifying gives the RMS duration of the example haversine shock as:

$$D_T = \sqrt{\frac{\frac{1}{4}A^2\left[\frac{1}{32} - \frac{15}{64pi^2}\right]T^3}{\frac{3}{8}A^2T}} = T\sqrt{\frac{1}{12} - \frac{5}{8\pi^2}} = 0.000707\,\text{s}. \tag{10.23}$$

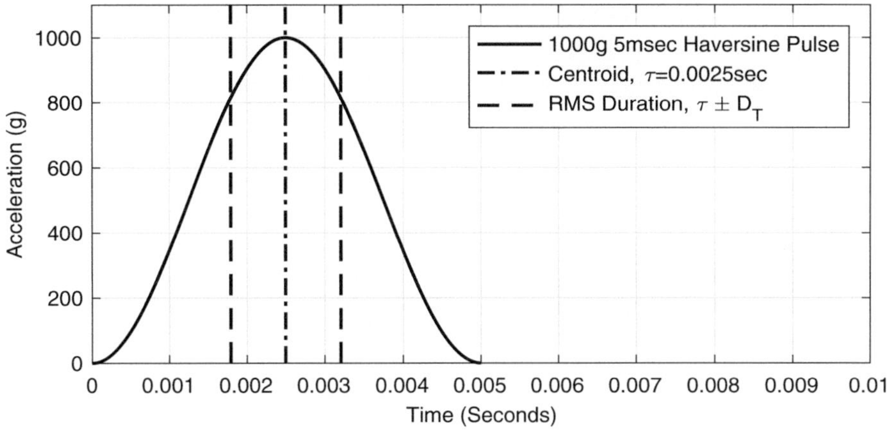

Fig. 10.7 Plot of a 1000 g 5 ms duration haversine shock pulse with the centroid and RMS duration marked

The example haversine shock time history along with the location of the centroid and the RMS duration are shown in Fig. 10.7. The centroid of this classical pulse is in the center of the shock pulse and the RMS duration represents the point on either side of the centroid that encompasses most of the shock energy. In this example, 53% of the area under the shock curve falls between the two RMS duration lines.

The third temporal moment for the haversine shock is given by Eq. 10.1 as:

$$m_3(\tau) = \int_0^T (t-\tau)^3 \left\{\frac{A}{2}\left[1-\cos(\alpha t)\right]\right\}^2 dt. \tag{10.24}$$

Expanding the polynomial gives:

$$m_3(\tau) = \int_0^T \left[t^3 - 3\tau t^2 + 3\tau^2 t - \tau^3\right] \left\{\frac{A}{2}\left[1-\cos(\alpha t)\right]\right\}^2 dt. \tag{10.25}$$

Similar to the equation for $m_2(\tau)$, three of the terms, m_0, m_1', and m_2' have already been calculated.

$$m_3(\tau) = \int_0^T t^3 \left\{\frac{A}{2}\left[1-\cos(\alpha t)\right]\right\}^2 dt - 3\tau m_2' + 3\tau^2 m_1' - \tau^3 m_0. \tag{10.26}$$

The remaining integral term will be designated m_3' and the solution is:

$$m_3' = \int_0^T t^3 \left\{\frac{A}{2}\left[1-\cos(\alpha t)\right]\right\}^2 dt = \frac{A^2}{4}\left[\frac{3}{8} - \frac{45}{32\pi^2}\right] T^4, \tag{10.27}$$

once all simplifications have been performed and the integration constant is added to ensure the integral is zero at $t = 0$. The skewness is the cube root of the normalized third temporal moment as:

$$S_T = \sqrt[3]{\frac{m_3(\tau)}{m_0}} = \sqrt[3]{\frac{m_3' - 3\tau m_2' + 3\tau^2 m_1' - \tau^3 m_0}{m_0}}. \tag{10.28}$$

Substituting the results from Eqs. 10.13, 10.15, 10.17, 10.21, and 10.27 into Eq. 10.28 and simplifying shows that the numerator under the radical is equal to zero. Therefore, the skewness of the example haversine is zero as would be expected. As defined earlier, the skewness of the shock is a measure of how much of the energy lies to the left or right of the shock centroid. Since a haversine shock is symmetric about the centroid the skewness should be zero. In contrast to the classical haversine shock shown in Fig. 10.7, Fig. 10.8 shows a theoretical shock pulse with positive skewness. Positive skewness is defined as a shock with the energy predisposed to occur near the beginning of the shock pulse whereas negative skewness is a shock with the energy predisposed to occur near the end of the shock pulse.

Finally, the fourth temporal moment for the haversine shock is given by Eq. 10.1 as:

$$m_4(\tau) = \int_0^T (t - \tau)^4 \left\{ \frac{A}{2} \left[1 - \cos(\alpha t)\right] \right\}^2 dt. \tag{10.29}$$

Expanding the polynomial gives:

$$m_4(\tau) = \int_0^T \left[t^4 - 4\tau t^3 + 6\tau^2 t^2 - 4\tau^3 t + \tau^4 \right] \left\{ \frac{A}{2} \left[1 - \cos(\alpha t)\right] \right\}^2 dt. \tag{10.30}$$

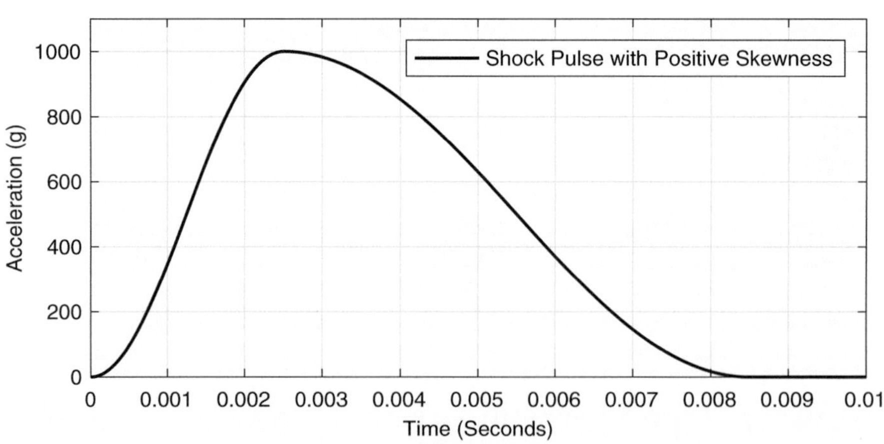

Fig. 10.8 Plot of a 1000 g 8.5 ms duration shock pulse with positive skewness

Here again, most of the terms in the integral have already been obtained while calculating the lower-order temporal moments. The only unsolved term, m'_4, is given by the expression

$$m'_4 = \int_0^T t^4 \left\{ \frac{A}{2} \left[1 - \cos(\alpha t)\right] \right\}^2 dt = \frac{A^2}{4} \left[\frac{189}{64\pi^4} - \frac{30}{16\pi^2} + \frac{3}{10} \right] T^5, \tag{10.31}$$

once all the simplifications have been performed. The kurtosis is the fourth root of the normalized fourth temporal moment as:

$$K_T = \sqrt[4]{\frac{m_4(\tau)}{m_0}} = \sqrt[4]{\frac{m'_4 - 4\tau m'_3 + 6\tau^2 m'_2 - 4\tau^3 m'_1 + \tau^4 m_0}{m_0}}. \tag{10.32}$$

Substituting the results from Eqs. 10.13, 10.15, 10.17, 10.21, 10.27, and 10.31 into Eq. 10.32 and simplifying yields:

$$\begin{aligned} K_T &= \sqrt[4]{\frac{\frac{1}{4} A^2 T^5 \left[\frac{189}{64\pi^4} - \frac{15}{32\pi^2} + \frac{3}{160} \right]}{\frac{3}{8} A^2 T}} \\ &= T \sqrt[4]{\frac{2}{3} \left[\frac{189}{64\pi^4} - \frac{15}{32\pi^2} + \frac{3}{160} \right]} \\ &= 0.0009 \text{ s}. \end{aligned} \tag{10.33}$$

While the preceding solution is very insightful from the standpoint of understanding temporal moments, it is not extremely practical for most shock applications. The example could be redone with any of the classical shock pulses for which a smooth, mathematically tractable description exists. However, what is far more common is to obtain a digital record of acceleration versus time with no obvious analytical conversion. In these cases, the obvious answer is to perform the integration numerically using a simple integrator.

Figure 10.9 shows a plot of the 18 May 1940 El Centro, California north-south ground response with the normalized centroid and RMS duration overlaid. The centroid and RMS duration were calculated by numerically integrating the temporal moment expressions given in Eqs. 10.7 and 10.8. In the case shown here, the signal centroid location is not as obvious as for the classical haversine shock. As stated earlier, the centroid is located at approximately the midpoint of the energy distribution and the RMS duration represents the time window which encompasses most of the shock energy. In Fig. 10.9 the bulk of the energy obviously occurs near the beginning of the shock pulse. By the end of the RMS duration, approximately 17.7 s, the shock pulse is still driving; however, its amplitude is about a fourth of its peak earlier in the record.

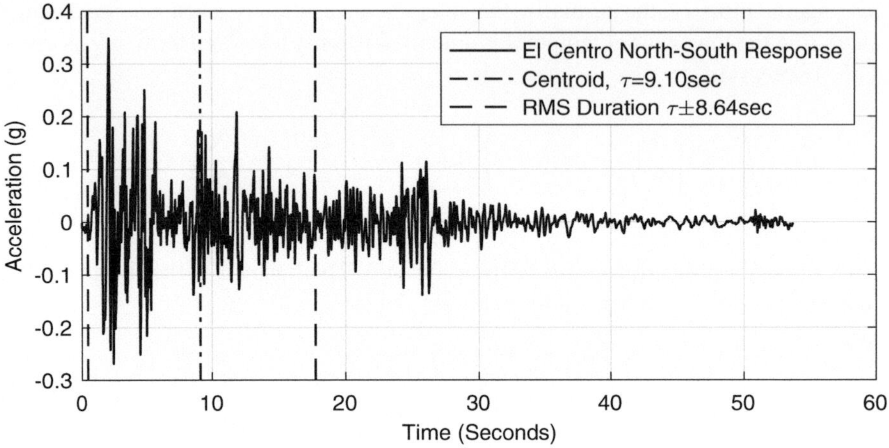

Fig. 10.9 El Centro, California 18 May 1940 north-south earthquake ground response with the centroid and RMS duration indicated

10.2.2 Limitations on the Use of Temporal Moments

Calculating temporal moments for ideal, classical shocks is straightforward and repeatable. However, calculating temporal moments on experimentally collected data can be a more subjective exercise. The reason for this is that classical shocks decay to zero and remain there. In contrast, experimentally obtained signals decay to small but finite noise-floor values. As a result, long duration shock histories will continue to accumulate energy in the temporal moments even if only noise is being recorded. The reason for this is seen in Eq. 10.1 where the time history is squared. As a result, the noise response does not cancel with nearly equal positive and negative excursions, but rather the noise squared is always positive so it is always additive. It has been previously noted that there is a need to exercise caution when calculating temporal moments from experimental data [4]. The warning is based on this idea of corrupting the temporal results by including too much of the noise floor; however, the warning has not been well quantified. A recent paper by the authors demonstrates some of the risks of including additional noise data in the temporal moment calculation [6].

Figure 10.10 shows an acceleration time history plot of a low-level shock from a drop table test. This time history is representative of one that could be reasonably used for calculating temporal moments. The signal includes some pre-shock data, the shock excitation nominally occurs at about $t = 0$, and some post-shock data were captured to ensure the response is completely characterized. The data shown in Fig. 10.10 appear to be of good quality, with the noise floor being almost imperceptible in the plot shown. The noise floor here is actually less than about 0.2 g, less than one-half percent of the acceleration peak. Test engineers would consider this very clean experimental data.

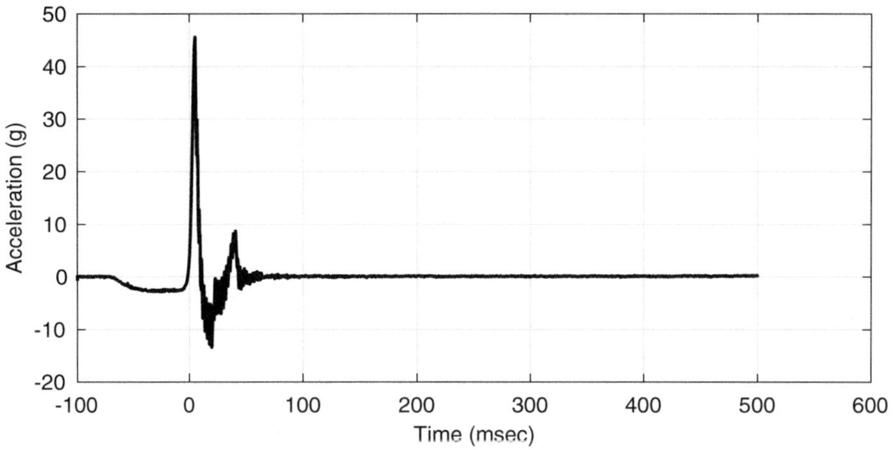

Fig. 10.10 Plot of the acceleration time history from a low-level drop table shock test

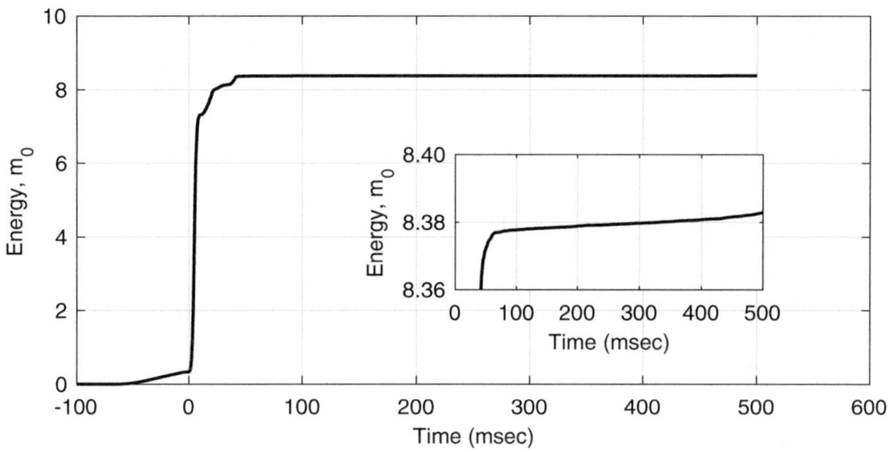

Fig. 10.11 Energy calculated from the measured drop table acceleration time history data

Normally, temporal moments are calculated and presented as scalar values. One calculates the RMS duration and records it as a number with no additional qualifiers. However, while the temporal moment equations can produce a single scalar result, it is also possible and insightful to calculate and plot the temporal moments as a function of time. Figure 10.11 shows a plot of how the zeroth order temporal moment, energy, changes with respect to time for the drop shock shown in Fig. 10.10. The energy starts near zero prior to the shock, rises slightly with the motion before the primary impact, rises sharply with the shock impact, and trends to a flat line as the shock decays to zero as expected. However, the inset plot in Fig. 10.11 gives a slightly different conclusion. The inset plot shows that the energy

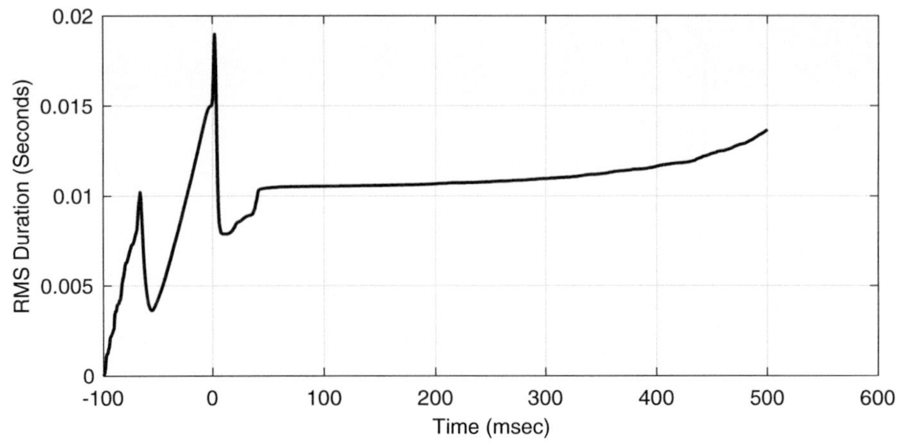

Fig. 10.12 RMS duration calculated from the measured drop table acceleration time history data

actually continues to increase after the shock until the record's end. The reason for this, as stated earlier, is that the noise floor is included in the calculation from immediately after the shock all the way to the end of the record. In this example, the energy at the record end is less than 0.1% of the energy immediately after the shock event—certainly a very minor difference.

Figure 10.12 shows a similar but more pronounced trend with the RMS duration. Even with this "clean" data, the instability in the RMS duration is obvious. The RMS duration that would normally be reported from a straightforward numerical integration is the value at the record's end, 13.7 ms, which is not trending to any particular fixed value. Neither is the value at the record's end the maximum value obtained by the calculation, 19.0 ms. Of course the maximum RMS duration value occurs before the shock transient has completely passed and as such would not be appropriate to report. The most correct RMS duration is probably the value occurring at about 50 ms, immediately after the shock transient. This value is 10.5 ms or about 77% of the value obtained if the full duration record is used for the calculation.

This example is not intended to discourage the use of temporal moments for shock analysis or specification definition, rather to highlight the need for careful definition of what exactly is being analyzed. Here again, the blind application of algorithms can lead to a result different than expected.

10.3 Summary

Shock analysis has a tendency to be focused almost exclusively on the SRS; however, the SRS is not sufficient to fully describe the shock event from which it was derived. Often the SRS is sufficient for defining test specifications or estimating damage. Other times, additional temporal information is necessary when the goal is to reproduce an SRS with a time signal that more closely mimics the one from which the original SRS was derived. How closely the laboratory experiment needs to match the field measurement is a topic for much debate, but in general, the test and field measurements should be relatively close. This similarity helps to ensure that the laboratory test is a reasonable representation of the anticipated environment that adequately simulates its damage potential.

Several definitions of shock duration were presented. The older MIL-STD-810 definitions for T_E and T_e are based on the time above a certain percentage of the maximum amplitude. The definition of T_e in the Change 1 update to MIL-STD-810G was shown to be somewhat subjective. The RMS duration defined by the second normalized temporal moment is a third measure of the shock duration. It is important to note that none of these definitions of shock duration will yield the same answer for a given shock. They are not differing ways of calculating the same numerical result, rather they are different ways of quantifying the shock duration. Since the methods are different, they are not interchangeable and there is no conversion from one method to another. Rather, a method should be selected and used consistently throughout a particular analysis.

Problems

10.1 Numerical calculation of temporal moments from experimental data was assumed to make use of the trapezoidal integration rule in this chapter. Calculate the temporal moments using a Simpson's rule integration routine or other more sophisticated numerical integrator and evaluate the difference between the methods. How do your answers change if the signal sample rate changes?

10.2 Calculate the RMS duration of a sample shock time history from data you have collected using different record lengths and record the differences in the results obtained. Compare the differences in the energy and kurtosis as well. Are these differences reasonable for the differences in the time history record length?

10.3 Calculate the temporal moments using closed-form techniques for an oscillatory shock excitation of the form:

$$\ddot{y}(t) = Ae^{-\zeta\omega t}\sin(\omega t) \tag{10.34}$$

How do the closed-form solution results compare with the numerically integrated results.

10.4 Two examples were provided in this chapter of disparate shocks with some identical temporal moments. Think about how you might accomplish this in conjunction with shock synthesis. Write some simple code to vary one time history until certain temporal moments match those of a reference time signal. You may find that it is helpful to converge some of the higher-order moments before the lower-order moments.

10.5 Calculate the shock duration for some of your test data using the different methods described here, and taken from different versions of MIL-STD-810. How do these results compare to what you would intuitively have selected for the shock duration? Is it any surprise that defining shock duration is a complicated task?

10.6 RMS duration has been proposed for use in conjunction with the SRS to better define a shock waveform. Compare the RMS durations from several shock time histories with the shock durations calculated using different versions of MIL-STD-810. While the numbers will be different because the methods differ, does any method stand-out as substantially better?

10.7 It was demonstrated that temporal moments can be heavily influenced by the inclusion of noise-floor data in the waveform. Can you think of a method to reduce the impact of noise-floor data during the calculation of temporal moments?

References

1. United States Department of Defense. (2014). Department of defense test method standard; environmental engineering considerations and laboratory tests, MIL-STD-810G (w/Change 1), 15 April 2014.
2. United States Department of Defense. (2008). Department of defense test method standard; environmental engineering considerations and laboratory tests, MIL-STD-810G, 31 October 2008.
3. United States Department of Defense. (1989). *Military standard; environmental test methods and engineering guidelines*, MIL-STD-810E, 14 July 1989.
4. Smallwood, D. O. (1994). Characterization and simulation of transient vibrations using band limited temporal moments. *Shock and Vibration, 1*(6), 507–527.
5. Smallwood, D. O. (1989). Characterizing transient vibrations using band limited moments. In *Proceedings of the 60th Shock and Vibration Symposium*, Virginia Beach, VA, November 1989. Portsmouth: Shock and Vibration Information Center.
6. Sisemore, C., Babuška, V., & Booher, J. (2017). Using temporal moments to detect interactions during simultaneous shock testing of multiple components. In *Proceedings of the 88th Shock and Vibration Symposium*, Jacksonville, FL, October 2017.

Chapter 11
Development of Shock Test Specifications

Many times, shock test specifications are provided by statute or contract and the requirement is simply to perform the test to the specifications. This is especially common when military specifications are levied on systems and components. Other times, it is the engineer's job to take environmental data from testing or analysis and derive the appropriate shock test for components. This happens most often for subcomponents or engineering development tests. This chapter is naturally focused on the later scenario.

The first precursor to developing appropriate shock test specifications is to define the goal of the test. This was discussed previously in Chap. 9. As discussed in that chapter, the fundamental goal of shock testing is to either replicate a field event with a high degree of fidelity or to demonstrate a particular level of robustness for the component being tested. While these goals may sound similar, they are distinctly different. If the goal is to reproduce a measured event, then great care must be taken to faithfully replicate nearly all aspects of the measured event to ensure that the same potential failure modes are being exercised. On the other hand, if the goal is simply to demonstrate robustness, then certain simplifications and approximations are rightfully permitted. In most cases, the goal of shock testing is to demonstrate component robustness. While this may not always be stated clearly, the stochastic nature of real-world shock events generally does not lend itself to exact replication of measured environments. Likewise, the measurement of an actual shock event is no guarantee that subsequent shocks will be identical to those measured.

The second area of consideration when developing shock test specifications is simplicity. One of the age-old engineering principles is that things should be made as simple as possible but no simpler. This is certainly true with shock testing. Simpler test specifications are easier to implement and are generally more consistent and repeatable. In addition, simpler test specifications are easier to justify—meaning that it is easier to relate them back to estimates of the field measured environments. For example, if the goal is to demonstrate that an item can survive being dropped

C. Sisemore, V. Babuška, *The Science and Engineering of Mechanical Shock*,
https://doi.org/10.1007/978-3-030-12103-7_11

by a person, then specifying a drop test from a 1 m height is both simple and easy to justify as being an accurate representation of the event.

How complicated should a shock test series be? Many shock test series apply shocks in three orthogonal axes. The purpose being to excite the unit under test in the standard Cartesian x-, y-, and z-coordinate frame. It is then assumed that the component could survive a single shock applied in any arbitrary direction. Whether or not this is true depends on the system being tested. In this case, the specification calls for three shocks to be applied in order to qualify to a single expected shock. The problem quickly grows if the environment requires more than one shock test to envelope the expected event. Thus, if the specification is written as two shocks per axis, one can expect six shocks to be applied to the unit under test. This multiplication gives rise to the complexity trade for shock testing. While two or more tests may be beneficial in some respects for testing, the benefits may be outweighed by the costs and risks associated with performing an inordinate number of shock tests on the same unit. For this reason, it is often desirable to perform a single, less representative test rather than two or three more representative tests.

This chapter presents basic information for developing appropriate shock test specifications. In this chapter, we discuss:

1. shock test environment definitions;
2. single axis versus multi-axis testing;
3. simple ways for specifying classical shocks with a defined velocity change;
4. matching tests equipment to the field environment;
5. developing drop shock test specifications;
6. developing resonant plate test specifications;
7. evaluation of complex shock specifications;
8. test specification error and tolerances;
9. margin specifications and test compression.

11.1 Shock Environment Definitions

One of the main objectives of shock testing is to provide data with which to verify that the component or system meets requirements and will operate successfully in the field. The test environment depends on when in the design cycle the test is performed. Early in the cycle development tests may be performed. The objectives of these early tests are to gather information about the design and determine engineering parameters. Once the design has matured, qualification tests are used to verify that the design meets requirements with margin and demonstrates adequate robustness. The qualification tests also validate acceptance and screening test procedures to which manufactured parts may be subjected. Acceptance tests are workmanship tests to demonstrate that a specific part is free from defects and meets functional and performance requirements. Customers who receive the parts may also perform their own acceptance or lot screening tests to verify that the received components are free from defects. This is standard practice for high consequence items.

All mechanical environments, especially shocks, are inherently stochastic with some amount of variability. No two earthquakes are alike; you do not drop your phone the same way every time; the intensity of pyrotechnic events depends on parameters that you do not control precisely. This means that the in-service environment is really a range of environments with some distribution. Shock test specifications are based on the *Maximum Predicted Environment*, or MPE, which is a conservative bound on in-service environments. For aerospace equipment, the shock MPE is defined as the P95/50 shock response spectrum or 4.9 dB above the log mean SRS, whichever is greater.

The P95/50 criterion means that the MPE will be greater than 95% of the in-service shock-spectra with a 50% confidence. An in-depth discussion of probability and confidence intervals is beyond the scope of this book; however, Fig. 11.1 illustrates the meaning of P95/50. This figure is sometimes called a horsetail plot. Each curve is an estimate of the cumulative distribution of N samples of X drawn from a population with a known or assumed distribution type. Since shock spectra are always positive, a log-normal distribution is typically used. A log-normal distribution is normal in logarithmic space, such that $\log_{10}(X)$ has a normal or Gaussian distribution. The mean and variance of the distribution are unknown and are estimated by the sample mean and variance. Typically, 3 dB is used as the standard deviation for aerospace shock environments [1, 2]. The value of X that is at the 95% level varies because each cumulative distribution is slightly different. The P95/50 value of X, 194, is shown as a diamond. It is the mid-point of all the

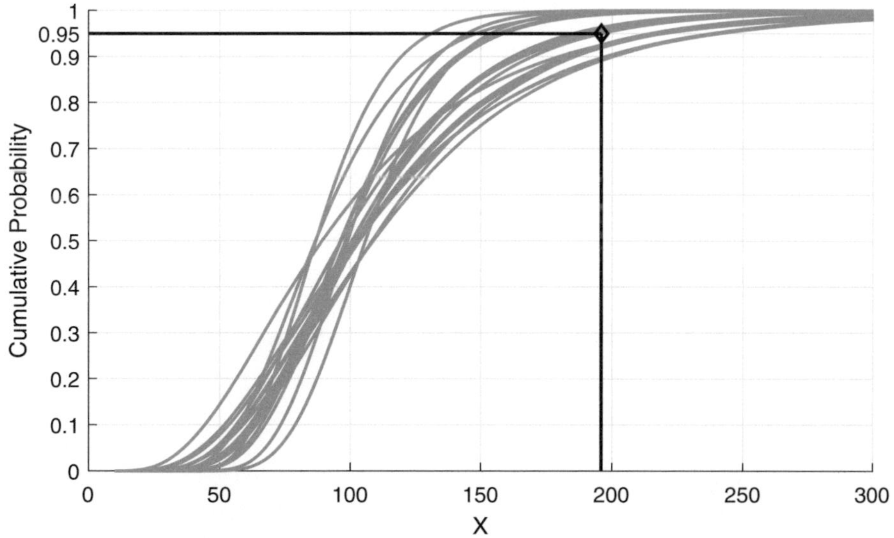

Fig. 11.1 Illustration of the definition of P95/50 environment with a horsetail plot

95% values of X. We do not need to create sample cumulative distributions because formulas exist to compute the probability/confidence values. If there is only one flight SRS on which to base the test specification, it is taken as the mean and the P95/50 SRS is +5 dB higher at all natural frequencies.

11.1.1 Qualification Test Specifications

The qualification level for military space systems is defined as the P99/90 shock spectrum. If we assume a log-normal distribution with a standard deviation of 3 dB, the P99/90 level will be 11 dB above the mean at each natural frequency, or equivalently 6 dB above the MPE at each natural frequency. NASA-STD-7003 defines the qualification environment for pyroshock slightly different, as +3 dB above the MPE.

Qualification tests are conducted at levels well above the in-service environment. While the units will have survived the qualification tests, the amount of remaining life is uncertain. Therefore, qualification tests are usually performed on prototype parts, and these parts are generally not put into service without refurbishment, if that is possible.

In addition to being at a level higher than the in-service environment, the qualification environment for shock involves exposure to multiple shocks. MIL-HDBK-340 specifies at least three shocks in both directions in each of the three axes for qualification of a component, subsystem, and system per shock event. This can be up to 18 shocks! NASA-STD-7003 on pyroshock specifies two shocks per axis and does not require shocks in both directions per axis because the environment is based on the maxi-max shock spectrum which is insensitive to direction [3].

We know that exposing a part to three shocks is not the same as exposing the part to one shock whose magnitude is three times as large. The reason for multiple shocks is not to increase the severity of the shock environment rather it is to expose latent damage caused by the first shock. If a part shows no damage after three shocks, then we can be fairly certain that it will not be damaged by the in-service shock environment.

The qualification environment should be selected based on the type of shock and the type of equipment. The military and NASA standards provide guidance and describe a philosophy that, if followed, will yield a robust design and reduce the risk of in-service failure. However, qualification test levels may be tailored and should be matched to the specific needs of the project.

11.1.2 Acceptance Test Specifications

The other type of test for which shock specifications are usually written is the *acceptance test*. The purpose of an acceptance test is to identify workmanship defects in specific parts or components. Because of this objective, there is usually less flexibility to tailor the test environment. One of the most important aspects of the acceptance test is that it must be severe enough to reveal defects but not so severe as to cause latent damage that may lead to premature failure in service. Acceptance tests are typically performed at or above the MPE, P95/50, level. Only one shock per axis is typically performed.

11.2 Shock Testing in Multiple Axes

How severe is it to test the same component three times in the three orthogonal axes? Many test specifications are written to require multiple shock tests on the same component in orthogonal axes—usually in the x-, y-, and z-axes. While this is a very common test series, many engineers think that their components are over-tested by this repeated shock exposure. There are a few different ways to deal with this conundrum. The first, of course, is to simply design the part with adequate strength to pass a full three-axis test series. The second approach is to essentially ignore the problem, which often results in the same design. The reason this works is that fatigue damage typically develops over a very large number of stress cycles. Shock testing a part three times is essentially the same as shock testing the part once from a fatigue standpoint so long as there is no inelastic material behavior. There are simply not enough stress reversal cycles in three, six, or even nine shocks, if the material response is linear, to make any appreciable difference in the test outcome from a fatigue standpoint.

In contrast, if material yielding is allowed, or happens, then the results can be quite different. Material yield results in permanent deformation of the part or internal parts of a components. As such, when the shock test is subsequently performed in another axis, the starting point for the yielded components is different from the expected starting point and their resulting displacements will differ accordingly. Likewise, if yielding occurs in the subsequent shocks, as would be expected, the final displacement will be shifted again. Furthermore, yielding and crack propagation can develop very rapidly in high stress environments meaning that a failure could be precipitated in only a few shocks if minor damage is allowed as a result of the shock test. For this reason, if some level of damage is permissible, then it is not advisable to repeatedly shock the same unit multiple times.

The final approach is to understand the likely ramifications of testing the same unit in multiple axes. For a simple example, consider the cantilever beam drawn in Fig. 11.2 with a fixed base and a lumped mass at the free end. The bending stress at the beam's fixed end is given by the common equation $\sigma = Mr/I$, where M is the bending moment, r is the beam's radius, and I is the beam's area moment of inertia If a shock is applied in the vertical, y-direction, then the bending stress is a maximum at point A, see inset picture, while the bending stress is zero at point C and approximately 70% of the maximum at the intermediate point, B. Likewise,

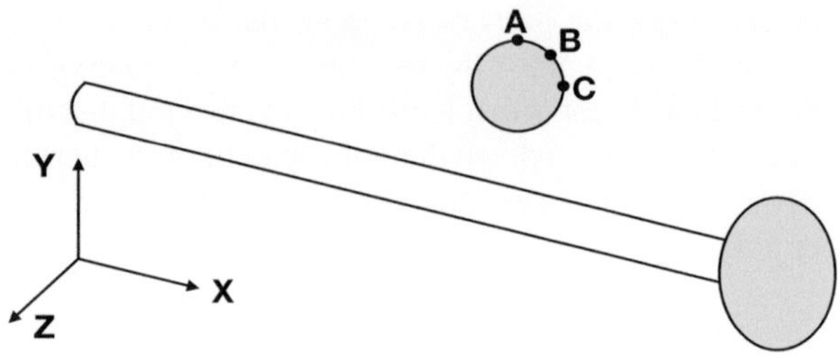

Fig. 11.2 Sketch of a cantilever beam with tip mass and beam cross-section

a lateral shock in the z-direction produces a maximum bending stress at point C, zero bending stress at point A, and the same 70% stress at point B. In this example, testing in the two orthogonal directions, y and z, does not over-stress the component because the material at points A and C have only been stressed once and the material at point B was stressed twice but at 70% of the maximum.

A longitudinal shock, along the x-axis will stress points A, B, and C equally; however, at a much lower level than the bending stress. The stress from an axial loading is given by the equation $\sigma = F/A$, where F is the applied force and A is the cross-sectional area. However, if the applied shock load, F, is the same in all three directions, then the bending moment is given by $M = FL$, where L is the beam length. The bending stress is then always larger than the axial stress, generally significantly larger, due to the additional factor of the beam length and the difference between the cross-sectional area, A, and the area moment of inertia, I. In other words, the x-direction shock produces equal stress across the beam cross-section but at a significantly lower level than the y- or z-direction shocks.

One interesting thing to note from this simple example is that point B on the beam was not tested to the full level during either test. It was tested twice to the 70% level. For the very simplistic case shown here, isotropic material and a symmetric component, one can infer that a shock applied through point B would survive if it had survived shocks through A and C. However, if a sensitive component is installed in a particular plane it would be prudent to test in that plane specifically rather than simply default to the standard x-, y-, and z-axes.

While most components are admittedly not cantilever beams, and we have ignored the stress from direct shear, the example is still applicable to many components. It is not uncommon for a loading in one direction to produce negligible stress in an orthogonal direction. As a result, testing the same unit in multiple axes is frequently not particularly damaging due to the differences in the induced stress. For this reason, if a part is designed not to yield under shock, one can frequently get away with ignoring the fact that their part will be tested in orthogonal axes.

11.3 Specifying Shocks with a Velocity Change

One of the simplest non-SRS methods of specifying a shock event is to simply specify the desired velocity change. It has been shown that velocity is closely related to stress, making it a good choice for specifying shock environments. It is entirely reasonable to simply specify a shock environment with a velocity change to occur over a specified time interval—indirectly specifying an acceleration as well with the time parameter. For example, a shock specification defined by a 5 m/s velocity change occurring in about 15 ms is not overly specific but is nonetheless well defined. A specification written this way naturally assumes a simple, classical shock pulse; but as has been shown previously, most classical shocks have similar SRS. This definition also assumes a largely one-sided shock since a shock arresting time is not defined.

Figure 11.3 shows a simple time history plot of how a defined velocity change specification could appear in test data. The plot meets the defined 5 m/s velocity change in 15 ms exactly. In a practical specification, a tolerance would be defined for the velocity change as opposed to the fixed values defined above. A typical specification might specify a velocity change between 5 and 5.5 m/s in 14–16 ms for example. Figure 11.3 does not show much of the time history beyond the peak as this is not defined by the specification. It is assumed that it would return to zero since the equipment under test will return to rest eventually; however, the manner of the return has not been specified and it is assumed to occur over a longer time than the rise. If the return occurred in less time than the rise, then the return portion could impart more damage than the defined shock specification.

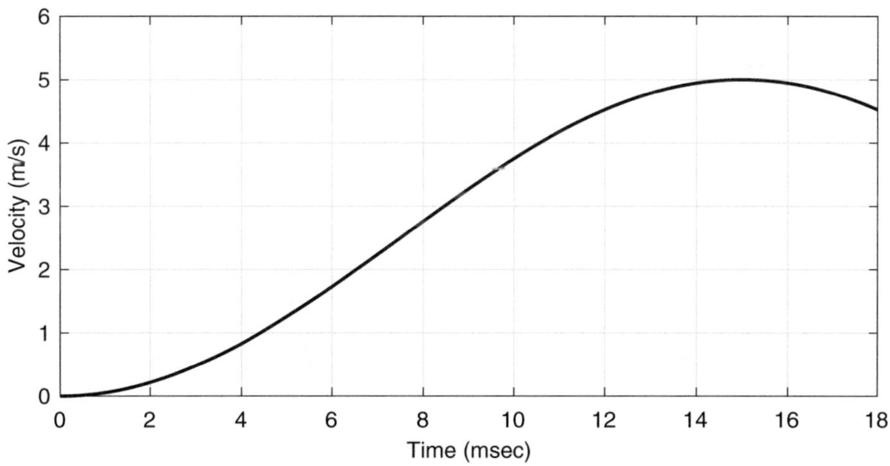

Fig. 11.3 Velocity time history plot showing how a simple velocity change specification might look in test data, 5 m/s velocity change in 15 ms

This simple method of specifying a shock test has been used with success on many systems. Simplicity in the test definition allows for testing flexibility and should not be construed as inferior to other methods. Rather test specifications should be selected based on the necessary rigor for the system under test.

11.4 Matching Shock Tests to the Environment

Shock testing should be matched to the expected environment if at all possible. If the shock environment is a drop, then the equipment should be tested on a drop table or package drop machine. If the environment is an oscillatory shock, then some form of resonating test fixture or shaker shock should be used. If the shock is a low-level excitation, then a shaker shock may be appropriate. A shock test specification matched to the environment will always be more representative.

An example of less than ideal shock test matching is shown in Figs. 11.4 and 11.5. Figure 11.4 shows the MMAA SRS measured from a pyroshock event overlaid with a best-fit haversine SRS. While the haversine SRS is a very good match to the test data above 5 kHz, the haversine SRS is significantly higher below 5 kHz. This is further highlighted with the pseudo-velocity SRS comparison shown in Fig. 11.5. In this plot, it is readily apparent that the haversine approximation is significantly more severe at the lower frequencies. This is a result of the essentially zero velocity change of the measured pyroshock event compared to the defined momentum change from a drop table test. While it may be acceptable to perform a

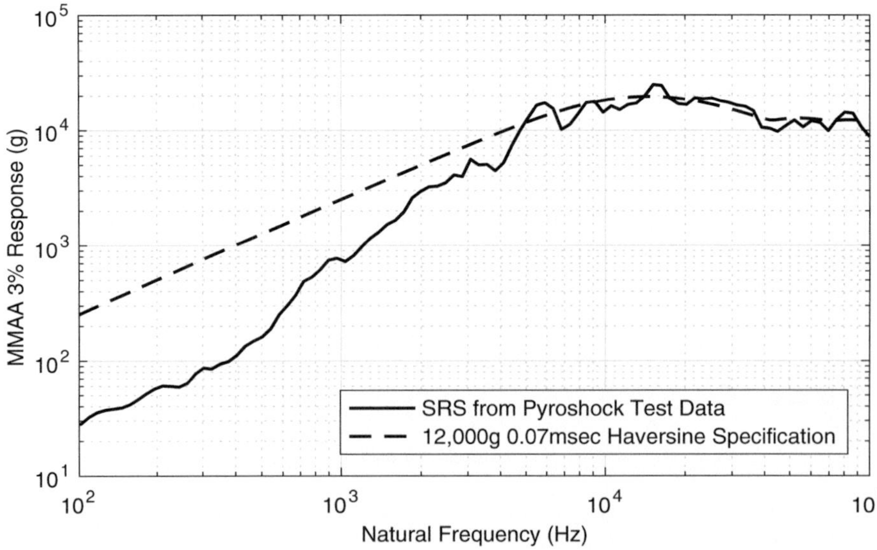

Fig. 11.4 Absolute acceleration haversine test specification plotted with pyroshock test data

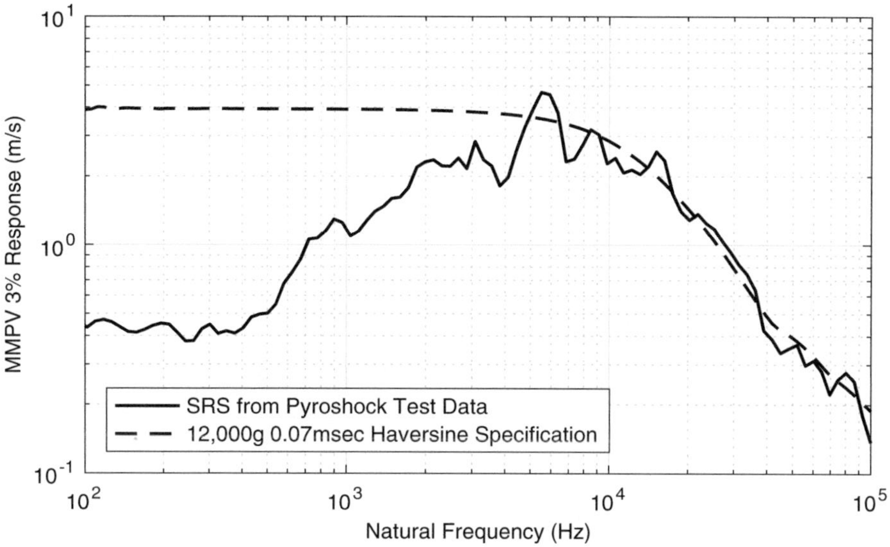

Fig. 11.5 Pseudo-velocity haversine test specification compared to pyroshock test data

test this way, and it may be a conservative test, it is not particularly representative of the expected environment.

Of course, from a practical standpoint, it is probably not possible to achieve the 0.07 ms haversine pulse duration shown in Fig. 11.4 on a drop table. The limitations of the equipment will also necessarily factor into deriving shock test specifications. If equipment availability or capability is limited, then it may be necessary to conduct a shock test in a less than ideal way. While this is certainly possible, the ramifications of this need to be fully understood and accepted by all parties involved. If equipment limitations are encountered, it is usually desirable to err on the side of a more severe test than the expected environment. This at least provides confidence that the component will survive the expected environment, albeit with a greater margin of safety than perhaps is desirable.

The example shown in Figs. 11.4 and 11.5 could equally well be reversed. If a drop shock with a defined momentum change is approximated with a pyroshock type event, then the momentum change is completely missing from the test. Likewise, many shaker shock test implementations have no momentum change associated with them due to the requirement that the armature returns to its original starting position. As such, a drop test approximated with a shaker shock or pyroshock substantially under-tests the low-frequency regime of the frequency spectrum. This can be a significant concern since the momentum change of a drop shock is often the most stressing to the structural components of a system or component.

11.5 Drop Shock Specification Development

The importance of matching a shock test to the expected environment has been shown. The question remains: how is this best accomplished? Figures 11.6 and 11.7 show the same MMAA SRS from a measured drop test of a component overlaid with two different approximating haversine shock SRS curves. These two plots show two common approaches to defining test specifications. The first approach, shown in Fig. 11.6, defines the haversine such that it generally fares through the test data. The second approach, shown in Fig. 11.7, defines the haversine such that it envelopes the test data. Which one is right? The answer of course depends on what the test designer is trying to accomplish.

The first approach, shown in Fig. 11.6, fares through the data, but also attempts to match the velocity change, defined by the low-frequency slope of the MMAA plot, and the maximum acceleration in the time history, defined by the flat tail at the high-frequency end of the spectrum. In many respects, this is a very good methodology for defining a test specification to match test data. While it does not fully encompass every blip of the SRS, it does achieve the spirit of the test by matching both the velocity and acceleration reasonably well. The peak acceleration in this case is actually a little high in this match, about 3600 g versus 3400 g but it is very close.

The test specification match shown in Fig. 11.6 was actually accomplished using a geometric simplex optimization routine. The geometric simplex is a method first proposed by Nelder and Mead [4] that can be applied to a wide range of problems. In this case, a simplex optimization algorithm was written that allowed the haversine

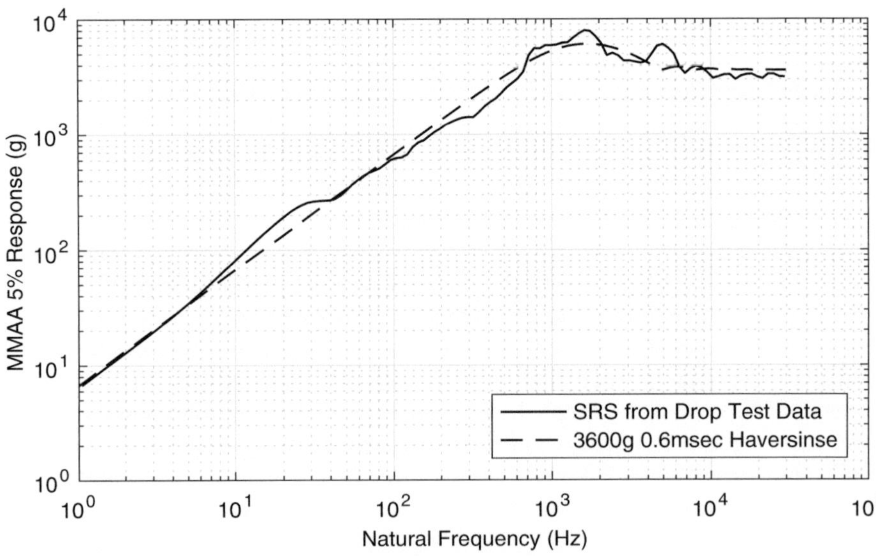

Fig. 11.6 Haversine test specification developed by matching velocity and faring through the test data

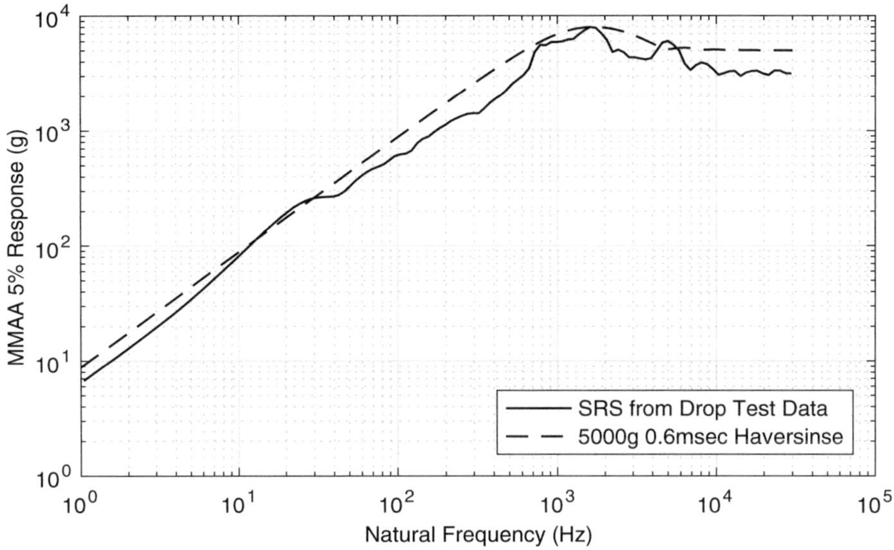

Fig. 11.7 Haversine test specification developed by matching velocity and enveloping the high-frequency test data

amplitude and duration to vary until the resulting MMAA SRS had the minimum total difference with the measured MMAA SRS [5]. Of course, a haversine match could also be performed by simple brute-force trial and error relatively quickly and there would be no harm in using the simpler approach. The answer would likely be the same with either method.

The alternative approach, shown in Fig. 11.7, envelopes the entire MMAA SRS spectrum. The consequence of this specification development method is that both the velocity and acceleration of the resulting laboratory test will be higher than the velocity and acceleration measured in the field environment. This difference in velocity is apparent by comparing the location of the low-frequency straight lines from the measured data and haversine SRS in Fig. 11.7. The difference in maximum acceleration in the time history is apparent by the difference in the SRS high-frequency tails. In this example, the test data have a peak acceleration of about 3400 g in the time domain where the test specification will have a peak acceleration of about 5000 g. Likewise, the velocity change with the second method is 14.7 m/s compared to 10.6 m/s from the first method. Since stress is proportional to velocity, the second approach yields a substantially more stressing test. If a more conservative approach to testing is desired, then the second approach may be a very appropriate and more desirable method. If a test more representative of the measured environment is the goal, then the first approach of faring through the data may be the most desirable.

When deriving test specifications from test data, one needs to also keep in mind the idea of significant figures in a test definition. The test laboratory will only have the ability to control the test to within certain tolerances. Therefore, while it is possible to define a test to a large number of significant figures, it is unlikely that the test laboratory will be able to actually perform a test with that precision. Specific test tolerances will be discussed further later in this chapter; however, your test laboratory will likely have their own range of acceptable tolerances for their test equipment. As such, many test specifications are typically defined with two significant figures to avoid over-specification. Other approaches may be more acceptable to your specific test lab and this should not be considered a hard rule of test specification development.

The example shown in Fig. 11.6 is relatively straightforward. The test was a drop test, the measured SRS had the general appearance of a haversine SRS, and a simple algorithm produced a good match to the data. Figures 11.8 and 11.9 present a more complicated example. The measured drop test data shown in Figs. 11.8 and 11.9 contain some additional high-frequency content shown in the SRS in addition to the primary drop excitation. In this example, there is an obvious drop in the measured SRS data at about 300 Hz. The question immediately arises; what does this mean? Does this dip indicate the existence of two separate shock pulses, a low-frequency shock and a high-frequency shock, or is it indicative of some sort of energy absorbing mechanism in the system at this frequency? How should the resulting test be specified to best represent the environment? Figure 11.8 shows one approach, assuming that the there are two substantial shock events, using a haversine to approximate the low-frequency excitation. This results in a 660 g 3.6 ms haversine test and no coverage of the high-frequency excitation. It is assumed at this point

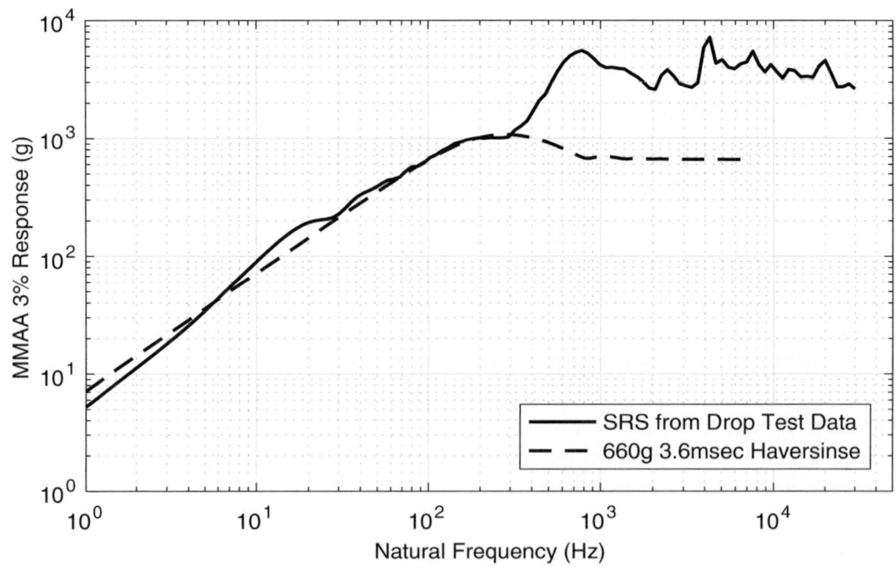

Fig. 11.8 Haversine test specification designed to match the first inflection point in the test data SRS

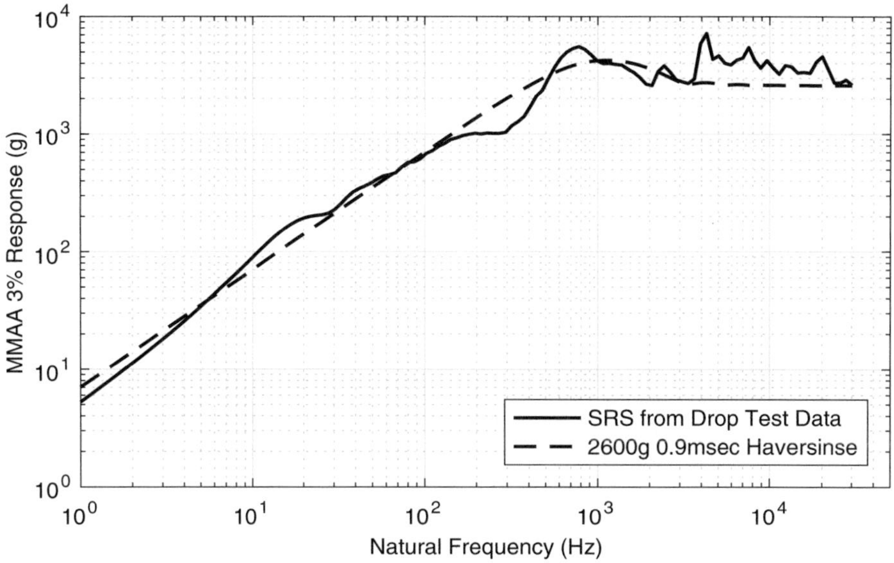

Fig. 11.9 Haversine test specification designed to match the second inflection point in the test data SRS

that a second test would be used to approximate the high-frequency portion of the MMAA SRS spectrum.

Figure 11.9 shows a second approach, assuming that the MMAA SRS represents one primary event and some secondary vibration. This approach fits a single haversine with a higher amplitude and shorter duration to the test data. In this case, the approximating haversine test is 2600 g 0.9 ms and includes good coverage of the measured SRS up to about 3.5 kHz. Which approach is more correct? Without knowledge of how the high-frequency content is to be covered in Fig. 11.8, it is not possible to make an appropriate judgment. In order to complete this example, we will assume for the moment that the intention is to cover the high-frequency portion of the MMAA SRS with a resonant plate test defined by the specification shown in Fig. 11.10. Fitting resonant plate specifications to test data will be discussed further in Sect. 11.6 but for now it is taken as a given.

The first concern with Fig. 11.10, two-shock test approach is the obvious requirement to hit the unit under test multiple times. If there is a strong desire to limit the number of tests performed, then the single shock approach shown in Fig. 11.9 may be a more appropriate choice. However, a more technical approach to consider is to plot the three test specifications in terms of pseudo-velocity as shown in Fig. 11.11. This plot shows immediately that all three of these test specifications have nearly identical peak velocity changes. In other words, the two haversine shocks will induce essentially identical stresses in the structural components with the higher level haversine having greater frequency bandwidth coverage. Furthermore, the defined resonant plate shock test does not have a significantly

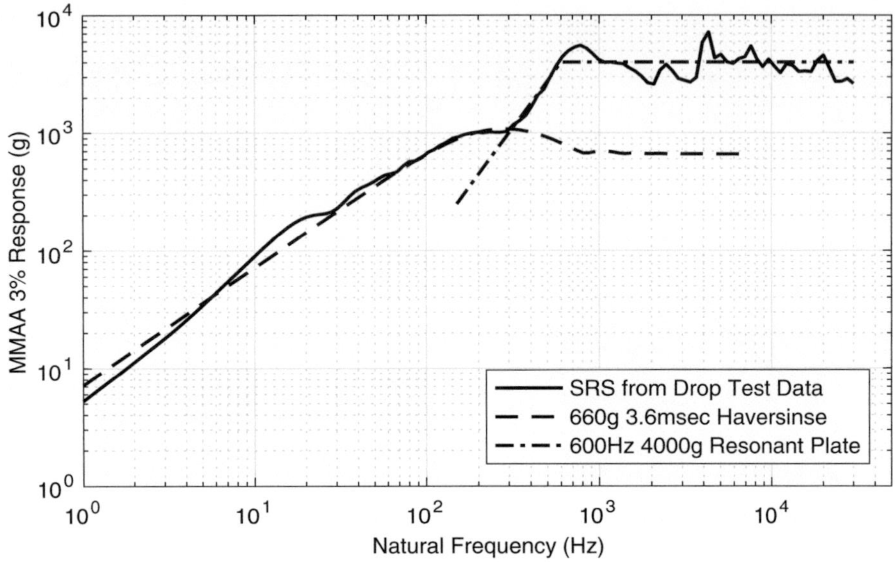

Fig. 11.10 Resonant plate test specification designed to match the highest inflection point in the test data SRS along with the lower-frequency haversine specification

different energy bandwidth than the higher-level haversine from Fig. 11.9. Some variation in the bandwidth could be obtained depending on how the test data are approximated—enveloping or faring through. From Fig. 11.11, it is apparent that the single haversine approach of Fig. 11.9 and the two-test approach of Fig. 11.10 are essentially identical from an energy bandwidth perspective making the single haversine test the more appealing option due to the fact that fewer tests are required. If the velocity plot had indicated a significant difference in energy bandwidth, then the two-test approach may have been more appropriate but still not necessarily advisable due to the added complexity.

As an aside, the same test was repeated at a later date with a new set of hardware and the SRS dip at 300 Hz was not apparent in the data. This development further highlights the limitations of deriving test specifications from a single data source and the risk of matching tests too closely to an SRS. In this case, the subsequent test shows that the single haversine approach was more correct than the two-test approach. In this case, one drop shock test specification to envelope one drop test event is the obvious conclusion. The fact that the high-frequency bumps in the MMAA SRS are not well covered by the haversine is most likely inconsequential since the energy at the high frequencies is generally low and the resulting strains are correspondingly low. While there may be some risk associated with ignoring some high-frequency content in the SRS, the additional energy could be covered by extending the haversine to higher frequencies if desired.

The discussion in the previous example also shows the need to consider unit-to-unit variability when defining test specifications. It is unfortunate but often

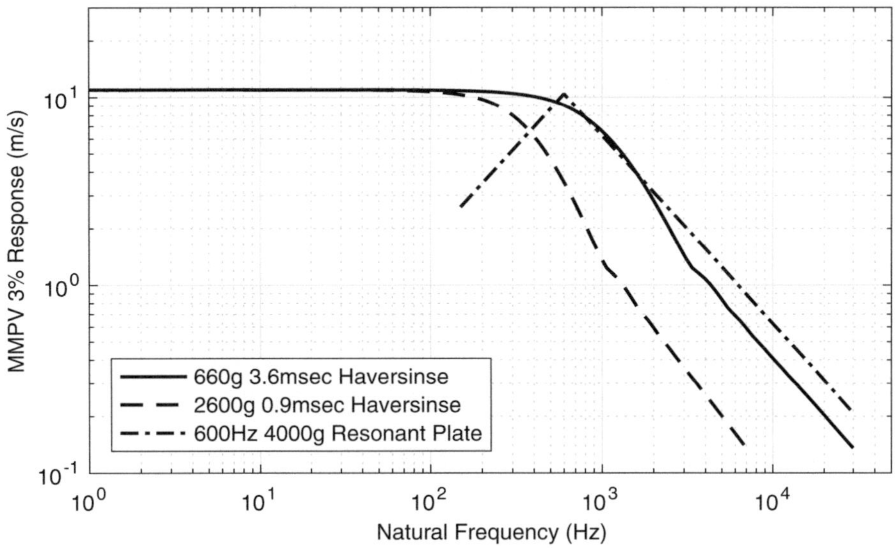

Fig. 11.11 Velocity comparison from the two haversine and the resonant plate test specifications matched to test data SRS

necessary that many test specifications are derived from a single test event. Wherever possible, multiple tests should be used to derive test specifications, especially with high-consequence systems or components. However, until ideal circumstances arrive, engineering judgment must be used to ensure that the defined test specifications best approximate not only the measured environment from the current test but the likely environment from future tests.

As a second example, consider the MMAA SRS shown in Fig. 11.12. This SRS shows the obvious characteristics of a haversine shock at the lower frequencies along with some additional high-frequency content. Like the previous example, the question here is how to properly define a test specification to adequately qualify a component to this SRS. Matching the low-frequency haversine is straightforward, but what about the additional high-frequency content? This example is contrived but serves to illustrate an important part of test specification development. The time history for this shock is shown in Fig. 11.13 in the upper subplot. The time history was actually made as the sum of the other two signals in Fig. 11.13. What this shows is that the primary haversine has an amplitude of 5000 g, while the oscillatory signal amplitude is less than 3000 g.

In contrast to the time histories, the MMAA SRS in Fig. 11.12 has a maximum of just over 10,000 g. If the high-frequency portion of the SRS were enveloped with a test specification, the test could easily have a 10,000 g input where the actual time history maximum was about 6400 g. The first question is how did the high-frequency bumps in the SRS rise above the asymptotic portion of the 5000 g haversine? The answer comes from Chap. 6 where it was demonstrated that the more oscillations

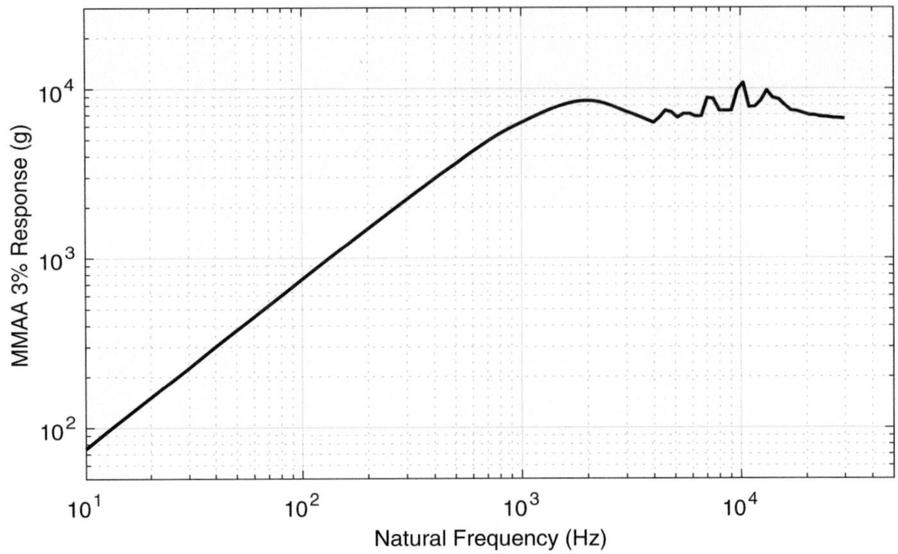

Fig. 11.12 MMAA SRS of a complex shock having both haversine and oscillatory components

occurring in the shock, the higher the final SRS. This is a feature of the mathematics of the SDOF oscillator. Of course, the SRS is not unique so a similar SRS could have been obtained with fewer oscillations and higher amplitudes on the secondary pulses.

Figure 11.14 shows the same MMAA SRS from Fig. 11.12 calculated separately for the two component parts of the complex shock—the classical haversine SRS along with the MMAA SRS of the oscillatory shock. Figure 11.14 shows that the SRS is largely additive in that the sum of the SRS of two signals is generally equal to the SRS of the sum of the signals. There are a few minor differences between the envelope of the two SRS curves in Fig. 11.14 and the single SRS curve in Fig. 11.12; however, the differences are negligible. The problem with this example is that the SRS alone indicates that a 10,000 g shock will be necessary to represent what is clearly a 3000 g shock in the time history. This should be a concern to any component designer or test engineer. This example clearly shows that we cannot blindly envelope a given SRS without some knowledge of the underlying source data and the behavior of the component under test. Certainly, the 5000 g haversine is likely to induce more bulk stress in the part than the high-frequency oscillatory excitation. However, the haversine has a lower bandwidth and will not excite the same frequency ranges as the oscillatory shock.

What should the final test be? If the fundamentally damaging modes of the system or component are all well excited by the haversine, then it may not be necessary to perform the high-frequency oscillatory shock at all. High-frequency shocks typically have much smaller displacements and consequently much less strain than the lower-frequency shocks. Therefore, if the primary concern is

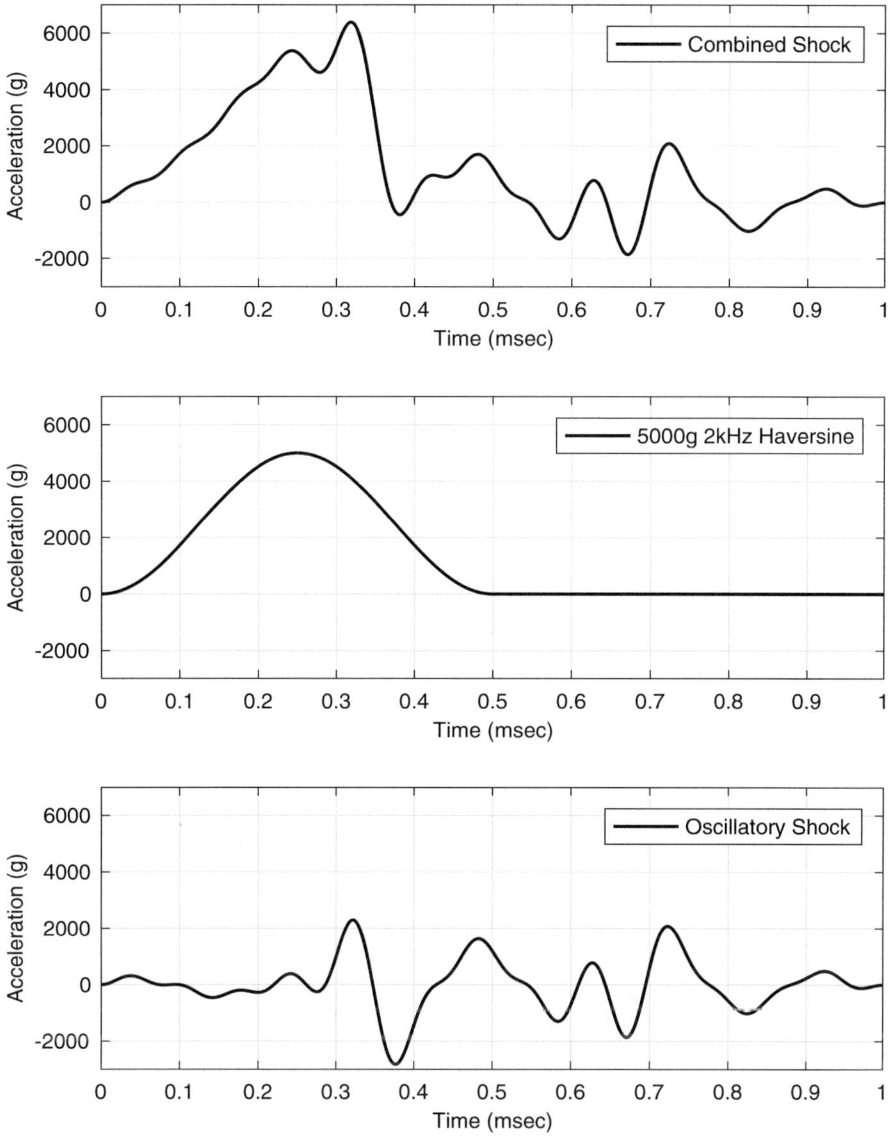

Fig. 11.13 Time history plot of the complex shock from Fig. 11.12 along with the five shock signals comprising the complex shock waveform

strain induced failure, the high-frequency oscillatory shock may be insignificant. Conversely, if the unit under test is sensitive to high-frequency excitations, then the oscillatory shock may be necessary for component qualification. Likewise, if the shock is necessary for qualification, then it is also of paramount importance to ensure that the exposed magnitude is comparable to the measured time history. In

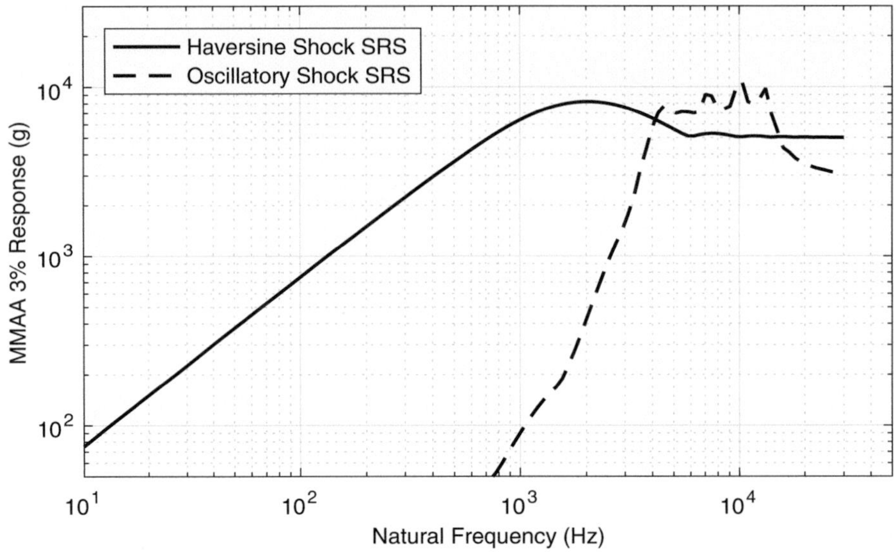

Fig. 11.14 MMAA SRS of the two parts of the complex shock calculated separately—the haversine shock and the oscillatory shock parts

other words, you should be looking for an excitation with a 3000 g time history excitation with several oscillations such that the SRS builds up to approximately the 10,000 g level seen in Figs. 11.12 and 11.14.

On final consideration when defining drop test specifications is the test laboratory's interpretation of pulse duration. A detailed discussion of pulse duration was presented in Sect. 9.2.1 but warrants reiterating. The shock laboratory will not be able to measure the theoretical pulse duration from experimentally collected data. The reason for this is that the theoretical haversine starts and ends with zero amplitude and zero slope. In contrast, experimental data start and end in the instrumentation noise floor, which is not zero. As a result, pulse duration in the laboratory is typically measured at some percent of the maximum amplitude, usually about 10% of the peak acceleration. Section 9.2.1 develops the simple relationship between this and the theoretical duration. The purpose in reiterating this point to ensure that the test specification and the test laboratory are using the same actual pulse duration for the applied shock. It is your job as the test specification developer to ensure that the test laboratory is performing the test as you intended. As such, you should be clear in your definition of pulse duration so that it is interpreted properly.

11.5.1 Drop Shock Scaling

Many times, drop shock data are measured in the field and then used to derive test specifications for a subsequent laboratory test. The previous section discussed this at length. However, it is also common to have data from one test but want to perform laboratory testing at a different level. This might occur if you want to evaluate margin to failure, or just to understand the component's response when it is dropped from different heights. But, what is the most appropriate way to scale the data. Obviously, one could simply multiply the SRS or the shock amplitude by a constant and generate a more or less severe test. While that approach will generate a different test, it is not representative of the system at a different drop height.

Conservation of energy can be used to derive the appropriate scaling for a drop shock event. For the case being considered here, an object in free fall, the energy in the system is given by the mass of the part and its drop height. The potential energy, U_h, is

$$U_h = mgh, \tag{11.1}$$

where m is the component mass, h is the original drop height, and g is the acceleration of gravity. Likewise, the kinetic energy, T, is given by:

$$T = \frac{1}{2}mv^2, \tag{11.2}$$

where v is the impact velocity. Setting Eqs. 11.1 and 11.2 equal and solving for the maximum velocity, it is apparent, and well known, that the impact velocity is independent of the component's mass and is given by:

$$v = \sqrt{2gh}. \tag{11.3}$$

Thus, a 5 m free fall results in an impact velocity of 9.9 m/s and a 10 m free fall generates a 14.0 m/s impact velocity as an example. The energy is linearly proportional to the drop height, as shown in Eq. 11.1, but the velocity is proportional to the square root of drop height. As a result, the 10 m drop contains twice the energy of the 5 m drop but the impact velocity is only 1.41 times greater. This implies that the scaling should be done in terms of impact velocity. However, referring back to the haversine velocity SRS plots, scaling in terms of velocity is a nonunique operation. While the velocity can be increased by a factor of 1.41, what should be done with the rest of the SRS?

It is intuitive that dropping an object from a greater height should result in greater compression of the impacting components. Likewise, higher drop heights typically result in greater damage. This can be examined by considering the stiffness and deformation of the component under test. Again, from conservation of energy, the potential energy stored in a spring is given by:

$$U_s = \frac{1}{2}ky^2, \tag{11.4}$$

where the displacement, y, is a combination of the deflection of both the falling component and the impacting surface. The spring rate, k, is a combination of the component stiffness and the impact surface stiffness. Since energy is always conserved, the potential energy prior to release, U_h, must equal the kinetic energy immediately prior to impact, T, and this in turn must equal the maximum energy stored in the spring deformation when the velocity is zero at the maximum impact depth. Thus:

$$mgh = \frac{1}{2}ky^2. \tag{11.5}$$

Rearranging Eq. 11.5 and solving for y give

$$y = \sqrt{\frac{2mgh}{k}} = \sqrt{2gh}\sqrt{\frac{m}{k}} = \frac{\sqrt{2g}}{\omega}\sqrt{h}. \tag{11.6}$$

The system natural frequency, ω, is substituted into Eq. 11.6, showing that impact depth is a function of the system frequency and the square root of drop height. As a result, the impact depth or impact deformation for the 10 m drop will be 1.41 times greater than for the 5 m drop height. However, greater impact depths will necessarily take a proportionately longer time to occur. Linear displacement is given as a function of acceleration and impact velocity by the well-known relationship:

$$y = y_0 + v_0 t + \frac{1}{2}at^2. \tag{11.7}$$

For the haversine shock, the velocity change is a function of the pulse duration and the peak acceleration:

$$v_0 = \frac{1}{2}at, \tag{11.8}$$

where the acceleration a is given in m/s^2. Substituting Eq. 11.8 into Eq. 11.7 and assuming that $y_0 = 0$ give

$$y = at^2. \tag{11.9}$$

Impact depth is a function of time squared, or reversing Eq. 11.9 shows that impact time is a function of the square root of the impact depth. Substituting Eq. 11.6

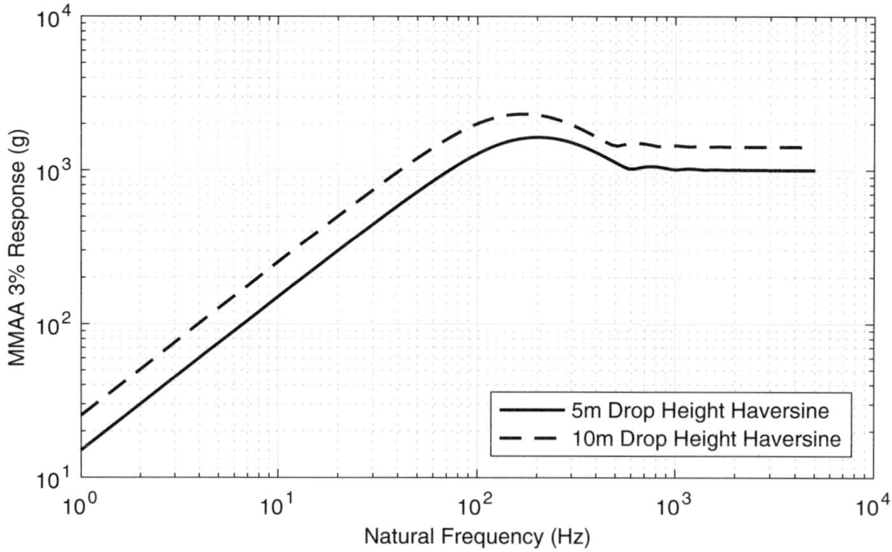

Fig. 11.15 Scaling of haversine shock SRS for drop height doubling

into Eq. 11.9 shows that impact time is proportional to the fourth-root of the drop height:

$$t = \sqrt{\frac{y}{a}} = \frac{\sqrt[4]{2g}}{\sqrt{a\omega}}\sqrt[4]{h}. \tag{11.10}$$

For the example above, if the drop height is doubled, then the velocity increases by a factor of $\sqrt{2} = 1.41$ and the impact time or pulse duration increases by a factor of $\sqrt[4]{2} = 1.189$. This is shown graphically in Fig. 11.15 for a theoretical doubling of the drop height. This SRS plot shows how the classical haversine shock increases in amplitude but moves down in frequency. Additional information about drop shock scaling can be found in [5].

11.6 Resonant Plate Specification Development

Resonant plate or beam test specifications are typically defined by three points on the MMAA SRS plot: a low-frequency point, the knee frequency, and a high-frequency point. A sample resonant plate shock test specification with a 2 kHz knee frequency and a 3000 g amplitude is shown in Fig. 11.16. The resonant plate test is supposed to represent the classical oscillatory shock discussed in Chap. 6; however, the defined specification does not necessarily follow the form of the SRS plots given in Chap. 6. The SRS from a theoretical oscillating shock has a low-frequency slope

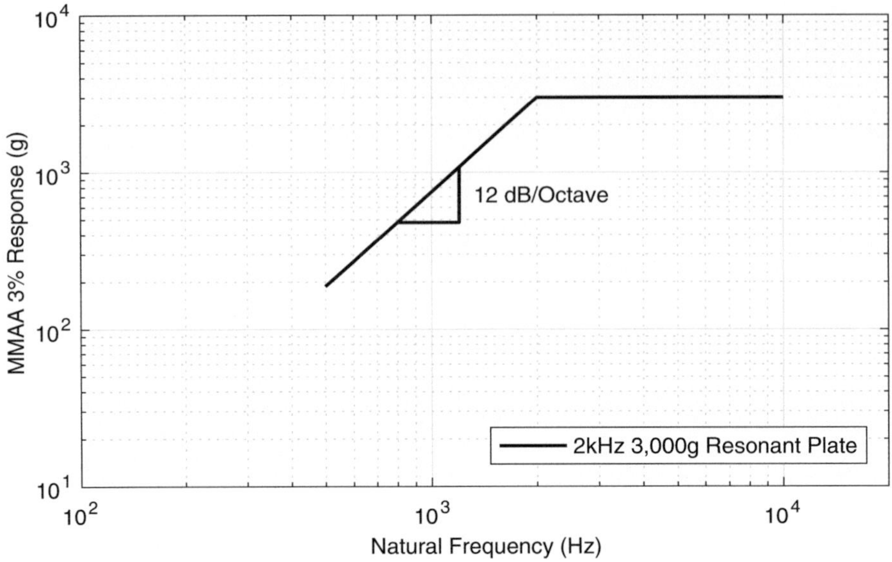

Fig. 11.16 Plot of a typical resonant plate test specification

that starts at 6 dB/octave, transitions to 9 dB/octave, then as high as 12–18 dB/octave or more near the peak before trending down to a flat at the high-frequency end of the spectrum. In contrast, a typical resonant plate specification has no peak, is flat beyond the knee frequency, and has a constant low-frequency slope of about 12 dB/octave. This admittedly sounds like a poor formula for success with resonant plate testing.

In actuality, resonant plate testing works quite well although it is a highly specialized area of shock testing that has often been likened more to an art than a science. In reality, the resonant plate response rarely mimics a pure oscillatory shock, nor does it reproduce the theoretical three-point SRS of Fig. 11.16. The practitioner of resonant plate testing needs to understand that what will be achieved in the laboratory is somewhat of a mix between the theoretical test specification SRS and the oscillatory shock SRS. While the result will be a hybrid between two SRS, the tests are known to be quite repeatable and tailorable. The equipment, on the other hand, is very specialized and only available at a few testing facilities.

Figure 11.17 shows a plot of some pyroshock test data overlaid with two potential resonant plate test specifications. The specifications do not look significantly different because of the nature of the logarithmic plot; however, the lower specification is a 15,000 g test where the higher is a 20,000 g test. Here again, the question is asked: is it better to envelope the test data or fare through the test data? As before, the answer is that it depends on the goal of the test. If the intention is to provide a bounding level test and there is good confidence in the component design, then perhaps enveloping the test data are both easy and best. If the goal is to be more

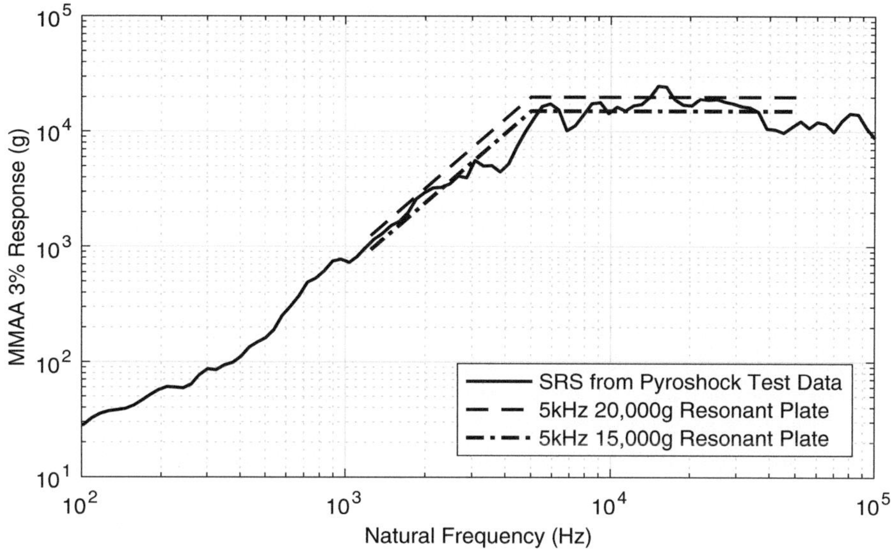

Fig. 11.17 Resonant plate test specification options for pyroshock test data

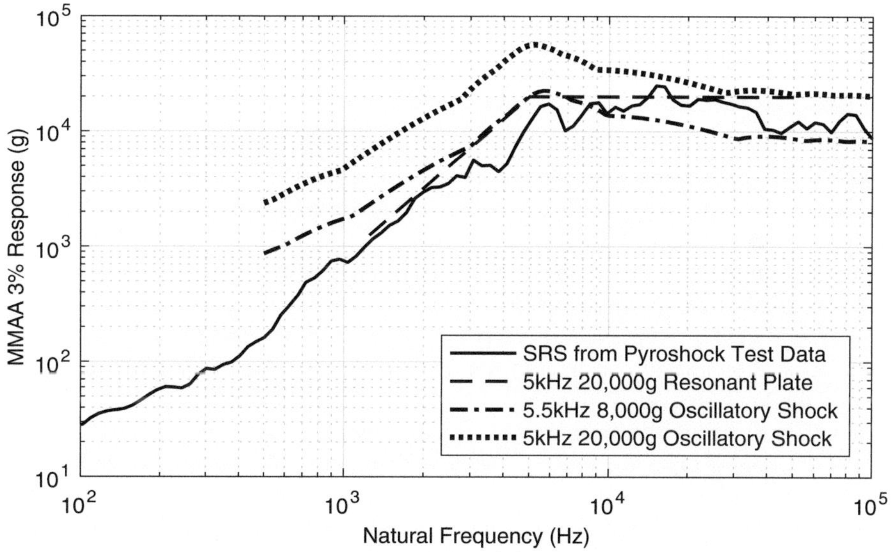

Fig. 11.18 Resonant plate test specification options for pyroshock test data

representative of the environment that generated the original SRS, then perhaps faring through the data are more appropriate. But why?

Figure 11.18 shows a plot of the same pyroshock data from Fig. 11.17 overlaid with the MMAA SRS from two theoretical oscillatory shocks. These shocks follow the form of those described in Chap. 6 defined by a single-frequency decaying sinusoid. Recall that the resulting MMAA SRS plots from Chap. 6 are not three-point SRS shapes, but rather have a peak at the sinusoid frequency. As such, if the test specification calls for a 20,000 g resonant plate shock, the unanswered question is whether that means that the 20,000 g should be the peak in the SRS, the peak in the time history, or something in between. As can be seen in Fig. 11.18, the 20,000 g SRS matches the test specification very well at the high frequencies but is a significant over-exposure at the peak and at the low-frequencies. On the other hand, the 8000 g shock SRS is a fairly good match around the peak but a little low at the high-frequency end of the spectrum.

It should always be remembered that the test laboratory will not have the benefit of studying the original field data from which the test specification was derived. Therefore, they will attempt to match the test specification to the best of their ability while blind to the original source. When a test engineer is confronted with a flat-line specification where their equipment is known to give a peak and tail, as shown in Fig. 11.18, they will usually try to split the difference with some of the specification above the line and some below the line so long as they remain within the defined test tolerances. Therefore, if a resonant plate test specification completely envelopes the test data, then it is likely that the resulting laboratory test will be high around the knee frequency of the specification and high at all frequencies below the knee.

11.6.1 Resonant Plate Frequency Range

When developing a resonant plate type test specification, the natural focus is on the location of the knee frequency and the acceleration amplitude. However, the high- and low-frequency extents of the test specification are very critical for a successful implementation of these tests. The test laboratory often spends more time and makes more compromises to keep the tails of the specification within tolerance than around the knee. As such, if compromises need to be made to bring the tails of the specification into tolerance, they are made at the expense of the frequency band around the knee frequency, the critical region. As such, the tails need to be defined judiciously to ensure that they are not adversely impacting the test.

At the high-frequency end of the test spectrum, the last point in the test specification should be limited to about one decade higher than the knee frequency or less. For example, if the test specification calls for a 2 kHz knee frequency, then the test specification should not be defined past 20 kHz. Likewise, if the specification calls for a 500 Hz knee frequency, then the test specification should only be defined to about 5 kHz or less. Most shock laboratories will also have a practical upper limit for test specification definitions and specifications should not be defined beyond that frequency if possible. This upper limit is usually around 10 kHz for mechanical shock.

The low-frequency point on the MMAA SRS is likely more critical than the high-frequency point as it is typically more difficult to control. The low-frequency point should likewise be limited to no more than one decade below the knee frequency and usually less. A more appropriate low-frequency point is typically between one-half and one-quarter of the knee frequency. The amplitude of the low-frequency should also be calculated to give a slope of 12 dB/octave or steeper. A slope shallower than 12 dB/octave is typically more difficult to control because it requires the plate to have more forward velocity at first impact, while the point of a pyroshock test is to have zero net velocity change. As a general rule for defining test specifications, the low-frequency point should be calculated assuming a 12 dB/octave slope and then the resulting number rounded down to the appropriate number of significant figures.

A quick formula for calculating the magnitude at the low-frequency point for a resonant plate test specification can be derived by rearranging Eq. 5.9 as:

$$S_1 = \frac{S_2}{(f_2/f_1)^m}, \tag{11.11}$$

where S_1 is the magnitude at the low-frequency point, f_1; S_2 is the magnitude at the knee frequency, f_2; and m is given by the equation:

$$m = \frac{\text{slope}}{20 \log_{10}(2)}, \tag{11.12}$$

where the slope is given in terms of dB/octave. For a slope of approximately 12 dB/octave, $m = 2$.

11.6.2 Pyroshock Attenuation

Resonant plate or beam test specifications are typically used to approximate a pyroshock event in the laboratory. While pyroshock events can be extremely harsh, they are also somewhat predictable since the location of the pyrotechnic device is known relative to the component of interest. The high-frequency stress wave propagation from the pyrotechnic detonation is also easily attenuated. The dominant contributor to shock attenuation is the structure's geometry. Two structural attenuation features are specifically addressed in NASA and military specification documents: distance (which means shock path distance) and joints. Neither takes into account sensitivity to shock magnitude.

NASA-STD-7003A [3] and NASA-HBDK-7005 [6] provide guidance on shock attenuation as a function of shock path distance from the source. In general, the attenuation of a shock due to distance is an exponential decay. Perhaps the most widely quoted shock attenuation curve in NASA-HBDK-7005 is Fig. 5.7, reproduced here in Fig 11.19. This figure recommends the amount of attenuation for the SRS ramp and the SRS peak (i.e., plateau) as a function of distance from the shock source. The attenuation of the SRS plateau is independent of natural frequency. This curve comes from Ref. [7], which described pyroshock

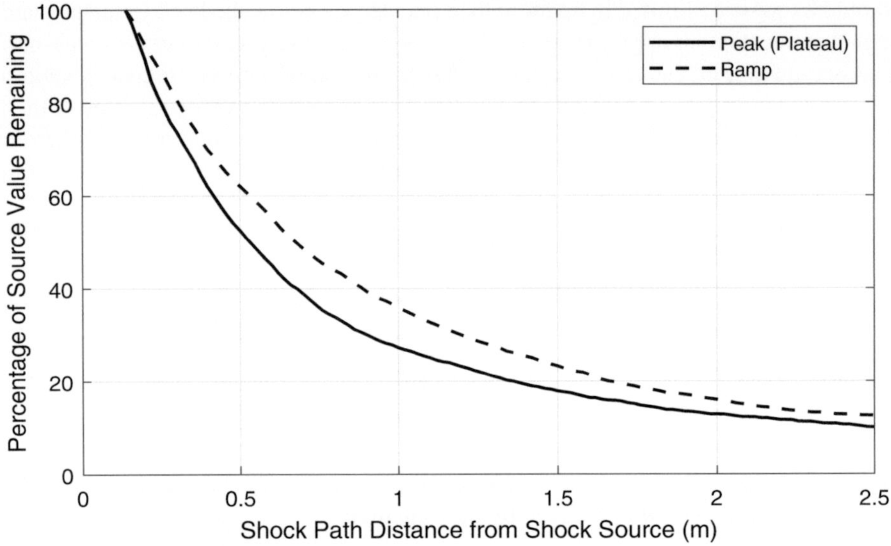

Fig. 11.19 Shock response spectrum attenuation vs. distance from pyroshock source NASA-HDBK-7005, Fig. 5.7 [6]

test requirements for Viking lander capsule components. No uncertainty data are provided nor is an explanation of how the curves were derived.

NASA-HBDK-7005 gives another distance attenuation model for complex structures subjected point source pryoshocks. It is an exponential decay model (Eq. 5.78 [6]):

$$\frac{\mathrm{SRS}(D_2)}{\mathrm{SRS}(D_1)} = \exp\left(-8 \times 10^{-4} f_n^{2.4 f_n^{-0.105}} [D_2 - D_1]\right). \tag{11.13}$$

Figure 11.20 is plot of the percentage of the SRS remaining as a function of distance from the shock source and natural frequency. This model assumes that the pyroshock is from a point source and it is more complex than the one shown in Fig. 11.19. It also predicts more attenuation. This is shown in Fig. 11.21, which shows the percentage of source remaining at the SRS natural frequency of 2 kHz. 2 kHz is a typical knee frequency for a pyroshock spectrum. This curve is lower than the SRS plateau curve in Fig. 11.19.

Since these curves were included in NASA-HDBK-7005, they have been used for all kinds of vehicles, presumably successfully, but the fact remains that they were derived for a specific vehicle so extrapolation to any other vehicle should be done with full knowledge of its origin. Data collected from a recent full-scale pyroshock separation test [8] suggest that the NASA-HDBK-7005 attenuation curves may not be conservative. Unpublished studies by the authors on launch vehicle stage separation pyroshock attenuation also have indicated that the NASA-HDBK-7005 attenuation curves are not conservative.

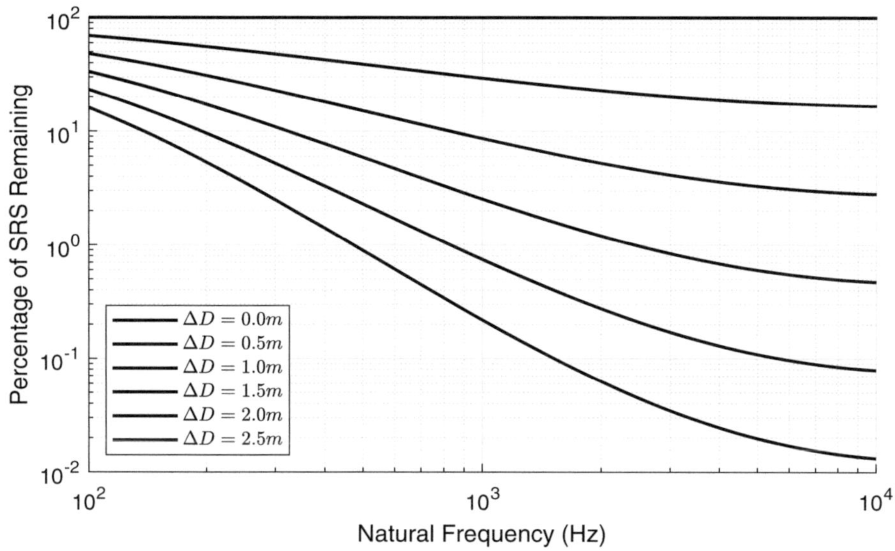

Fig. 11.20 Shock response spectrum as a function of distance from the pyroshock source NASA-HDBK-7005, Eq. 5.78 [6]

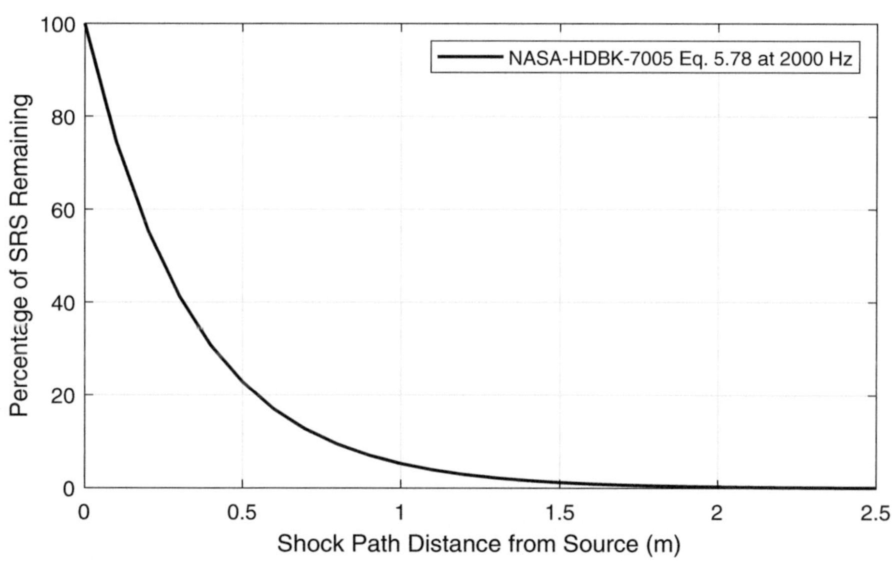

Fig. 11.21 Attenuation of the SRS at 2 kHz vs. distance from the pyroshock source NASA-HDBK-7005, Eq. 5.78 [6]

References [9] and [10] presented an exponential model of peak acceleration attenuation for truss and cylindrical shell structures from 456 shock test measurements. The attenuation model is

$$\frac{\mathrm{SRS}(D_2)}{\mathrm{SRS}(D_1)} = \exp(-\alpha[D_2 - D_1]). \tag{11.14}$$

The overall mean attenuation coefficient, α, is 0.033 and the associated coefficient of variation is 27%. The attenuation coefficients vary from 0.023 to 0.044. References [9] and [10] do not provide enough information to determine if there are joints in the shock path or the truss materials. The amount of attenuation versus distance predicted by this model is in between the two NASA-HDBK-7005 models.

NASA-HDBK-7005 also provides guidance regarding the amount of attenuation across joints. The guidance comes from a paragraph in Ref. [7] that describes pyroshock test requirements for the Viking lander. It reads

> ...to ensure adequate conservatism it was elected to use the following approach to joint attenuation: Assume a peak attenuation of 40% for each joint, up to a maximum of three joints, with no attenuation applied to the ramp of the spectrum.

This is the so-called three-joint rule that is specified in various other documents, such as Refs. [11] and [12]. This thee-joint rule essentially limits the amount of attenuation from joints in the shock path to 13 dB, regardless of the types of joints, shock amplitude, or number of joints if there are more than three. The three-joint rule is not unreasonable because some types of joints such as bolted joints lose their attenuating properties as excitation amplitudes decrease. The three-joint rule would be better interpreted as a "three-joint guideline" and applied with additional, design-specific information.

More nuanced information is provided in Refs. [9] and [10]. In these references, attenuation ranges are assigned to specific interface types. Solid joints and riveted butt joints provide no attenuation. A matched angle joint reduces the shock pulse by 30–60% and a solid joint with interface material provides a reduction of 0–30%. Load path bifurcations are discussed briefly also. The amplitude of a shock may be reduced by 20–70% when there is a corner in the shock path.

Shock attenuation effects due to shock path distance, joints, and structural features are complicated and not well understood. All information in the military standards and NASA handbooks is quite crude and should be used as a guideline rather than as a rigid design rule. Modeling shock attenuation across structural features is still a relatively new field of research, so vehicle-specific testing is still the best way to reduce uncertainty and avoid the risks associated with over- or under-conservatism of pyroshock environments. The information provided here is intended as a guide for developing test specifications for a component subject to a nearby pyroshock event. Since the component location is known, along with the source location, appropriate consideration of the intervening structure will help to ensure test specifications are appropriate to the expected field environments.

11.7 Test Specification Error

In addition to defining a test specification for a given environment, the allowable test error must also be defined. No matter how good the test equipment or the test engineers, the actual test event will differ somewhat from the requested shock excitation. Sometimes, the differences are small and sometimes more significant, but they will be present. Shock testing on a shaker system typically provides a shock event much closer to the test definition due to the closed-loop feedback controller. Drop testing or other shock machine testing usually has larger deviations from the defined test specification due to the open-loop nature of the shock machine configuration.

The exact nature of the allowable test specification error will need to be discussed and agreed upon by the specification designers and the test personnel so a hard rule is not appropriate. Rather some general guidelines will be provided here for consideration. Typically, there are two general types of error limits defined for shock test specifications: tolerance bands and an average error estimate. Both of these methods will be presented in more detail. Your test specifications may make use of either or both methods or you may find a more appropriate method applicable to your hardware.

11.7.1 Tolerance Bands

Tolerance bands are ranges within which the experimentally obtained SRS is intended to fall. Typically, these will be scaled versions of the original test specification both higher and lower. Figure 11.22 shows an arbitrary resonant plate test specification with ±6 and ±9 dB tolerance bands plotted for comparison. For resonant plate testing, a common tolerance is to require a high percentage of the experimentally measured SRS to fall within the ±6 dB tolerance band and all of the experimentally measured SRS to fall within the ±9 dB tolerance band. This, of course, is not a universal requirement. Some tests are performed with ±3 dB tolerance bands. Tests with tolerance bands larger than ±9 dB would be rare since the allowable excitation range would be getting quite large. Even ±6 dB tolerance bands are large in many respects. After all, ±6 dB means that the SRS could fall anywhere between a low of one-half of the defined test specification up to double the defined specification. Wider tolerance bands are often necessary for resonant plate and beam type tests due to the larger variability seen with this test method.

In contrast, shaker shocks are often extremely repeatable due to the nature of the closed-loop feedback controller. For this reason, much tighter tolerances can be applied to shaker shocks with no adverse concerns from the test laboratory. Most shaker shocks can be easily controlled within considerably less than a ±3 dB tolerance band. Here again, the equipment limitations must be considered with defining test specifications tolerances.

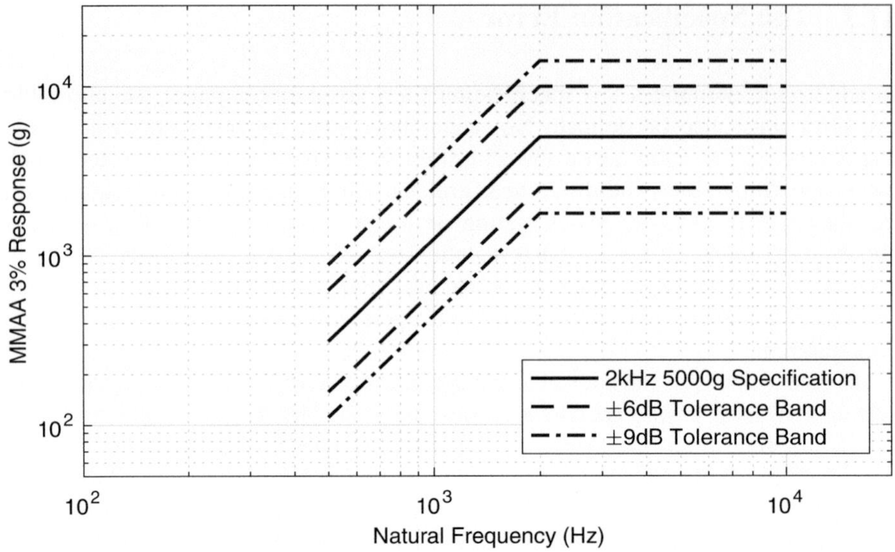

Fig. 11.22 Resonant plate test specification shown with typical test tolerance bounds

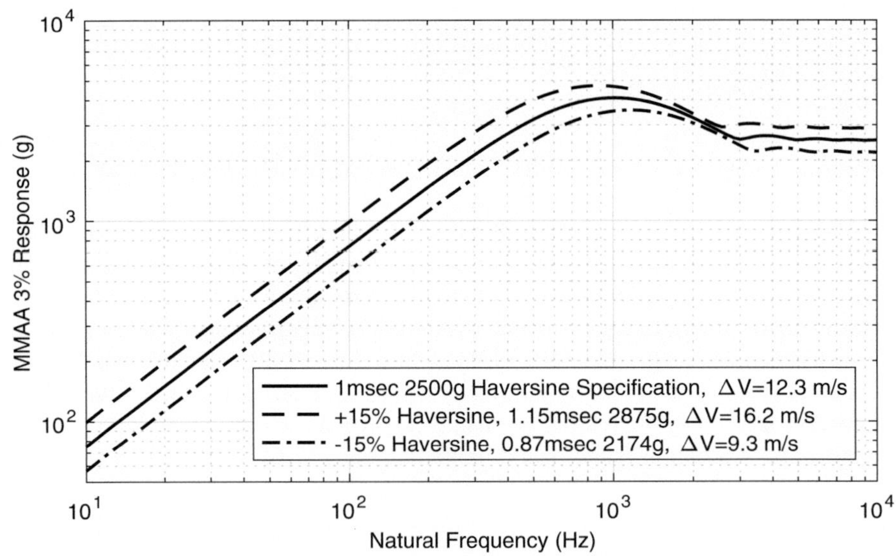

Fig. 11.23 HaversineDrop test specification shown with typical tolerance bounds

Drop shocks are usually not defined by their SRS but by an amplitude and duration. Since the test specification is defined by an amplitude and duration, the test tolerances should be defined in the same manner. For this reason, a common drop table tolerance is $\pm 15\%$ on the amplitude and $\pm 15\%$ on the duration, although tighter or broader tolerances are certainly permissible. Figure 11.23 shows an

Table 11.1 MIL-STD-810 pyroshock test tolerances

Environment	Test device	Bandwidth	Test tolerances
Near field	Pyrotechnic device	100 Hz to 20 kHz	−3 dB to +6 dB at 80% or more natural frequencies −6 dB to +9 dB at 100% natural frequencies At least 50% of SRS values exceed the nominal level
Mid field	Mechanical device	100 Hz to 10 kHz	−3 dB to +6 dB at 90% or more natural frequencies −6 dB to +9 dB at 100% natural frequencies At least 50% of SRS values exceed the nominal level
Far field	Mechanical device	100 Hz to 3 kHz	−3 dB to +6 dB at 90% or more natural frequencies −6 dB to +9 dB at 100% natural frequencies At least 50% of SRS values exceed the nominal level
Far field	Shaker	100 Hz to 10 kHz	−1.5 dB to +3 dB at 90% or more natural frequencies −3 dB to +6 dB at 100% natural frequencies At least 50% of SRS values exceed the nominal level

example of the range that a haversine MMAA SRS might fall within if ±15% amplitude and duration tolerances were applied. While such a plot would usually not be made, it is instructive for this application to understand how an out-of-tolerance drop shock might alter the defined test specification. The first observation is that the ±15% tolerances for a drop shock are considerably tighter than the ±6 dB tolerance bands of the resonant plate tests, about 2.4 dB at the maximum difference. The tolerance on the duration implies that the peak in the MMAA SRS can shift left or right on the frequency scale, thus ensuring that the tolerance band will not be a constant scaling from the baseline test specification as it was for the resonant plate test specifications.

The specific tolerance bands that are used to evaluate whether a shock test met its intent are usually specified in discipline-specific standards. For example, Table 11.1 lists SRS test tolerances for simulations of pyroshock events given in MIL-STD-810G. The mechanical device used for mid-field and far-field pyroshocks is typically a resonant beam or plate. The MIL-STD-810G pyroshock test tolerances are often adopted for other environments that are simulated with the same device and similar bandwidths. However, information in standards and handbooks is not always completely consistent. For example, MIL-HDBK-340 gives slightly different allowable test tolerances. It calls out ±6 dB for natural frequencies less than or equal to 3 kHz and more generous tolerances of +9 dB/−6 dB for natural frequencies above 3 kHz.

Figure 11.24 shows a typical mid-field pyroshock test SRS. The slope of the SRS is approximately 13 dB/octave. The SRS high-frequency asymptote, 2000 g, is the specified amplitude of the shock. The break-point natural frequency is 2 kHz indicating the resonant frequency content of the shock starts at that frequency. Test tolerances for this test specification are obtained from NASA-STD-7003A.

A spacecraft component was tested to this environment with an axial resonant beam test as described in Sect. 9.3.2. The acceleration time history applied to part

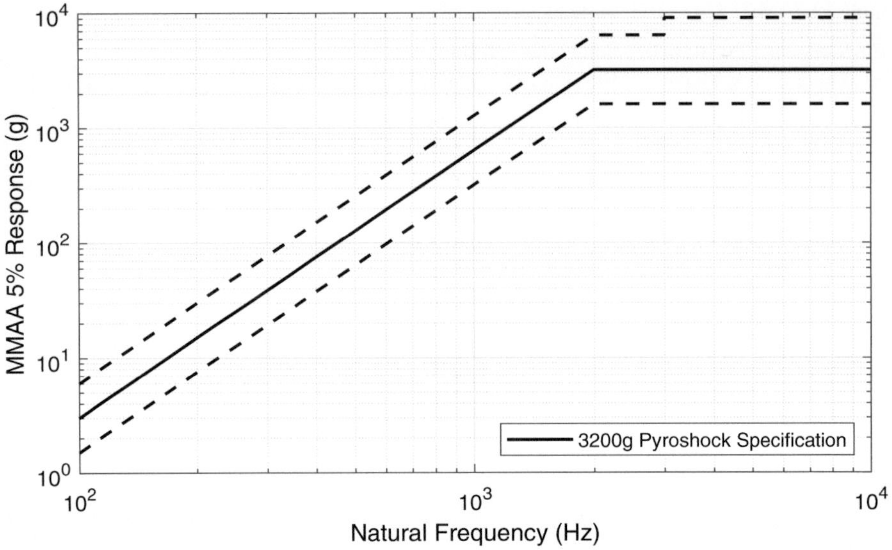

Fig. 11.24 Typical spacecraft component pyroshock test SRS with NASA-STD-7003 [3] recommended tolerance bounds

of the component was shown in Fig. 9.15. The MMAA SRS from the input was overlaid with the shock specification and the test tolerance bands in Fig. 9.16. As was presented previously, the test was out-of-tolerance; however, the component was sufficiently robust that the test was still considered a success. While the NASA-STD-7003A test tolerance bounds are considered to be fairly generous, it is often difficult to perform a laboratory test that fully lies within the given tolerance bands.

11.7.2 Average dB Error

Another useful option for defining test error is with the average dB error measurement. The average dB error is a measure of the difference between the measured SRS from the test and the defined test specification. The equation for the average dB error is give as:

$$\text{Avg dB Error} = \frac{1}{N}\sum_{n=1}^{N} 20\log_{10}\left(\frac{S(n)_{Measured}}{S(n)_{Specification}}\right). \tag{11.15}$$

In this equation, $S(n)_{Measured}$ is the measured SRS from the test and $S(n)_{Specification}$ is the test specification SRS. The difference between the two SRS is calculated for each frequency in terms of dB, implying that the two SRS must be sampled at the same frequencies. An average of the frequency-by-frequency differences yields the average dB error.

The goal of a shock test should be to end up with 0 dB error over the defined test specification frequency range. This rarely happens. The average dB error is a useful tool to indicate whether a test was low, high, or on target. If the average dB error is positive, then the test was overall more severe than the test specification. Likewise, if the average dB error is negative, the test was overall less severe than the defined test specification. The average dB error can also be useful in conjunction with defined tolerance bands to help ensure that an in-tolerance test is performed in the spirit of the test specification.

One disadvantage to the average dB error methodology is that it averages over the entire frequency range of the test specification. Most test specifications are defined over a relatively wide range but are considered critical over a fairly narrow range. For example, Fig. 11.23 showed the MMAA SRS plot of a haversine. The low-frequency portion of the SRS is a straight line that extends all the way down to 0 Hz. Likewise, the high-frequency portion of the SRS is also a straight line extending out to ∞ Hz. The critical portion of the SRS is in the vicinity of the hump at the haversine's primary frequency. By averaging over the entire frequency range, it may be possible to have a measured SRS that matches well in the frequency regions of little interest and poorly in the regions of greatest concern. If this happens, the average dB error might look good while the test could be a relatively poor match. For this reason, the average dB error is a good measure of a shock test's match to the specification but not necessarily the definitive measure.

11.8 Margin Test Specifications

Margin testing is a way to determine how much capability a component has beyond the requirement. In other words, how close is the requirement to the likely failure level. The difference is margin. There are two types of margin that could be considered, one is amplitude margin and the other is duration margin. Amplitude margin is essentially defined as how much harder can the component be hit before the onset of failure. In contrast, duration margin is equivalent to how many more times can the component be hit at the same level before the onset of failure. Duration margin is a fatigue problem whereas amplitude margin is a strain problem.

Whether to test amplitude margin or duration margin depends on the expected loading scenario. If the tested environment is a single shock load, such as an underwater mine explosion or an accidental crane drop, then amplitude margin is the appropriate choice. To test amplitude margin, the component should first be shock tested to the qualification test specification. The qualification specification in this context is often referred to as the 0 dB level or baseline level. The shock test is then repeated with successively increasing amplitudes. Usually, stepping up about 3 dB at each step, until the onset of failure. The exact incremental step is arbitrary but too large of an increment risks catastrophic failure instead of detecting the onset of failure. Too small of an increment may require an excessive number of tests and risks the possible accumulation of low-cycle fatigue damage.

If the expected environment is a repeated shock exposure, such as transportation shock where the number of shocks is proportional to the distance driven, then duration margin is more appropriate. In this case, the baseline shock test should ideally be repeated until the onset of failure. This can admittedly be a long process. For example, MIL-SST-810G suggests a total of 66 terminal peak saw-tooth shocks for every 5000 km of on-road transportation [13]. To estimate the margin for shock transportation, this sequence of 66 shocks would be repeated until failure occurred and the margin number calculated accordingly.

The goal of margin testing should always be to discover the onset of failure in a component. Ideally, multiple units would be tested so that an understanding of the range and variation of failures can be quantified. Failure is always a stochastic phenomenon with a high degree of variability. One test to failure is interesting and important, but is by no means definitive. Furthermore, in some circumstances it may not be possible to initiate a failure with the equipment available. In these cases, the engineer has demonstrated some margin for the component, but has not found the actual margin value. Demonstrated margin tests are still very useful to the component and system designer.

11.9 Test Compression and Repeated Shocks

In vibration testing, it is quite common to perform accelerated testing to save time in the laboratory. For example, transportation vibration testing is frequently time compressed. After all, if you want to make sure a component can survive a 5000 km cross-country trip in a truck at 100 km/h, do you really want to test your part for 50 h in the laboratory? Generally not. Time compression is a way to increase the vibration test levels to increase the induced stress in the part at each cycle. This moves the exposure to a shorter duration along the fatigue curve. In the same way, a similar process can be used to perform accelerated shock testing.

There are two contrary opinions with respect to accelerated shock testing. The first opinion is that accelerated testing does not matter for low numbers of shocks. The logic is that if a part is not yielding on the first shock, then an additional two or three shocks of the same level will not accumulate any significant level of fatigue damage and hence one shock or a few shocks gives the same result. This logic should only be applied when the number of shocks is very low, perhaps ten or less. This would not hold true for a system exposed to a large number of repeated shocks or systems where yielding is anticipated during the first shock.

The second approach to shock test compression is to apply the same formulas for accelerated testing used for accelerated life sinusoidal vibration testing [14]. These formulas are based on Miner's rule and the fact that failure is based on the number of stress reversals at a given amplitude. If the stress amplitude increases, then the number of cycles to failure will decrease. The accelerated testing equivalence ratio is given by the expression:

$$\frac{S_2(f)}{S_1(f)} = \left(\frac{N_1}{N_2}\right)^b, \qquad (11.16)$$

where $S_1(f)$ and $S_2(f)$ are the original and accelerated SRS test specifications, N_1 and N_2 are the number of applied shocks at each of the corresponding SRS test levels, and b is the fatigue damage coefficient. Fatigue damage coefficients are experimentally determined and specific to a particular material. However, generic values for average materials are frequently used. For example, $b = 0.15$ is a popular choice for generic structural metals and $b = 0.25$ is recommended by NASA for electronic components where solder failure is the primary concern [6].

It should be noted here that the formula for accelerating shock testing is usually only applied in one direction. In other words, test are almost always accelerated (shortened duration) by using Eq. 11.16. Tests are almost never decompressed (lengthened duration) using this formula. The first reason is that there is always a desire to shorten time in the laboratory and rarely ever a desire to test for longer. More importantly, the fatigue damage exponent is frequently not the same when going up the *S–N* curve, in the direction of increasing stress, as compared to going down the *S–N* curve, decreasing stress direction.

As a final note on test compression, as a general rule the compressed test specification should not increase the stress by more than a factor of two. The goal of test compression is to reduce the time in the test laboratory while keeping the same failure mechanisms. If the component stresses are increased such that the failure transitions from a high-cycle fatigue mode to a low-cycle fatigue mode, then the results may not be representative. Worse yet, it could be possible to transition from a fatigue failure mode to a first-passage yielding or buckling. For this reason, care should be taken when compressing tests to ensure that unrepresentative failure modes are not inadvertently introduced to the system.

11.10 Summary

Derivation of shock test specifications is a lot more than simply drawing lines on paper around test data. To be a good shock test specification designer, one must understand the source of the shock in order to match the test specification and shock test machine to the field environment. One must also understand the capabilities of the shock test machines that are available for use. It is quite frustrating to everyone when a test specification is defined in such a way that it cannot be performed in the laboratory.

Furthermore, it is incumbent on the test specification designer to understand what part of the field data are critical to match and what part can be ignored or minimized. For example, with a drop test, the impact velocity and the primary frequency are critical. All of the secondary fluctuations in the SRS may not be that important. It is also important to understand that it is likely impossible to exactly recreate a measured field environment in the laboratory. Even if you succeeded in recreating the environment, that measured event was likely only one of an infinite number of stochastically similar events. Ideally, a good test specification designer should spend time in the shock laboratory, should understand the performance and capability of the test equipment, and witness tests to understand the installations, fixtures, processes, and results seen during testing.

The development of shock test specifications can often be a balance between an art and a science. Of course, engineering often straddles the boundary between art and science because of the need to develop a solution when the science of the problem may not be fully explored. Shock testing is also notorious for producing unanticipated results. For this reason, the test specification developer should always strive to be present in the laboratory when testing is performed. There is nothing more enlightening to the engineer than to witness the damage first hand when a failure occurs. This attitude of "let's go see" is an important quality for the engineer to develop in all their work [15]. It is especially important with test work.

Personally witnessing a shock test gives the specification developer insight into the shock test levels. A 100 g test sounds like a number. Witnessing the same 100 g test on actual hardware will give the engineer an immediately obvious indication of whether the test was benign or severe. Did the part look like it performed well or did you have to go find pieces and parts that were ejected during the test? Did the instrumentation remain in place or come loose? Was the part installed correctly on the shock machine?

One of the more subtle problems with engineering tests is that the people conducting the tests do not always have an intuitive feel for what the test should look like. If you, as the specification developer, hand off a test specification and never follow-up to ensure it is being performed as you intended, you may find that it is not. Usually, the laboratory or the component owners will only come back to the specification developer if the part fails or the test cannot be performed. If the part passes the test, there is frequently no information passed back to the specification developer. This can be especially problematic if the component passes the test but the test was performed incorrectly. How would you ever know? For this reason, you should always strive to go and see.

The authors have had numerous experiences where shock tests either went bad or almost went bad because things were installed improperly or machines were setup incorrectly. Orientation-specific components can be installed in the wrong orientation if not positively keyed to a specific orientation. Drop test machines or hammer impact machines can be raised to the wrong height. The author even had one unfortunate shock test where the internal shock isolation system was inadvertently locked out by the installation technicians resulting in a catastrophic failure of the unit under test simply because they did not understand the hardware.

Problems

11.1 Derive the actual relationship between the axial and lateral stresses for the cantilever beam from Fig. 11.2. The circular cross-section cantilever beam is a special case where the outer fiber is stressed in one direction and not in the other. A beam with a square cross-section will have a different result since the corner fiber will be stressed by bending in both directions. Should this be concerning? Why or why not? How do the results change if the beam is built from standard structural sections such as L-sections or I-beams?

11.2 Section 11.3 discusses a very simple method for specifying a shock test with a required velocity change and rise time. Generate a few sample time histories meeting the 5 m/s velocity change occurring in approximately 15 ms. Compare the resulting SRS profiles from the various time histories to see if they are similar. Are they similar to any of the classical shock pulses described in Chap. 5? Can you synthesize a classical shock to match your velocity change test specifications?

11.3 When developing test specifications from experimental data, is it better to envelope the data or to fare through the data? Enveloping the data are more conservative but perhaps unnecessarily so. Use some sample test data to develop test specifications both ways and make a determination of the nominal increase in severity for enveloping versus faring through the test data.

11.4 Two shock test specifications may appear to be a better match to measured test data than a single test specification but they come with added complexity. Take the set of test specifications shown in Figs. 11.9 and 11.10 and overlay them with reasonable error bounds. Make a statement about whether or not the two methods are significantly different given the range of possible test errors. Can minor adjustments be made to make the two methods equivalent?

11.5 Test compression is a challenging problem since many assumptions are usually made about the failure mechanisms of the unit under test. Calculate test compression ratios for various values of the fatigue damage coefficient, b, ranging from 0.1 to 0.5 and compare the results. Look-up values for common metals such as steel and stainless steel and compare the variation in test compression results.

11.6 Use the relationship provided in Sect. 11.9 along with the dB relations given in Sect.5.2.1 to determine the accelerated testing factor in dB required to reduce the number of shocks by a factor of ten for various generic values of the fatigue damage coefficient.

11.7 Average dB error is a reasonable measure of test accuracy; however, it can be skewed by inclusion of SRS data away from the region of greatest interest. Use some existing test data and the corresponding test specification to calculate the average dB error for the test. See how the average error changes as more or less of the test specification frequency range is included in the calculation. Can you envision a scenario where the average dB error is good but the test is bad? What about a good test with a poor average dB error?

11.8 Collect drop shock test data on the same system from two different drop heights. Demonstrate the drop shock scaling method with real test data. Does the data agree with the scaling methodology detailed in this chapter?

References

1. United States Department of Defense. (1999). *Test requirements for launch, upper-stage and space vehicles*. Department of Defense Handbook, MIL-HDBK-340A.
2. Yunis, I. (2005). The standard deviation of launch vehicle environments. In *Proceedings of the 46th Conference on AIAA/ASME/ASCE/AHS/ASC Structures, Structural Dynamics, and Material*, Austin, TX.
3. National Aeronautics and Space Administration. (2011). *Pyroshock test criteria*. NASA Technical Standard, NASA-STD-7003A.
4. Nelder, J. A., & Mead, R. (1965). A simplex method for function minimization. *Computer Journal, 7*, 308–313.
5. Sisemore, C., & Skousen, T. (2014) A method for extrapolating haversine shock test levels. In *Proceedings of the 86th Symposium on Shock and Vibration*, Orlando.
6. National Aeronautics and Space Administration. (2001). *Dynamic environmental criteria*, NASA-HDBK-7005.
7. Barrett, S. (1975). The development of pyro shock test requirements for Viking Lander capsule components. In *Proceedings of the Institute of Environmental Sciences 21st Annual Technical Meeting* (pp. 5–10).
8. Ott R. J., & Folkman, S. (2017). Full-scale pyroshock separation test: Attenuation with distance. *AIAA Journal of Spacecraft and Rockets, 54*(3), 602–608. https://doi.org/10.2514/1.A33705
9. Kacena, W. J., McGrath, M. B., & Rader, W. P. (1970). *Aerospace systems pyrotechnic shock data (ground test and flight)* (Vol. 1). Summary and Analysis, NASA-CR-116437.
10. Kacena, W. J., McGrath, M. B., & Rader, W. P. (1970). *Aerospace systems pyrotechnic shock data (ground test and flight)* (Vol. 6). Pyrotechnic Shock Design Guidelines Manual, NASA-CR-116406.
11. National Aeronautics and Space Administration. (2013). *General environmental verification standard (GEVS) for GSFC flight programs and projects*, GSFC-STD-7000A.
12. Sarafin, T. P. (2003) *Spacecraft structures and mechanisms: From concept to launch. Space technology series*. El Segundo, CA/Dordrecht: Microcosm, Inc./Kluwer Academic Publishers.
13. United States Department of Defense. (2014). Department of defense test method standard; environmental engineering considerations and laboratory tests, MIL-STD-810G (w/Change 1).
14. Fackler, W. C. (1972). *Equivalence techniques for vibration testing*, SVM-9, The Shock and Vibration Information Center, United States Department of Defense.
15. Skakoon, J. G., & King, W. J. (2001). *The unwritten laws of engineering*. New York: ASME Press.

Chapter 12
Energy Spectra Methods

The shock response spectrum (SRS), in one of the several forms, is considered to be the standard for characterizing mechanical shock events. The popularity of the SRS is due largely to its ability to represent damage potential reasonably well. In general, two time histories with similar SRS are considered to have similar damage potential. While the SRS is the standard for shock specifications and describing shock events, the SRS is not the only method for characterizing shock environments. Several non-SRS methods have been proposed along with methods used to supplement the SRS with additional information, such as temporal moments, which were covered in Chap. 10. While all of the non-SRS methods have merit, not all of the methods aggregate the same information about the shock transient.

One method that is often used to evaluate the severity of a shock is the energy response spectrum (ERS). The energy response spectrum falls into the group of analysis approaches called *energy methods*. The basis for energy methods is conservation of energy, which requires that all of the energy input to a structure be absorbed, dissipated, or converted to kinetic energy. The civil engineering community, specifically the earthquake engineering community, uses energy methods for analyzing the damage potential to structures from earthquakes. However, energy methods can provide insight into how any structure might fail when subjected to shock or vibration loading.

The energy response spectrum is a simplified way of evaluating the distribution of energy in a structure. It is a plot of an energy quantity of interest as a function of the natural frequency of a single degree-of-freedom oscillator. In this sense it is analogous to the shock response spectrum, but the energy response spectrum uses quantities more directly related to failure mechanisms of interest such as potential energy, kinetic energy, and input energy, rather than displacement type quantities. While energy response spectra hold some advantages over traditional shock response spectra, an energy response spectrum is still a spectral method. As a result, the energy response spectra suffer from the same non-uniqueness limitations and accuracy issues associated with the SRS. Of course, energy response spectra

C. Sisemore, V. Babuška, *The Science and Engineering of Mechanical Shock*,
https://doi.org/10.1007/978-3-030-12103-7_12

also have some advantages over the SRS. One of these is that energy accumulates. The SRS of two identical shocks applied sequentially is the same as the SRS of the single shock, if the system is allowed to come to rest between the shocks. An energy response spectrum, on the other hand, is cumulative so the energy from two shocks will be greater than the energy from just one shock.

The concept of energy response spectra is nearly as old as the SRS. Hudson [1] and Housner [2] are generally credited with proposing energy quantities for characterizing transient base excitations resulting from earthquakes. Further fundamental analytical developments were made by Zahrah and Hall [3, 4] with the derivation of the various energy terms. Uang and Bertero [5] and Kalkan and Kunnath [6] made significant contributions by showing that energy imparted to a structure can be expressed in terms of relative input energy or absolute input energy.

Many researchers have explored various aspects and applications of energy methods. For example, Takewaki and Fujita [7] looked at both time domain and frequency domain energy formulations of earthquake input energies in tall buildings. Inelastic energy dissipation, which includes hysteretic and elastic plastic mechanisms, has been a popular area of research. Bruneau and Wang [8] did early work on using energy methods for characterizing the inelastic response of SDOF structures. Nicknam et al. [9] proposed hysteretic energy capacity as a cumulative damage parameter for SDOF systems, and Segal and Val [10] modeled inelastic energy dissipation with a Ramberg–Osgood hysteresis model to assess structural response to seismic loading. Others have extended energy-based analyses to general vibration problems. Edwards [11] presented a framework for the use of energy methods with vibration loads. The references cited here are by no means an exhaustive list; that would be impossible. They are some of the more significant contributions to the development and dissemination of energy methods for evaluating structural performance. The interested reader can use these citations as starting points for locating references for his or her specific problem.

This chapter presents an overview of energy response spectra. We take the approach that energy quantities are specific types of outputs, expressed through the output equations so all the mechanics developed for the SRS are applicable. Specifically, we explain:

1. the main energy quantities of interest—input energy, dissipated energy, kinetic energy, and absorbed energy;
2. absolute and relative energy methods;
3. energy balance in the absolute and relative energy methods;
4. energy response spectra with the main energy quantities of interest.

12.1 Energy Quantities

To compute the response of a structure we need an equation of motion, initial conditions, and an output equation. In Chap. 3 we presented equations of motion for a single degree-of-freedom oscillator (Fig. 3.3) and we discussed various response quantities such as absolute acceleration, relative velocity, and pseudo-velocity. In Chap. 8 we extended those ideas to multi-degree-of-freedom systems. Energy methods use the same equations of motion, but the output quantities are *specific energy* quantities. The term *specific energy* means energy per unit mass, and we use these quantities because response spectra are parameterized by the modal parameters ω_n and ζ.

Almost all papers and references on energy methods start with an energy balance equation, which is a statement of the conservation of energy. We present the energy quantities as output quantities for an equation of motion. We take this approach because in practice, we solve an equation of motion for the state variables and then use them to compute the energy terms. We do not solve for energies directly with the energy balance equation. The energy balance equation explains the relationships between the energy quantities.

The six energy response quantities are:

1. Absolute kinetic energy:

$$E_{KA}(t) = \frac{1}{2}\left[\dot{x}^2(t) - \dot{x}^2(0)\right]. \tag{12.1}$$

2. Relative kinetic energy:

$$E_{KR}(t) = \frac{1}{2}\left[\dot{z}^2(t) - \dot{z}^2(0)\right], \tag{12.2}$$

   where $\dot{z}(t)$ is the relative velocity of the SDOF oscillator mass.
3. Dissipated energy:

$$E_D(t) = 2\zeta\omega_n \int_0^t \dot{z}^2(\tau)d\tau. \tag{12.3}$$

4. Absorbed or strain energy:

$$E_S(t) = \frac{1}{2}\omega_n^2\left[z^2(t) - z^2(0)\right]. \tag{12.4}$$

5. Absolute input energy:

$$E_{IA}(t) = \int_0^t \ddot{x}(\tau)\dot{w}(\tau)d\tau, \tag{12.5}$$

   where $\dot{w}(t)$ is the prescribed velocity at SDOF oscillator base.

6. Relative input energy:

$$E_{IR}(t) = \int_0^t \ddot{w}(\tau)\dot{z}(\tau)d\tau, \tag{12.6}$$

where $\ddot{w}(t)$ is the prescribed acceleration at the SDOF oscillator base.

These quantities are discussed in the following sections.

12.1.1 Relative Energy Method

The relative energy method gets its name from the form of the equation of motion used to describe structure excited by a base acceleration. The fundamental equation of motion for shock analysis introduced in Chap. 3 is:

$$\ddot{z}(t) + 2\zeta\omega_n\dot{z}(t) + \omega_n^2 z(t) = -\ddot{w}(t), \tag{12.7}$$

where $z(t)$, $\dot{z}(t)$, and $\ddot{z}(t)$ are the displacement, velocity, and acceleration of the SDOF oscillator mass relative to the base, respectively, and $\ddot{w}(t)$ is the prescribed base acceleration.

As discussed previously, this equation can be viewed as the equation of motion of a fixed-based SDOF oscillator excited by an inertial specific force induced by the base acceleration. Figure 12.1 introduced in Chap. 3 shows the fixed base single degree-of-freedom oscillator. It is repeated here, with only prescribed base acceleration loading and $p(t) = 0$.

Since work is defined as force times displacement, the specific work done by each of the terms in Eq. 12.7 is obtained by integrating with respect to the relative

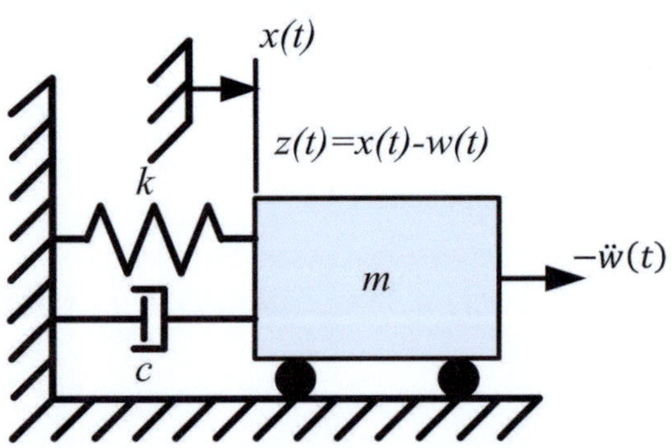

Fig. 12.1 Base excitation of a single degree-of-freedom oscillator

displacement, z. Integrating Eq. 12.7 gives an energy balance equation,

$$\int_{z(0)}^{z(t)} \ddot{z} dz + 2\zeta\omega_n \int_{z(0)}^{z(t)} \dot{z} dz + \omega_n^2 \int_{z(0)}^{z(t)} z dz = -\int_{z(z(0))}^{z(t)} \ddot{w} dz. \tag{12.8}$$

Recognizing from differential calculus that $dz = \dot{z}dt$ and changing the limits of integration appropriately gives the energy balance equation integrated with respect to time:

$$\int_0^t \ddot{z}(\tau)\dot{z}(\tau)d\tau + 2\zeta\omega_n \int_0^t \dot{z}^2(\tau)d\tau + \omega_n^2 \int_0^t z(\tau)\dot{z}(\tau)d\tau = -\int_0^t \ddot{w}(\tau)\dot{z}(\tau)d\tau. \tag{12.9}$$

This equation can be written compactly as

$$E_{KR}(t) + E_D(t) + E_S(t) = E_{IR}(t). \tag{12.10}$$

The first term is the specific relative kinetic energy. Integrating by parts yields Eq. 12.2

$$E_{KR}(t) = \int_0^t \ddot{z}(\tau)\dot{z}(\tau)d\tau = \frac{1}{2}\left[\dot{z}^2(t) - \dot{z}^2(0)\right]. \tag{12.11}$$

The second term on the left-hand side of Eq. 12.8 is the viscously dissipated, or damping, energy defined in Eq. 12.3. It is always positive and remains as an integral.

The absorbed or strain energy is the third term on the left-hand side of Eq. 12.8. The integral is easily evaluated yielding Eq. 12.4

$$E_S(t) = \omega_n^2 \int_{z(0)}^{z(t)} z dz = \frac{1}{2}\omega_n^2\left[z^2(t) - z^2(0)\right]. \tag{12.12}$$

Typically, energy associated with inelastic (e.g., hysteretic) mechanisms is grouped in with the absorbed energy. However, grouping all dissipative energy together in the dissipated energy term, $E_D(t)$, keeps the strain energy conservative. If there is no permanent deformation in the structure when it comes to rest, the final strain energy must be zero, but dissipated energy, no matter the mechanism, will be positive. In the case of a viscously damped SDOF oscillator, the absorbed energy is just the strain energy, $E_S(t)$.

The right-hand side of Eq. 12.9 is the specific relative input energy, Eq. 12.6. It is the work done by the inertial load (acceleration) on the fixed base SDOF oscillator.

The energy balance equation, Eq. 12.10, says that all of the energy imparted to the system through the base motion is distributed between kinetic energy, strain energy, and dissipated energy. Because the strain energy and kinetic energy are conservative, if the SDOF oscillator starts at rest and returns to rest with no residual deformation, then all energy input into the system must have been dissipated, $E_{IR}(\infty) = E_D(\infty)$.

This is the main idea behind energy methods. If a structure cannot dissipate all of the energy input to it, it will fail. If the maximum strain energy exceeds the capability of the structure, it will also fail. The first type of failure, when the system cannot dissipate all of the input energy, is a fatigue type of failure, whereas the second type is an overstress failure.

12.1.2 Absolute Energy Method

The relative energy method uses the relative motion model of a system, so it's not surprising that the absolute energy method uses the absolute motion model. In Chap. 3, we showed that the equation of motion of the SDOF oscillator with no load applied directly at the mass can be written as:

$$\ddot{x}(t) + 2\zeta\omega_n\dot{z}(t) + \omega_n^2 z(t) = 0, \tag{12.13}$$

where $\ddot{x}(t)$ is the absolute acceleration of the mass.

Following the same approach that we used for the relative energy balance equation, the absolute energy balance expression is

$$\int_0^t \ddot{x}(\tau)\dot{z}(\tau)d\tau + 2\zeta\omega_n \int_0^t \dot{z}(\tau)^2 d\tau + \omega_n^2 \int_0^t z(\tau)\dot{z}(\tau)d\tau = 0. \tag{12.14}$$

The second and third terms are the same dissipative and strain energy that we obtained previously. The strain energy and dissipated energy terms are the same because the damping and elastic forces are functions of the relative motion. The first term includes both absolute and relative motion terms. It is not particularly useful in this form, but may be decomposed into two terms by substituting $\dot{x} - \dot{w}$ for $\dot{z}$

$$\int_0^t \ddot{x}(\tau)\dot{z}(\tau)d\tau = \int_0^t \ddot{x}(\tau)\dot{x}(\tau)d\tau - \int_0^t \ddot{x}(\tau)\dot{w}(\tau)d\tau. \tag{12.15}$$

The absolute energy balance equation becomes

$$\int_0^t \ddot{x}(\tau)\dot{x}(\tau)d\tau + 2\zeta\omega_n \int_0^t \dot{z}(\tau)^2 d\tau + \omega_n^2 \int_0^t z(\tau)\dot{z}(\tau)d\tau = \int_0^t \ddot{x}(\tau)\dot{w}(\tau)d\tau, \tag{12.16}$$

where the absolute kinetic energy is

$$E_{KA}(t) = \int_0^t \ddot{x}(\tau)\dot{x}(\tau)d\tau = \frac{1}{2}\left[\dot{x}^2(t) - \dot{x}^2(0)\right], \tag{12.17}$$

and the term on the right-hand side is the specific absolute input energy given in Eq. 12.5. The specific absolute input energy is the total work done by the inertial motion of the mass. It includes rigid body motion as well as vibratory motion. The specific absolute input energy and the specific relative input energy are related by

$$E_{IA}(t) = \frac{1}{2}\dot{w}^2(t) + \dot{w}(t)\dot{z}(t) - E_{IR}(t). \tag{12.18}$$

The energy balance equation for the absolute energy method can be written compactly as

$$E_{KA}(t) + E_D(t) + E_S(t) = E_{IA}(t). \tag{12.19}$$

In most cases, the relative energy formulation is more intuitive than the absolute energy formulation and is more applicable to mechanical shock problems.

12.2 Energy Response Spectra

Any of the energy quantities can be used to generate an energy response spectrum (ERS). The energy spectra are calculated in the same manner as the SRS, on a frequency by frequency basis. As with the SRS, the selection of frequencies is arbitrary but traditionally they are logarithmically spaced to optimize the calculation time for a log–log plot presentation.

Recall that peak response is what is used in the SRS. With energy methods, we can use peak energies to make an ERS, but we can also use final time energy, where the final time is the time that the system has returned to a vibration free state. Just as the oscillator starts from rest, it must return to a vibration free state when the damping is non-zero. This final time can be $t_f = \infty$ for notational convenience.

The peak energy response spectrum is formed from the infinity norms of the energy terms for individual SDOF oscillators parameterized by natural frequency, ω_n, and damping ratio, ζ.

$$\mathrm{ERS}_X = \max_t |E_X(t, \omega_n, \zeta)| \forall \omega_n, \zeta; 0 < t, \tag{12.20}$$

where the subscript X identifies one of the six energy types.

The strain energy and the relative kinetic energy will reach their peak values during or just after the forced vibration era. Once the SDOF oscillator stops vibrating, the strain energy, being a function of relative displacement, must be zero. Similarly, the relative kinetic energy must be zero when the vibration stops. Absolute kinetic energy will also reach its peak value during or just after forced vibration era, but its final value may not be zero if the shock imparts a velocity change.

Let T_E be the time that the base excitation ceases, i.e., the duration of the forced vibration era. The peak value of the specific relative input energy may occur at or before the end of the forced vibration era.

The dissipated energy reaches its peak value when the SDOF oscillator vibration ceases. The peak dissipated energy is just the final relative input energy. For $t > T_E$, $\ddot{w}(t) = 0$ so the final value of the relative input energy is

$$E_{IR}(\infty) = E_{IR}(T_E) = \int_0^{T_E} \ddot{w}(\tau)\dot{z}(\tau)d\tau. \tag{12.21}$$

Because $E_S(\infty) = E_{KR}(\infty) = 0$, the relative energy balance equation 12.10 is

$$E_D(\infty) = E_{IR}(T_E). \tag{12.22}$$

The total energy dissipated by the structure is the relative input energy evaluated at $t = T_E$. This makes the relative input energy such a useful quantity because we only need to consider motion during the forced vibration era.

The peak and final time energy quantities are illustrated with two examples. In the first example, the excitation is a haversine shock, which is the representative of an impact or a shock applied on a drop table or shock machine. This type of shock imparts a velocity change to the system. The second example is a random transient. This type of shock does not impart a velocity change to the system.

12.2.1 Example 1: Impact Shock

Figure 12.2 shows a plot of the calculated input energy spectrum overlaid with thc dissipated energy spectrum for the 100 g 10 ms classical haversine shock shown in Fig. 2.7. In this plot, the dissipated energy spectrum exactly overlays the input energy spectrum. One should immediately notice the similarity of this plot to the classical haversine relative velocity SRS plots shown in Chap. 4. The reason for the similarity was derived in Sect. 12.1.1.

Figure 12.2 shows the energy spectrum—a combination of the energy results at each of the selected calculation frequencies. Figure 12.3 shows the time history energy results for the 100 g 10 ms haversine shock pulse evaluated at 50 Hz. At this frequency, which is lower than the 100 Hz pulse frequency, the specific input energy rises from zero to a constant value at the end of the shock pulse, 10 ms in this example. The specific dissipated energy lags behind and approaches the specific input energy magnitude. If the time history were continued further, the dissipated energy will reach the same amplitude as the input energy. This was derived in Sect. 12.1.1 and is also shown in the spectral plot in Fig. 12.2 where the dissipated and input energy curves at 50 Hz are on top of one another. Also notice in Fig. 12.3 how the dissipated energy has a monotonically increasing oscillation with time. This

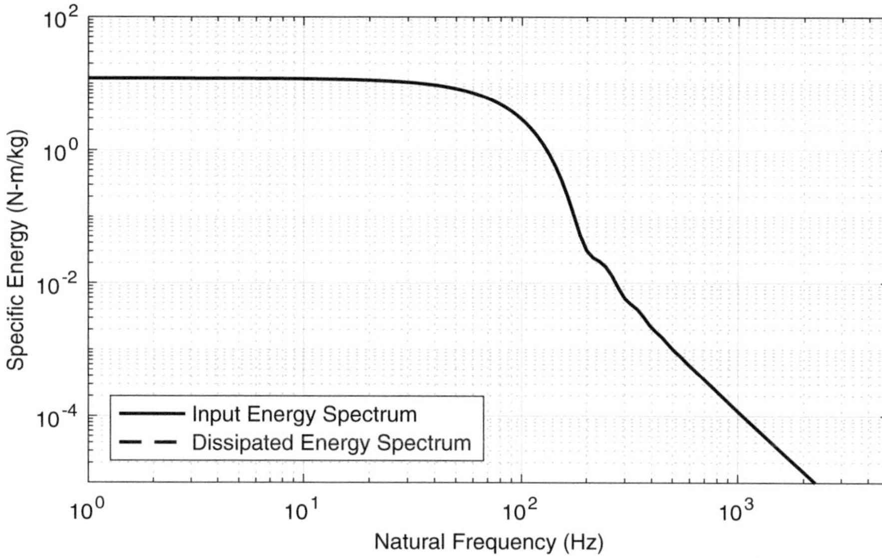

Fig. 12.2 Plot of the input and dissipated energy spectra for a 100 g 10 ms haversine shock pulse

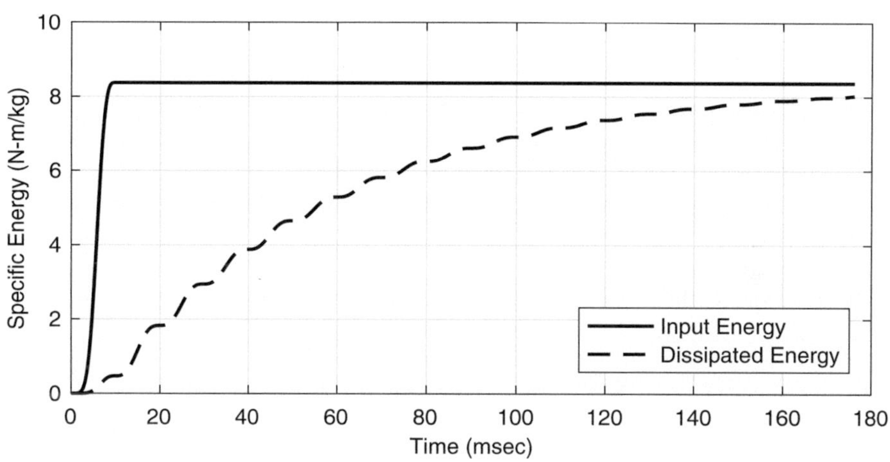

Fig. 12.3 Plot of the input and dissipated energy with time for the 100 g 10 ms haversine shock pulse through a 50 Hz SDOF oscillator

oscillation corresponds to the positive and negative oscillations of the 50 Hz SDOF oscillator.

To extract the energy spectrum value at each frequency, the time histories are calculated, as shown in Fig. 12.3. As with SRS, time histories are only required for a little more than three-quarters of one complete cycle of the SDOF oscillatory response after the shock transient has ended because the maximum specific strain

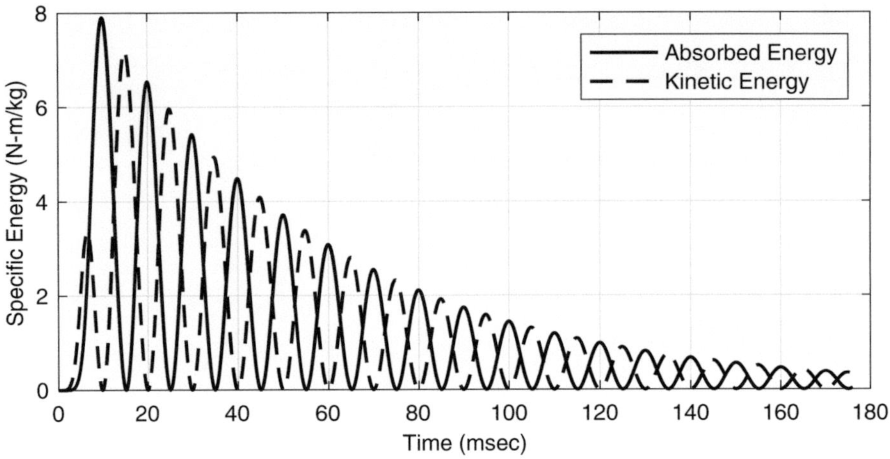

Fig. 12.4 Plot of the absorbed and kinetic energy with time for the 100 g 10 ms haversine shock pulse through a 50 Hz SDOF oscillator

energy and maximum specific relative kinetic energy occur when the relative displacement and relative velocity reach their maxima, respectively. For the specific input energy, the value at $t = \infty$ is the same as the value at $t = T_E$ because no more energy can be imparted to the system once the shock transient has passed. In contrast, energy from the shock pulse may continue to be dissipated long after the pulse has passed. This is clearly shown with the continued increase in the dissipated energy with time, asymptotically approaching the input energy value. However, the maximum dissipated energy is the final time specific energy, so we do not need to compute the response past the end of the forced vibration era to get the dissipated energy response spectrum.

In addition to the specific input and dissipated energy, Fig. 12.4 shows time history plots of the specific strain and relative kinetic energies for the 100 g 10 ms haversine shock pulse evaluated at 50 Hz. The maximum strain energy is frequently used to create an ERS and is directly relatable to the maximum pseudo-velocity SRS. This plot shows the out-of-phase oscillation between the strain energy and the relative kinetic energy after the initial transient has passed. The out-of-phase nature of the oscillation comes from the motion of the SDOF oscillator. When the displacement of the mass is greatest, the spring is stretched to its maximum but the velocity of the mass is zero. Zero velocity implies zero kinetic energy. Likewise, as the mass passes back through its neutral position the velocity is highest, implying that the kinetic energy is maximum. At this point the spring is un-stretched and thus there is no strain energy. This trade-off between strain energy and kinetic energy continues with each cycle until the motion stops. The decay seen in Fig. 12.4 matches the increase in dissipated energy from Fig. 12.3.

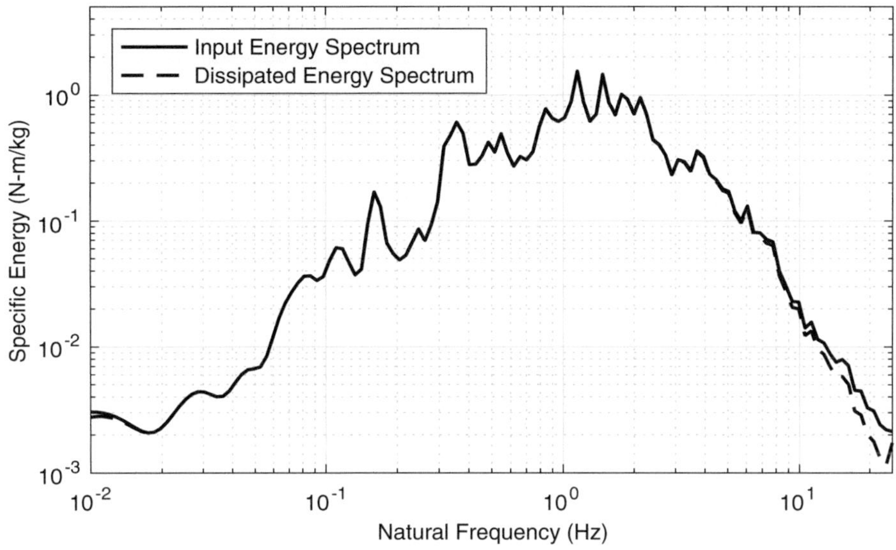

Fig. 12.5 Plot of the input and dissipated energy spectra for the May 1940 El Centro earthquake shock through a 5 Hz SDOF oscillator

12.2.2 Example 2: Random Transient Base Acceleration

Figure 12.5 shows the specific input and dissipated energy spectra for the May 1940 El Centro earthquake time history from Fig. 2.12. The energy spectra shown here are much more featured than those of the classical shock pulse but show similar trends. In this example, the dissipated energy spectrum is similar to the relative input energy spectrum but not identical at all points. While it is not obvious from the spectral plot, the dissipated energy is slightly lower than the input energy across all frequencies.

The difference is more clearly seen in Fig. 12.6. This plot shows the specific relative input and dissipated energy time history responses for a 5 Hz SDOF oscillator exposed to the earthquake shock. In Fig. 12.6 it can be clearly seen that the dissipated energy is a smoothed version of the input energy at a consistently lower amplitude. The difference is more pronounced on the linear amplitude scale here than the logarithmic amplitude scale in Fig. 12.5.

Figure 12.6 also shows that the relative input energy is not monotonically increasing for this real-world shock event. The classical haversine passed through the lower-frequency SDOF oscillator shown in Fig. 12.3 appeared to be monotonically increasing. The relatively higher frequency (compared to the fundamental frequencies of the shock event) coupled with the oscillatory nature of the earthquake shock results in a system where the SDOF oscillator does not necessarily accept all of the energy that the shock can provide. Thus, the final value of the specific input energy can be equal to or less than the maximum value of the specific input

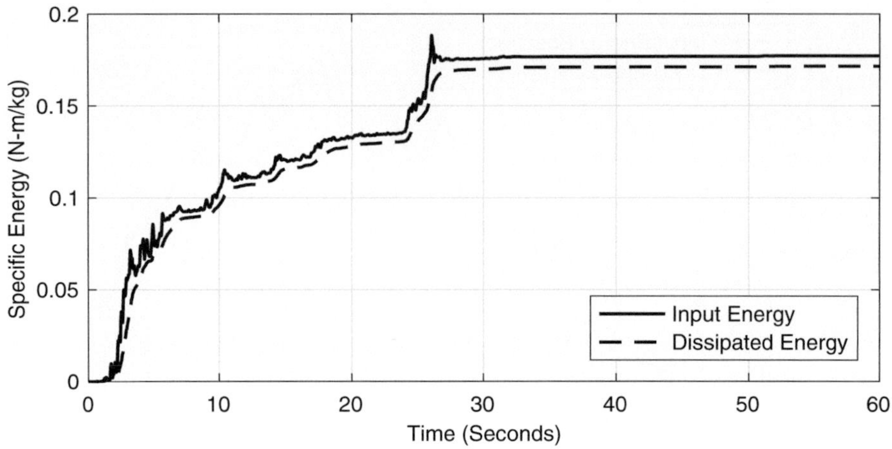

Fig. 12.6 Plot of the input and dissipated energy with time for the May 1940 El Centro earthquake shock through a 5 Hz SDOF oscillator

energy. That is why we create energy response spectra with the maximum relative input energy. The dissipated energy spectrum is the final time relative input energy spectrum.

The dissipated energy example shown in Fig. 12.6 reaches its asymptote almost immediately after the shock transient. This is due to the relatively high frequency of the SDOF oscillator compared to the earthquake frequency. If the energy terms were calculated for a much lower-frequency SDOF oscillator, then dissipated energy time history would rise much more slowly, similar to the manner shown in Fig. 12.3.

12.3 Energy Response Spectrum Relation to the SRS

The specific strain energy expression in Eq. 12.4 is a function of the relative displacement of the SDOF oscillator squared. The strain energy response spectrum is just the peak value of the specific strain energy of a single degree-of-freedom oscillator, so

$$ERS_S = \frac{1}{2} SRS_{PV}^2, \tag{12.23}$$

where SRS_{PV}^2 is the square of the maxi-max pseudo-velocity SRS.

Similarly, the specific relative kinetic energy in Eq. 12.2 is a function of the relative velocity squared, so

$$\mathrm{ERS}_{KR} = \frac{1}{2} \mathrm{SRS}_{RV}^2, \tag{12.24}$$

where SRS_{RV} is the maxi-max relative velocity SRS.

12.4 Calculation of Energy Terms

Calculating an energy response spectrum is analogous to calculating the SRS because we use the same variables from the equation of motion. The only difference is that for the input energies we must integrate the product of the base excitation with the response. This is one additional computational step. The ISO standard for calculating the SRS makes use of digital recursive filters to resolve the SDOF oscillator response at a given frequency. This was discussed extensively in Chap. 4. Likewise, the ISO standard filter weights and methodology can be used for calculating the various energy spectrum terms. It was already shown in Sect. 12.3 that specific strain energy was related to the pseudo-velocity SRS, so once we have computed the pseudo-velocity SRS, we have the strain energy response spectrum.

The specific kinetic energy is similar to the strain energy but is a measure of the relative velocity between the SDOF oscillator mass and ground. As with the strain energy response spectrum, once we have computed the relative velocity SRS, we have the relative kinetic energy response spectrum.

The absolute kinetic energy response spectrum requires different, but analogous calculations can also be done with the digital filter method and ISO standard filter weights. The relative input energy spectrum is also computed with the digital filter method, but the relative velocity must be multiplied by the base acceleration and the result numerically integrated. These additional steps are not time intensive because we only need to compute the response during the forced vibration era.

The use of the ISO standard filter weights in the above expressions greatly simplifies the calculation of the various energy response spectra. The coding of an energy spectrum routine is straightforward and essentially the same as an SRS function. The additional numerical integration steps do not add a significant computational burden.

12.5 Summary

This chapter presented the energy response spectrum as a way to describe a shock excitation and its damage potential on a structure. We described the relative energy method and the absolute energy method. These methods yield different input energies and kinetic energies. Through two illustrative examples we showed that the relative energy approach is more intuitive for mechanical shock problems.

This chapter presented the energy response spectrum as a way to describe a shock excitation and its damage potential on a structure. We described the relative energy method and the absolute energy method. These methods yield different input energies and kinetic energies. Through two illustrative examples we showed that the relative energy approach is more intuitive for mechanical shock problems.

Problem

12.1 Use a readily available shock time history and calculate the time history for the specific input, dissipated, absorbed, and kinetic energies for several SDOF oscillator frequencies ranging from very low to very high, relative to your shock pulse frequency. How do the different energy metrics change as the SDOF oscillator frequency changes? How long do you have to propagate the solution in time to obtain a stable result?

References

1. Hudson, D. E. (1956). Response spectrum techniques in engineering seismology. In *Proceedings of the World Conference on Earthquake Engineering* (pp. 4.1–4.12). Berkeley, CA: University of California.
2. Housner, G. W. (1959). Behavior of structures during earthquake. *Journal of the Engineering Mechanics Division (ASCE), 85*(4), 109–129.
3. Zahrah, T. F., & Hall, W. J. (1982). *Seismic energy absorption in simple structures*. Civil engineering studies. Structural research series no. 501. Urbana, IL: University of Illinois.
4. Zahrah, T. F., & Hall, W. J. (1984). Earthquake energy absorption in SDOF structures. *Journal of Structural Engineering, 110*(8), 1757–1772.
5. Uang, C.-M., & Bertero, V. V. (1990). Evaluation of seismic energy in structures. *Earthquake Engineering & Structural Dynamics, 19*(1), 77–90.
6. Kalkan, E., & Kunnath, S. K. (2008). Relevance of absolute and relative energy content in seismic evaluation of structures. *Advances in Structural Engineering, 11*(1), 17–34.
7. Takewaki, I., & Fujita, K. (2009). Earthquake input energy to tall and base-isolated buildings in the time and frequency dual domains. *The Structural Design of Tall and Special Buildings, 18*, 589–606.
8. Bruneau, M., & Wang, N. (1996). Some aspects of energy methods for the inelastic seismic response of ductile SDOF structures. *Engineering Structures, 18*(1), 1–12.
9. Nicknam, A., Shahbazian A. A., & Sabeti, M. R. (1999). The assessment of hysteretic energy capacity in energy-based damage model design. In *Developments in analysis and design using finite element methods* (pp. 215–221). Edinburgh: Civil-Comp Press.
10. Segal, F., & Val, D. V. (2006). Energy evaluation for Ramberg–Osgood hysteretic model. *Journal of Engineering Mechanics, 132*(9), 907–913.
11. Edwards, T. S. (2007). *Using work and energy to characterize mechanical shock*, SAND2007-0851J. Albuquerque, NM: Sandia National Laboratories.

Appendix A
Units

Units are one of the most critical components of any engineering problem. If the units are not right, the answer is not right, no matter how good the numbers look. Most engineers and scientists have a good working knowledge of the international system of units or Système International (SI). The SI unit system is the modern form of the metric system initiated in France in the 1790s. In addition to the SI unit system, the United States Customary System (USCS) is still in common use throughout the USA. The USCS is perhaps even more common within the US military to which much of the mechanical shock field owes its origins. For that reason, a brief review of units, and especially the differences between the SI and USCS systems, is appropriate.

In structural dynamics, the fundamental units can be traced back to Newton's law,

$$\mathbf{F} = m\mathbf{a}, \tag{A.1}$$

or force equals mass times acceleration. It is precisely at this point that the two unit systems diverge. The SI unit system is fundamentally a mass, length, and time system. The three basic units being the kilogram (kg), meter (m), and second (s). As such, acceleration is a derived quantity, being length divided by time squared or meters per second squared. The force is also a derived quantity, mass times acceleration, and given the name Newton (N) in honor of Isaac Newton and the law that he proposed. Thus, force in the SI system is defined as

$$1\,\mathrm{N} = (1\,\mathrm{kg})\left(1\,\frac{\mathrm{m}}{\mathrm{s}^2}\right). \tag{A.2}$$

In contrast to the SI system, the USCS is a force, length, and time system. The fundamental units being the pound or pound-force ($\mathrm{lb_f}$), foot (ft), and second (s). As a result, mass is a derived unit. This is fundamentally different from the SI system and likely the cause of many misunderstandings between the two systems.

C. Sisemore, V. Babuška, *The Science and Engineering of Mechanical Shock*,
https://doi.org/10.1007/978-3-030-12103-7

The problem is admittedly further exasperated by the unfortunate use of the word "pound" to refer to weight in the context of both force and mass. If that were not enough, the use of the questionable term "pound-mass" only further complicates people's perception of the unit system. The pound is a measure of force. Weight is also a measure of force. For this reason, the use of the pound in this text always refers to the proper pound-force and abbreviated $\mathrm{lb_f}$, and never to mass. The USCS mass, given the name slug, is defined as:

$$1\,\mathrm{slug} = \frac{1\,\mathrm{lb_f}}{1\,\mathrm{ft/s^2}} = 1\frac{\mathrm{lb_f\,s^2}}{\mathrm{ft}}. \tag{A.3}$$

Thus, one slug is the mass accelerated by one foot per second squared when a force of one pound is applied. Since standard gravity at the Earth's surface is $32.174\,\mathrm{ft/s^2}$ it is often stated that one slug of mass weighs 32.174 pounds. Similarly, one slug is approximately equal to 14.59 kg based on standard gravity.

Both unit systems are frequently complicated by the fact that the primary length units are not convenient to many structural dynamics problems. Many systems are designed in terms of millimeters or inches as opposed to the primary units of meters or feet. Consequently the mass units must be adjusted accordingly. In terms of SI units, grams is frequently the mass unit of choice when millimeters are used although this means that the forces are no longer in Newtons. Likewise, in USCS units an inch version of the slug must be used although a universally agreed upon name does not exist. The mass in terms of inches is obtained by dividing the force in pounds by the standard acceleration of gravity in inches, $386.09\,\mathrm{in/s^2}$. Colloquially this is sometimes referred to as a blob, slinch, or sometimes snail.

One further unit should be mentioned here because it appears frequently in US systems, the kip. The kip is not an official USCS unit but is commonly defined as

$$1\,\mathrm{kip} = 1000\,\mathrm{lb_f}. \tag{A.4}$$

The origin of this unit is unclear but it is generally believed that the word kip is a shortened version of kilo-pound which is an admittedly peculiar mixing of unit system naming conventions. Nevertheless, it is frequently used for systems with large weights.

Appendix B
Recommended References

This appendix summarizes some of the most important and useful references about all aspects of shock engineering. The references cover specific topics discussed in this book in greater depth. Most of these references were cited in the chapters.

Textbooks

- Scavuzzo, R. J. and Pusey, H. C. (2007) *Principles and Techniques of Shock Data Analysis, Second Edition,* SVM-16, The Shock and Vibration Analysis Center, HI-TEST Laboratories, Inc., Arvonia, Virginia.
  - This book is an update and significant rewrite of the first edition of *Principles and Techniques of Shock Data Analysis* written by R. D. Kelly and G. Richman, originally published in 1969 by the Shock and Vibration Information Center (SVIC). The focus of this book is on the collection and analysis of mechanical shock data. The book discusses fundamental tools like the Fourier transform and the shock spectrum, which can be found in other references. The book has an extensive discussion on instrumentation and practical analyses of acceleration measurements from shock tests. It also has a chapter on ground shock that is unique and not covered in other mechanical shock texts.
- Scavuzzo, R. J. and Pusey, H. C. (2000) *Naval Shock Analysis and Design*, SVM-17, The Shock and Vibration Information Analysis Center, Booz-Allen and Hamilton, Inc., Falls Church, Virginia.
  - This book is important because it is perhaps the only textbook devoted to the study of mechanical shock as it applies specifically to naval vessels. The focus is on the underwater shock phenomena although considerable attention is paid to the underlying theory of shock response spectra and modal theory. It discusses the dynamic design analysis method (DDAM), which is the

C. Sisemore, V. Babuška, *The Science and Engineering of Mechanical Shock*,
https://doi.org/10.1007/978-3-030-12103-7

U.S. Navy's approach to the analysis of ship shock response to underwater explosions.

- Lalanne, C. (2014) *Mechanical Shock, Mechanical Vibration and Shock Analysis, Third Edition, Vol. 2,* John Wiley & Sons, 2014.
  - This book is volume two of a five volume set written by the author dedicated to sinusoidal vibration, mechanical shock, random vibration, fatigue damage, and specification development. It provides a comprehensive treatment of mechanical shock with an emphasis on the shock response spectrum. There is a heavy focus on the theory of the shock response spectra and on the use of shakers for shock testing. There is some discussion of shock test specification development and a very brief discussion of some common shock test machines. It includes some of the same topics covered in this book.
- Steinberg, D. S. (2000) *Vibration Analysis for Electronic Equipment, 3rd Edition,* John Wiley & Sons, 2000.
  - This is a highly cited and useful book covering shock, vibration, and acoustic environments with an emphasis on simple design approaches for electronic components. The book includes practical formulas for estimating the damage potential of mechanical environments, including shock. While the context is electronic components much of the material is applicable to general structures. Chapter 11 of this book is devoted to designing electronics for shock environments. The book does not specifically address the analysis of shock data or test methods but it does supply some pertinent design information for this particular class of problem.
- Gupta, A.K. (1992) *Response Spectrum Method In Seismic Analysis and Design of Structures,* CRC Press, Inc., Boca Raton, FL, 1992 ISBN 978-0849386282.
  - This book describes the use of response spectra in earthquake engineering. The context is on the response of buildings to earthquake shocks, but the material is applicable to mechanical and aerospace systems too. It contains an excellent chapter on using response spectra for analyzing the shock response of multi-degree-of-freedom systems including multi-directional loading.

Standards

- United States Department of Defense (2017), *Requirements for Shock Tests, H.I. (High-Impact) Shipboard Machinery, Equipment, and Systems,* MIL-DTG-901E, July 2017.
  - This military specification describes high-impact shock testing requirements for machinery, equipment, systems, and structures on all type of ships. These requirements are the basis for verification of shipboard installations' capability to survive shocks from both normal and hostile sources.

- United States Department of Defense (2014), *Department of Defense Test Method Standard; Environmental Engineering Considerations and Laboratory Tests*, MIL-STD-810G (w/Change 1), 15 April 2014.
  - This military standard is divided into three parts. Part one contains general program guidelines and guidelines for laboratory tests. Part two describes environmental laboratory test methods for a wide variety of tests that many people use as starting points for specific tests. It contains tailoring information, environmental stress data, and laboratory test methods. The methods are divided into "procedures" which are subtopics for different types of environments. General shock is described in Method 516.6, pyroshock is covered in Method 517.1, gunfire shock is covered in Method 519.6, ballistic shock is covered in Method 522.1, and rail impact (railroad car impact) is in Method 526. Part three addresses testing in various climatic conditions such as extreme heat or cold found in different regions of the world, and their affects on materials.
- Air Force Space Command (2014) *Test Requirements for Launch, Upper-stage and Space Vehicles*, Space and Missile System Standard SMC-S-016, 2014.
  - This SMC Standard replaced MIL-STD-1540. It describes environmental testing requirements for launch vehicles, upper-stage vehicles, space vehicles, and their subsystems and units. Two very useful sections are Section 3—definitions and Section 4—general requirements. These sections contain technical descriptions of terms that are widely used in all aerospace programs. As with all military standards, it is a starting point for tailoring test requirements for specific applications. The information in this standard is invaluable when designing tests and defining test requirements. Often the requirements in this standard are used directly, particularly in the early stages of a project.
- National Aeronautics and Space Administration (2011) *Pyroshock Test Criteria*, NASA Technical Standard, NASA-STD-7003A, 2011.
  - This standard describes NASA's processes for pyroshock testing and verification of NASA vehicles' ability to withstand pyroshock events. It covers many aspects of pyroshock testing at a relatively high level, including test environments, prediction and analysis for NASA vehicles including spacecraft, payloads, and launch vehicles. It does not contain much detailed technical information. This can be found in NASA-HDBK-7005.
- National Aeronautics and Space Administration (2013) *General Environmental Verification Standard (GEVS) for GSFC Flight Programs and Projects,* GSFC-STD-7000A, April 22, 2013.
  - NASA's Goddard Space Flight Center (GSFC) administers many of NASA's satellite programs. This standard describes environment verification requirements and processes for satellite and spacecraft instrument programs managed by the GSFC. It contains information similar to that in SMC-S-16. Section 2.4

covers structural and mechanical verification requirements. Mechanical shock qualification is addressed in Section 2.4.4. While it is a GSFC standard, it is widely used and referenced by industry for spacecraft instruments and satellites sponsored by other NASA centers and non-NASA entities.

- ISO 18431-4 (2007) *International Standard, Mechanical vibration and shock — Signal processing — Part 4: Shock response spectrum analysis*
  - This is Part 4 of ISO Standard 18431 in mechanical vibration and shock. The other parts are: Part 1, which covers general information, and Part 2 which covers windowing of data in Fourier transform processing. Part 4 addresses the numerical algorithms for calculating the shock response spectrum with the digital filter approach, specifically with the ramp invariant filter. It defines expressions for the coefficients of ramp invariant filters for various responses of a single degree-of-freedom system, including absolute acceleration, relative velocity, relative displacement, and pseudo-velocity. While a shock is a transient signal, the digital filters and the coefficients in this standard can be used for computing the response of a single degree-of-freedom oscillator any type of excitation.
- National Institute of Building Sciences Building Seismic Safety Council (2010), *Earthquake-Resistant Design Concepts: An Introduction to the NEHRP Recommended Seismic Provisions for New Buildings and Other Structures*, FEMA P-749, December 2010.
  - This is a design guide for earthquake resistant buildings put out by the National Institute of Building Sciences Building Seismic Safety Council. It has a lot of information about earthquakes and seismic hazards associated with them. It discusses design and construction features that should minimize damage to structures from strong earthquakes. The guide generally covers these topics at a conceptual level.

Handbooks

- Piersol, A.J. and Paez, T.J. (2010) *Harris's Shock and Vibration Handbook, 6th Edition*, McGraw Hill, New York, 2010.
  - This is a comprehensive book with a wealth of information on mechanical vibration and shock. It covers basic concepts in vibration and shock as well as specific topics like testing, data acquisition, pyroshock, and human response to shock and vibration. Most of the handbook is about vibration but important aspects of mechanical shock, such as shock instrumentation, shock response analyses, shock testing, and standards are covered. It is not a textbook in that it does not include derivation and background for the equations and methods so it is more of a comprehensive reference for readers who possess some familiarity with mechanical vibration and shock.

- National Aeronautics and Space Administration (2001) *Dynamic Environmental Criteria* NASA-HDBK-7005, 2001.
  - As the title suggests, this NASA handbook covers much more than shock. It covers almost everything related to dynamic environments of spacecraft, including the environments themselves, methods for predicting the loads from the dynamic environments, methods for computing responses of the spacecraft to the dynamic loads, as well as design procedures. This handbook has a lot of detailed information presented in an accessible way.
- United States Department of Defense (1999), *Test Requirements for Launch, Upper-stage and Space Vehicles*, Department of Defense Handbook, MIL-HDBK-340A, 1999.
  - This DOD handbook is similar to NASA-HDBK-7005. It is testing centric, covering test procedures for components, subsystems, and space systems so that the space vehicle will meet requirements and ensure a successful launch and performance. It is a companion to DOD Standard SMC-S-16, providing more in-depth treatment of the material in SMC-S-16.
- European Space Agency (2015), *Space engineering: Mechanical Shock Design and Verification Handbook,* ESA European Cooperation for Space Standardization Handbook,ECSS-E-HB32-25A, July 2015.
  - It is a comprehensive collection of material on the spacecraft shock environment. It includes guidelines, recommendations, and best practices for spacecraft shock design and verification. In that sense, it covers similar material that is in the NASA and DOD standards and handbooks, but all in one place. It has a unique section on the use of statistical energy analysis (SEA) for modeling shock response that is not covered in any reference book on mechanical shock.

Internet References

- Irvine, T. (2018) www.vibrationdata.com "Shock, vibration, and acoustic analysis resource"
  - This website, created and maintained by Mr. Tom Irvine, contains a wide range of information about shock, vibration, and acoustic environments. Various pages provide explanations and derivations of many concepts in the shock and vibration fields.

Glossary

Conservative Test A conservative test is a test that is more severe than the requirements over most or all of the defined frequency range. The term conservative in this context is used to indicate that a higher safety factor to the expected environment has been demonstrated.

Damping An energy dissipation mechanism for a system or structure. Viscous damping, which is proportional to velocity, is usually assumed because of its simpler mathematical formulation. However, hysteresis damping, proportional to displacement, is also quite common and frequently more realistic, although it is more mathematically difficult.

Damping Ratio The damping ratio, ζ, is defined as the percentage of critical damping in a system. This is given mathematically as $\zeta = c/c_{cr}$, where c_{cr} is the critical viscous damping for the system and c is the viscous damping inherent in the system.

Drop Shock A shock event or shock test that results from dropping an item and allowing it to impact the ground or other relevant surface of interest.

Fourier Transform A reversible integral transform of a continuous function from the time domain to the frequency domain. The Fourier transform has the effect of translating a function of time into a complex function in terms of frequencies. Likewise, the inverse Fourier transform can convert a complex frequency function back into the original, real-valued time function.

Maxi-Max An abbreviation indicating that the shock response spectrum is calculated by extracting the maximum absolute value of the response quantity of interest from both the primary and residual portions of the shock transient.

Pseudo-Velocity The pseudo-velocity is a response quantity defined as the frequency times the displacement as: $PV(t) = \omega x(t)$. As a result, the pseudo-velocity has the units of velocity but is in phase with displacement and therefore not the true velocity. Pseudo-velocity correlates well with shock damage potential and as a result is a very popular quantity of interest.

C. Sisemore, V. Babuška, *The Science and Engineering of Mechanical Shock*,
https://doi.org/10.1007/978-3-030-12103-7

Pyroshock A high-intensity, high-frequency shock resulting from the detonation of a pyrotechnic or explosive device. Pyroshocks are usually characterized by high accelerations, short durations, and little or no velocity change.

Response Spectrum A measure of the responses to a specific excitation of a series of idea single degree-of-freedom oscillators (spring-mass-viscous damper system). The response spectrum is a measure of the response quantity of interest in terms of the natural frequency of the theoretical SDOF oscillator.

SDOF Oscillator The single degree-of-freedom oscillator is a mathematical model consisting of a lumped mass, m, connected to ground with a spring of spring stiffness, k, and a viscous damper with damping, c. This simple system can often be used to represent more complicated systems and is the fundamental basis of the various response spectra.

Shock A transient excitation characterized by a high amplitude and a short time duration. The transient duration is ideally much shorter than the fundamental period of the system or structure being excited.

Shock Bandwidth The shock bandwidth is typically defined as the range of frequencies where the energy contained in the shock pulse is above one-half of the maximum energy at any frequency.

Shock Response Spectrum The shock response spectrum (SRS) is a plot of an external response quantity of interest from a series of SDOF oscillators to a transient excitation. The external quantity of interest can be any measurable quantity such as maximum acceleration, velocity, or displacement. The SRS is a good tool to assess the potential severity of a shock excitation on a structure.

Test Specification The definition of how a shock test should be performed in the laboratory. For example, a drop shock test specification will define the type of shock, the amplitude of the shock, and the shock pulse duration. A test specification can take the form of a description of requirements, a table or requirements, a plot, or all three.

Time History The raw data collected from an instrumented test. Usually a list of numerical data given in terms of time and acceleration although velocity or displacement may be used for some applications.

Index

C. Sisemore, V. Babuška, *The Science and Engineering of Mechanical Shock*,
https://doi.org/10.1007/978-3-030-12103-7

MIX
Papier aus verantwortungsvollen Quellen
Paper from responsible sources
FSC® C105338

If you have any concerns about our products,
you can contact us on
ProductSafety@springernature.com

In case Publisher is established outside the EU,
the EU authorized representative is:
Springer Nature Customer Service Center GmbH
Europaplatz 3, 69115 Heidelberg, Germany

Printed by Libri Plureos GmbH
in Hamburg, Germany